Safety, Quality and Processing of Fruits and Vegetables

Safety, Quality and Processing of Fruits and Vegetables

Special Issue Editors

Urszula Tylewicz
Silvia Tappi
Malgorzata Nowacka
Artur Wiktor

MDPI • Basel • Beijing • Wuhan • Barcelona • Belgrade

Special Issue Editors

Urszula Tylewicz
University of Bologna
Italy

Silvia Tappi
University of Bologna
Italy

Malgorzata Nowacka
Warsaw University of Life Sciences
Poland

Artur Wiktor
Warsaw University of Life Sciences
Poland

Editorial Office
MDPI
St. Alban-Anlage 66
4052 Basel, Switzerland

This is a reprint of articles from the Special Issue published online in the open access journal *Foods* (ISSN 2304-8158) in 2019 (available at: https://www.mdpi.com/journal/foods/special_issues/Fruits_Vegetables).

For citation purposes, cite each article independently as indicated on the article page online and as indicated below:

LastName, A.A.; LastName, B.B.; LastName, C.C. Article Title. *Journal Name* **Year**, *Article Number*, Page Range.

ISBN 978-3-03928-086-5 (Hbk)
ISBN 978-3-03928-087-2 (PDF)

© 2020 by the authors. Articles in this book are Open Access and distributed under the Creative Commons Attribution (CC BY) license, which allows users to download, copy and build upon published articles, as long as the author and publisher are properly credited, which ensures maximum dissemination and a wider impact of our publications.

The book as a whole is distributed by MDPI under the terms and conditions of the Creative Commons license CC BY-NC-ND.

Contents

About the Special Issue Editors vii

Urszula Tylewicz, Silvia Tappi, Malgorzata Nowacka and Artur Wiktor
Safety, Quality, and Processing of Fruits and Vegetables
Reprinted from: *Foods* **2019**, *8*, 569, doi:10.3390/foods8110569 1

Artur Wiktor, Ronit Mandal, Anika Singh and Anubhav Pratap Singh
Pulsed Light treatment below a Critical Fluence (3.82 J/cm^2) minimizes photo-degradation and browning of a model Phenolic (Gallic Acid) Solution
Reprinted from: *Foods* **2019**, *8*, 380, doi:10.3390/foods8090380 5

Juan A. Tomas-Egea, Pedro J. Fito and Marta Castro-Giraldez
Analysis of Apple Candying by Microwave Spectroscopy
Reprinted from: *Foods* **2019**, *8*, 316, doi:10.3390/foods8080316 18

Abdul Ghani Dars, Kai Hu, Qiudou Liu, Aqleem Abbas, Bijun Xie and Zhida Sun
Effect of Thermo-Sonication and Ultra-High Pressure on the Quality and Phenolic Profile of Mango Juice
Reprinted from: *Foods* **2019**, *8*, 298, doi:10.3390/foods8080298 32

Malgorzata Nowacka, Artur Wiktor, Magdalena Dadan, Katarzyna Rybak, Aleksandra Anuszewska, Lukasz Materek and Dorota Witrowa-Rajchert
The Application of Combined Pre-Treatment with Utilization of Sonication and Reduced Pressure to Accelerate the Osmotic Dehydration Process and Modify the Selected Properties of Cranberries
Reprinted from: *Foods* **2019**, *8*, 283, doi:10.3390/foods8080283 45

Maria Paciulli, Ilce Gabriela Medina Meza, Massimiliano Rinaldi, Tommaso Ganino, Alessandro Pugliese, Margherita Rodolfi, Davide Barbanti, Michele Morbarigazzi and Emma Chiavaro
Improved Physicochemical and Structural Properties of Blueberries by High Hydrostatic Pressure Processing
Reprinted from: *Foods* **2019**, *8*, 272, doi:10.3390/foods8070272 61

Zoran Stamenković, Ivan Pavkov, Milivoj Radojčin, Aleksandra Tepić Horecki, Krstan Kešelj, Danijela Bursać Kovačević and Predrag Putnik
Convective Drying of Fresh and Frozen Raspberries and Change of Their Physical and Nutritive Properties
Reprinted from: *Foods* **2019**, *8*, 251, doi:10.3390/foods8070251 76

Irena Žuntar, Predrag Putnik, Danijela Bursać Kovačević, Marinela Nutrizio, Filip Šupljika, Andreja Poljanec, Igor Dubrović, Francisco J. Barba and Anet Režek Jambrak
Phenolic and Antioxidant Analysis of Olive Leaves Extracts (*Olea europaea* L.) Obtained by High Voltage Electrical Discharges (HVED)
Reprinted from: *Foods* **2019**, *8*, 248, doi:10.3390/foods8070248 90

Malgorzata Nowacka, Silvia Tappi, Artur Wiktor, Katarzyna Rybak, Agnieszka Miszczykowska, Jakub Czyzewski, Kinga Drozdzal, Dorota Witrowa-Rajchert and Urszula Tylewicz
The Impact of Pulsed Electric Field on the Extraction of Bioactive Compounds from Beetroot
Reprinted from: *Foods* **2019**, *8*, 244, doi:10.3390/foods8070244 119

Trond Løvdal, Bart Van Droogenbroeck, Evren Caglar Eroglu, Stanislaw Kaniszewski, Giovanni Agati, Michel Verheul and Dagbjørn Skipnes
Valorization of Tomato Surplus and Waste Fractions: A Case Study Using Norway, Belgium, Poland, and Turkey as Examples
Reprinted from: *Foods* **2019**, *8*, 229, doi:10.3390/foods8070229 . . . . **131**

Setya B.M. Abduh, Sze Ying Leong, Dominic Agyei and Indrawati Oey
Understanding the Properties of Starch in Potatoes (*Solanum tuberosum* var. Agria) after Being Treated with Pulsed Electric Field Processing
Reprinted from: *Foods* **2019**, *8*, 159, doi:10.3390/foods8050159 . . . . **152**

Benjamin Opuko Wayumba, Hyung Sic Choi and Lim Young Seok
Selection and Evaluation of 21 Potato (*Solanum Tuberosum*) Breeding Clones for Cold Chip Processing
Reprinted from: *Foods* **2019**, *8*, 98, doi:10.3390/foods8030098 . . . . **173**

About the Special Issue Editors

Urszula Tylewicz is Junior Assistant Professor at the Department of Agricultural and Food Sciences, University of Bologna (Italy). She obtained her M.Sc. degree in Food Technology at the Warsaw University of Life Science, Poland, in 2006, and her Ph.D. degree in Food Science and Biotechnology at the University of Bologna, Italy, in 2011. Her research activity is mainly focused on the application of nonthermal technologies for fruit and vegetable processing (pulsed electric field, ultrasound, vacuum impregnation, osmotic dehydration). She is member of the ISEKI Food Association (IFA) "Integrating Food Science and Engineering Knowledge into the Food Chain" and the International Society for Electroporation-Based Technologies and Treatments (ISEBTT). She was a chair of the 6th PEF School (3–7.06.2019, Cesena, Italy). She has authored and co-authored more than 50 publications in peer-viewed international journals and 4 book chapters.

Silvia Tappi is a food technologist with a Ph.D. in Food Science and Biotechnology, obtained at the University of Bologna in 2016. She is currently Research Fellow at the Interdepartmental Centre for Industrial Research (CIRI-Agro) of the University of Bologna, where she is mainly focused on the optimization and scaling up of emerging mild/nonthermal technologies for food processing for the stabilization and quality improvement of food products and for byproduct valorisation, with a particular focus on process/product sustainability and product innovation. She is author of more than 35 publications in peer-reviewed international journals and 2 book chapters.

Malgorzata Nowacka is Associate Professor at the Institute of Food Science at the Warsaw University of Life Sciences (WULS-SGGW). Her research area is connected with water removal from plant products using innovative technologies (novel drying methods, osmotic dehydration) and innovative methods of pretreatment such as ultrasound processing, pulsed electric field, etc., prior to technological processes occurring in food technology, i.e., mainly processes of mass transfer. In 2009, she received a PhD degree in agricultural sciences in the discipline of food technology and nutrition, and in 2018, she completed habilitation. She is an author or co-author of more than 45 publications in peer-reviewed international journals and numerous book chapters.

Artur Wiktor is Assistant Professor at the Warsaw University of Life Sciences (WULS-SGGW). His main research activity is on the utilization of nonthermal technologies (such as pulsed electric field, ultrasound, cold plasma) in order to enhance mass and heat transfer-based unit operations or to improve the quality of final products. A big part of his scientific activity concerns drying, osmotic dehydration, freezing of plant tissue, and extraction of bioactive compounds from food matrices. He is a member of the International Society for Electroporation-Based Technologies and Treatments (ISEBTT) and Polish Society of Food Technologists. He has authored or co-authored more than 40 publications in peer-reviewed international journals and numerous book chapters.

Editorial

Safety, Quality, and Processing of Fruits and Vegetables

Urszula Tylewicz [1,2,*], Silvia Tappi [2], Malgorzata Nowacka [3] and Artur Wiktor [3]

[1] Department of Agricultural and Food Sciences, University of Bologna, P.zza Goidanich 60, 47521 Cesena, Italy
[2] Interdepartmental Centre for Agri-Food Industrial Research, University of Bologna, Via Quinto Bucci 336, 47521 Cesena, Italy; silvia.tappi2@unibo.it
[3] Department of Food Engineering and Process Management, Faculty of Food Sciences, Warsaw University of Life Sciences, Nowoursynowska 159c, 02-776 Warsaw, Poland; malgorzata_nowacka@sggw.pl (M.N.); artur_wiktor@sggw.pl (A.W.)
* Correspondence: urszula.tylewicz@unibo.it; Tel.: +39-0547-338-120

Received: 8 October 2019; Accepted: 11 November 2019; Published: 13 November 2019

Abstract: Nowadays, one of the main objectives of the fruit and vegetable industry is to develop innovative novel products with high quality, safety, and optimal nutritional characteristics in order to respond with efficiency to the increasing consumer expectations. Various emerging, unconventional technologies (e.g., pulsed electric field, pulsed light, ultrasound, high pressure, and microwave drying) enable the processing of fruits and vegetables, increasing their stability while preserving their thermolabile nutrients, flavour, texture, and overall quality. Some of these technologies can also be used for waste and by-product valorisation. The application of fast noninvasive methods for process control is of great importance for the fruit and vegetable industry. The following Special Issue "Safety, Quality, and Processing of Fruits and Vegetables" consists of 11 papers, which provide a high-value contribution to the existing knowledge on safety aspects, quality evaluation, and emerging processing technologies for fruits and vegetables.

Keywords: fruit; vegetable; safety; quality; emerging technologies; unconventional processing

---

In the last few years, consumers have become more exigent and demand high-quality and convenient food products with natural flavours and taste, free from additives and preservatives [1]. Therefore, the challenge for the fruit and vegetable industry is to develop such products, taking into account quality and safety aspects along with consumer acceptance. Emerging, unconventional processing of fruit and vegetables is more and more studied in order to develop products rich in bioactive compounds, paying attention at the same time to waste and by-product valorisation [2–4]. This Special Issue "Safety, Quality, and Processing of Fruits and Vegetables" gives an overview of the application of emerging, unconventional technologies to obtain high-quality fruit juice, semi-dried and dried products, waste valorisation, and process control. It also provides some insights into principles and fundamentals of nonthermal technologies.

The importance of the quality standards for potatoes intended for the processing industry is explained by Wayumba et al. [5]. This study was designed with the purpose of identifying specialized potato clones with acceptable qualities for processing chips, as compared to two selected control varieties, Dubaek and Superior. From this study, the authors concluded that for quality processing of potato chips, clones with combined traits of high dry matter, low levels of glycoalkaloids and reducing sugars, should be used as raw materials along with the acceptable chip colour [5]. Starch is the major component in potato, that contributes to its nutritional and technological quality. Different food processing techniques including boiling, cooling, reheating, conventional frying, and air frying have been shown to change the digestibility of starch. In the paper by Abduh et al. [6], the effect of emerging processing using pulsed electric field (PEF)—usually used for structure modification in fruit and vegetables—on the properties of starch in potatoes was investigated, showing that PEF did not change the properties of starch within the potatoes, but it narrowed the temperature range of gelatinisation and reduced the digestibility of starch collected from the processing medium. Therefore, this study confirms that, when used as a processing aid for potato, PEF does not result in detrimental effects on the properties of potato starch [6]. PEF has been shown to be effective in the extraction of bioactive compounds (mainly betalains) from beetroot, increasing the extraction yield [7]. The greatest increase in the content of betalain compounds in the red beet extract was noted when an electric field at 4.38 kV/cm was applied [7]. The increase in the extraction rate of polyphenols from olive leaves was also observed by using high-voltage electrical discharges (HVED) as a green technology [8]. HVED parameters included different green solvents (water, ethanol), treatment times (3 and 9 min), gases (nitrogen, argon), and voltages (15, 20, 25 kV). The highest yield of phenolic compounds was obtained for the sample treated with argon/9 min/20 kV/50% (3.2 times higher as compared to conventional extraction (CE)). In general, HVED presents an excellent potential for phenolic compound extraction for further application in functional food manufacturing [8].

Valorisation of waste and by-products is the topic of the paper by Løvdal et al. [9]. It provides an overview of tomato production in Europe and the strategies employed for processing and valorisation of tomato side streams and waste fractions. Special emphasis was put on the four tomato-producing countries Norway, Belgium, Poland, and Turkey. These countries are very different with regard to their climatic preconditions for tomato production and volumes produced and represent the extremes among European tomato producing countries.

Osmotic dehydration and drying of berries were the objective of papers by Nowacka et al. [10] and Stamenković et al. [11]. In the paper by Nowacka et al. [10], osmotic dehydration of cranberries was combined with blanching, ultrasound, and vacuum application. Unconventional pretreatment of cranberries caused a significant increase of osmotic dehydration effectiveness. Cranberries subjected to combined treatment, in particular to ultrasounds, had comparable or higher polyphenolic, anthocyanin, and flavonoid content than a blanched tissue subjected to osmotic dehydration alone. Taking into account the evaluated physical and chemical properties of dehydrated cranberries and the osmotic dehydration process, it has been concluded that the best combined pretreatment method was a 20 min sonication followed by a 10 min lowered pressure treatment. In the paper by Stamenković et al. [11], the effectiveness of convective drying of Polana raspberries was compared to freeze-drying, which allows producers to obtain products of high quality but also with high cost. The authors concluded that convective drying of Polana raspberry with air temperature of 60 °C and air velocity of 1.5 m·s^{-1}, may be considered as a sufficient alternative to freeze-drying [11].

Another emerging nonthermal technology studied on fruits and vegetables is high-pressure processing, with the aim of better preserving nutritional and organoleptic properties. In fact, the results presented in the paper by Paciulli et al. [12] revealed the mild impact of high-pressure treatments on the organoleptic properties of blueberries along with better texture and colour maintenance. The effects of ultra-high pressure (UHP) and thermo-sonication (TS) were also tested on quality of mango juice [13]. Both treatments had minimal effects on the total soluble solids, pH, and titratable acidity of mango juice. The residual activities of three enzymes (polyphenol oxidase, peroxidase, and pectin methylesterase), antioxidant compounds (vitamin C, total phenolics, mangiferin derivatives, gallotannins, and quercetin derivatives) and antioxidant activity sharply decreased with the increase in the temperature of the TS treatment. Nevertheless, the UHP treatment retained antioxidants and antioxidant activity at a high level. The UHP process is apparently superior to TS in bioactive compound and antioxidant activity preservation. Therefore, the mango juice products obtained by ultra-high-pressure processing might be more beneficial to health [13].

In the paper by Wiktor et al. [14], instead, the effect of pulsed light treatment with different fluence was studied on a gallic acid aqueous solution—as a model system of phenolic abundant liquid food matrices. It was demonstrated that pulsed light can modify the optical properties of gallic acid and cause reactions and degradation of gallic acid. However, application of pulsed light did not significantly alter the overall quality of the model gallic acid solution at low fluence levels. Cluster analysis revealed that below 3.82 J/cm^2, changes in gallic acid were minimal, and this fluence level could be used as the critical level for food process design aiming to minimize nutrient loss.

Finally, Tomas-Egea et al. [15] studied the importance of process control in the industry, which requires fast, safe, and easily applicable methods. In this sense, the use of dielectric spectroscopy in the microwave range can be a great opportunity to monitor processes in which the mobility and quantity of water is the main property to produce a high-quality and safe product, such as candying of fruits. They demonstrated that the use of dielectric properties in γ-dispersion at relaxation frequency allowed us not only to monitor the osmotic drying and hot-air-drying processes of the apple candying, but also to predict the supersaturation state of the liquid phase until vitrification.

Conflicts of Interest: The authors declare no conflict of interest.

References

1. Tylewicz, U.; Tappi, S.; Mannozzi, C.; Romani, S.; Dellarosa, N.; Laghi, L.; Ragni, L.; Rocculi, P.; Dalla Rosa, M. Effect of pulsed electric field (PEF) pre-treatment coupled with osmotic dehydration on physico-chemical characteristics of organic strawberries. *J. Food Eng.* **2017**, *213*, 2–9. [CrossRef]
2. Barba, F.J.; Parniakov, O.; Pereira, S.A.; Wiktor, A.; Grimi, N.; Boussetta, N.; Saraiva, J.; Raso, J.; Martin-Belloso, O.; Witrowa-Rajchert, D.; et al. Current applications and new opportunities for the use of pulsed electric fields in food science and industry. *Food Res. Int.* **2015**, *77*, 773–798. [CrossRef]
3. Deng, L.-Z.; Mujumdar, A.S.; Zhang, Q.; Yang, X.-H.; Wang, J.; Zheng, Z.-A.; Gao, Z.-J.; Xiao, H.-W. Chemical and physical pretreatments of fruits and vegetables: Effects on drying characteristics and quality attributes–a comprehensive review. *Crit. Rev. Food Sci.* **2019**, *59*, 1408–1432. [CrossRef] [PubMed]
4. Putnik, P.; Lorenzo, J.M.; Barba, F.J.; Roohinejad, S.; Režek Jambrak, A.; Granato, D.; Montesano, D.; Kovačević, D.B. Novel food processing and extraction technologies of high-added value compounds from plant materials. *Foods* **2018**, *7*, 106. [CrossRef] [PubMed]
5. Wayumba, B.O.; Choi, H.S.; Seok, L.Y. Selection and Evaluation of 21 Potato (Solanum Tuberosum) Breeding Clones for Cold Chip Processing. *Foods* **2019**, *8*, 98. [CrossRef] [PubMed]
6. Abduh, S.B.; Leong, S.Y.; Agyei, D.; Oey, I. Understanding the Properties of Starch in Potatoes (*Solanum tuberosum* var. Agria) after Being Treated with Pulsed Electric Field Processing. *Foods* **2019**, *8*, 159. [CrossRef] [PubMed]
7. Nowacka, M.; Tappi, S.; Wiktor, A.; Rybak, K.; Miszczykowska, A.; Czyzewski, J.; Drozdzal, K.; Witrowa-Rajchert, D.; Tylewicz, U. The Impact of Pulsed Electric Field on the Extraction of Bioactive Compounds from Beetroot. *Foods* **2019**, *8*, 244. [CrossRef] [PubMed]
8. Žuntar, I.; Putnik, P.; Bursać Kovačević, D.; Nutrizio, M.; Šupljika, F.; Poljanec, A.; Dubrović, I.; Barba, F.J.; Režek Jambrak, A. Phenolic and Antioxidant Analysis of Olive Leaves Extracts (*Olea europaea* L.) Obtained by High Voltage Electrical Discharges (HVED). *Foods* **2019**, *8*, 248. [CrossRef] [PubMed]
9. Løvdal, T.; Van Droogenbroeck, B.; Eroglu, E.C.; Kaniszewski, S.; Agati, G.; Verheul, M.; Skipnes, D. Valorization of Tomato Surplus and Waste Fractions: A Case Study Using Norway, Belgium, Poland, and Turkey as Examples. *Foods* **2019**, *8*, 229. [CrossRef] [PubMed]
10. Nowacka, M.; Wiktor, A.; Dadan, M.; Rybak, K.; Anuszewska, A.; Materek, L.; Witrowa-Rajchert, D. The Application of Combined Pre-Treatment with Utilization of Sonication and Reduced Pressure to Accelerate the Osmotic Dehydration Process and Modify the Selected Properties of Cranberries. *Foods* **2019**, *8*, 283. [CrossRef] [PubMed]
11. Stamenković, Z.; Pavkov, I.; Radojčin, M.; Tepić Horecki, A.; Kešelj, K.; Bursać Kovačević, D.; Putnik, P. Convective Drying of Fresh and Frozen Raspberries and Change of Their Physical and Nutritive Properties. *Foods* **2019**, *8*, 251. [CrossRef] [PubMed]
12. Paciulli, M.; Medina Meza, I.G.; Rinaldi, M.; Ganino, T.; Pugliese, A.; Rodolfi, M.; Barbanti, D.; Morbarigazzi, M.; Chiavaro, E. Improved Physicochemical and Structural Properties of Blueberries by High Hydrostatic Pressure Processing. *Foods* **2019**, *8*, 272. [CrossRef] [PubMed]

13. Dars, A.G.; Hu, K.; Liu, Q.; Abbas, A.; Xie, B.; Sun, Z. Effect of Thermo-Sonication and Ultra-High Pressure on the Quality and Phenolic Profile of Mango Juice. *Foods* **2019**, *8*, 298. [CrossRef] [PubMed]
14. Wiktor, A.; Mandal, R.; Singh, A.; Pratap Singh, A. Pulsed Light treatment below a Critical Fluence (3.82 J/cm^2) minimizes photo-degradation and browning of a model Phenolic (Gallic Acid) Solution. *Foods* **2019**, *8*, 380. [CrossRef] [PubMed]
15. Tomas-Egea, J.A.; Fito, P.J.; Castro-Giraldez, M. Analysis of Apple Candying by Microwave Spectroscopy. *Foods* **2019**, *8*, 316. [CrossRef] [PubMed]

© 2019 by the authors. Licensee MDPI, Basel, Switzerland. This article is an open access article distributed under the terms and conditions of the Creative Commons Attribution (CC BY) license (http://creativecommons.org/licenses/by/4.0/).

Article

Pulsed Light treatment below a Critical Fluence (3.82 J/cm^2) minimizes photo-degradation and browning of a model Phenolic (Gallic Acid) Solution

Artur Wiktor [1,2], Ronit Mandal [1], Anika Singh [1] and Anubhav Pratap Singh [1,*]

1. Food, Nutrition and Health, University of British Columbia, 2205, East Mall, Vancouver, BC V6T 1Z4, Canada
2. Faculty of Food Sciences, Department of Food Engineering and Process Management, Warsaw University of Life Sciences (WULS-SGGW), Nowoursynowska 159c, 02-776 Warsaw, Poland
* Correspondence: anubhav.singh@ubc.ca; Tel.: +1-604-822-5944

Received: 7 July 2019; Accepted: 30 August 2019; Published: 1 September 2019

Abstract: Pulsed light (PL) is one of the most promising non-thermal technologies used in food preservation and processing. Its application results in reduction of microbial load as well as influences the quality of food. The data about the impact of PL on bioactive compounds is ambiguous, therefore the aim of this study was to analyze the effect of PL treatment of a gallic acid aqueous solution—as a model system of phenolic abundant liquid food matrices. The effect of PL treatment was evaluated based on colour, phenolic content concentration and antioxidant activity measured by DPPH assay using a design of experiments approach. The PL fluence (which is the cumulative energy input) was varied by varying the pulse frequency and time. Using Response Surface Methodology, prediction models were developed for the effect of fluence on gallic acid properties. It was demonstrated that PL can modify the optical properties of gallic acid and cause reactions and degradation of gallic acid. However, application of PL did not significantly alter the overall quality of the model gallic acid solution at low fluence levels. Cluster analysis revealed that below 3.82 J/cm^2, changes in gallic acid were minimal, and this fluence level could be used as the critical level for food process design aiming to minimize nutrient loss.

Keywords: pulsed light; fluence; gallic acid; non-thermal treatment

1. Introduction

For a long time, the food industry has been using thermal methods like pasteurization, sterilization, etc. for preservation of foods and extension of their shelf life. However, thermal processing operations have drawbacks associated with them. For instance, due to the high processing temperature, the nutrients may be destroyed. Also, the sensory characteristics may be affected [1]. Modern day consumers are more aware than ever before. They continuously demand food which is safe, of good eating quality and nutritionally sound. This has led the food processing scientists to seek and research food processing methods which can make the food safe, while keeping the nutritional properties intact.

Pulsed light (PL) technology has now been widely explored as a novel non-thermal food preservation method that uses a form of energy other than heat for achieving food preservation. PL uses high-intensity short duration white light (wavelength of 200–1100 nm) for microbial inactivation [2]. The electrical energy is stored in capacitors and discharged in short bursts or pulses of high intensity. The ultra-violet (UV) fraction of the spectrum is associated with microbial inactivation as well as other chemical changes in food products.

A multitude of phenolic compounds like ellagic acid, ferulic acid, gallic acid, etc. are synthesized by plants' fruits, vegetables, as part of their secondary metabolism. In the early 1960s these phenolic compounds were considered as by-products of the plant metabolism, which were present in the vacuoles of cells. These compounds act as complex constituents of pigments, antioxidants, flavoring agents, in plants and plant-based foods. Thus, they form a major part of our diet. Apart from that, they are also bioactive compounds that are anti-inflammatory, anticarcinogenic, can decrease blood sugar levels, reduce body weight and ageing [3].

The light sensitivity of phenolics is a topic of utmost importance for studying the processing of foods using light. The photoinduced degradation of gallic acid (a model system representing plant phenolics) was reported earlier in [4,5]. Benitez et al. [5] demonstrated that gallic acid subjected to UV radiation degrades following first-order kinetics reaction. Thus, UV radiation could be used in the wastewater treatment process after cork production. However, the authors pointed out that UV- provoked photolysis of gallic acid was a rather slow process—after 90 min of radiation the concentration of gallic acid decreased from 50 to 10–40 ppm, depending on pH. Also, the progress of the process may be different depending on the wavelength spectrum of the light used. A photoinduced decrease in total phenolic content was also observed for real food systems, i.e., pineapple juice subjected to UV-C treatment with a dose 10.76 mJ/cm^2 [6] or pumelo juice treated with a UV-C dose of 15.45–27.63 mJ/cm^2 [7]. It should be emphasized therefore that the data about the impact of UV light treatment of food on its bioactive compounds is ambiguous. There are articles which report no significant changes of phenolics after exposure of juices or solid food matrices to UV light or PL [8–10]. Similar statements can be made regarding the antioxidant activity. Thus, the data concerning the impact of PL on model systems becomes even more important to understand the basic mechanisms of effects and in designing proper PL processing systems for foods rich in phenolics.

PL technology is still in its infancy and therefore there are a limited number of studies that have been carried out on the effect of PL on nutrient attributes, while most available studies focus on the microbiological safety aspect of PL [11]. It is a matter of immense importance that this novel process ensures food safety while retaining the bioactive compounds in food and keeping the sensory properties intact. To the best of our knowledge, there is no literature available on the effect of PL on model polyphenolic solutions like gallic acid. Also, the literature on the effect of PL on liquid foods is scarce. In accordance, the aim of this study was to evaluate the effect of PL processing on the physicochemical properties of a model gallic acid solution using an experimental setup designed for thin-profile treatment of liquid foods.

2. Materials and Methods

2.1. Material

Gallic Acid (3,4,5-trihydroxybenzoic acid; Sigma Aldrich Co., Oakville, ON, Canada) was used to prepare a model solution with a concentration of 0.5 mg/mL. Methanol ($\geq$99.8%) and Folin-Ciocalteu reagents were purchased from Merck KGaA (Darmstadt, Germany). Ethanol, 2,2-diphenyl-1-picrylhydrazyl (DPPH) free radical and Na_2CO_3 were purchased from Alfa Aesar, Thermo Fisher Scientific Chemicals, Inc. (Ward Hill, MA, USA). ($\pm$)-6-Hydroxy-2,5,7,8-tetramethylchromane-2-carboxylic acid (Trolox^{TM}) was purchased from Sigma-Aldrich Co. (Oakville, ON, Canada).

2.2. Pulsed Light Equipment

The experiments were carried out in a bench-top pulsed light equipment designed at the Faculty of Land and Food Systems, University of British Columbia in collaboration with Solaris Disinfection Inc. (Mississauga, ON, Canada). The equipment consisted of two parts: (1) a cylindrical annular chamber built of quartz glass for thin profile liquid treatment. The chamber has an inlet and an outlet for flowing liquid in and out of the chamber; (2) a xenon flashlamp placed at the axial center of the cylindrical chamber that emits light pulses for liquid treatment. The annular volume of the treatment chamber was 75 mL and its average distance from the lamp axis was 2 cm. The liquid thickness in the chamber was around 1 mm. Pulsed light lamps emitted 30 J of light energy per pulse (comprising wavelengths ranging from far UV to near IR in the electromagnetic spectrum) on a chamber area (impact surface) of 675 cm². A schematic diagram of the equipment is given in Figure 1.

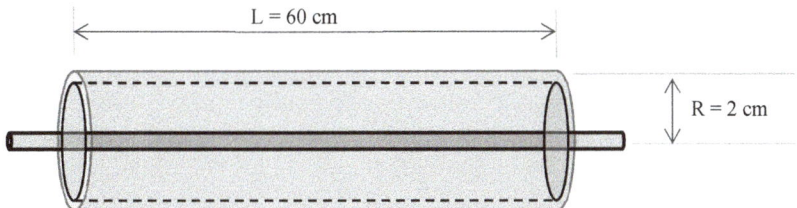

Figure 1. PL processing chamber. The chamber has an annular design inside which the liquid is housed and treated by PL lamps placed at the axis of chamber. Suitable for batch or continuous processing.

2.3. Design of Experiment

Table 1 shows the plan of the experiments. The response surface methodology approach was used for the experimental design to evaluate the effect of pulsed light treatment parameters on the physicochemical properties of gallic acid solutions. The central composite rotatable design (CCRD) used ($\alpha = k^{1/2}$ with two numeric factors k = 2 – frequency of pulses and treatment time) was prepared utilizing Statistica 13 Design (Statsoft Inc., Tulsa, OK, USA). It was composed of 10 experimental trials with two separate replicates in the central point (run 9 (C) and 10 (C)). Frequency varied from 1 to 10 Hz whereas time ranged from 5 to 50 s.

Energy input (in terms of fluence) depended on the parameters pulse frequency and treatment time and it was equal to 1.07–17.2 J/cm², since the device applied 30 J with one single pulse regardless of the frequency. The boundary conditions of treatment were selected based on literature data considering the energy necessary to inactivate microorganisms [11,12]. Each run was performed in two separate replicates, which means that central point was repeated four times in total. The responses described in subsequent sections were evaluated for these treatments and also for the untreated sample.

Table 1. The set-up of the performed experiment with fluence during PL treatment.

Run *	Factor A		Factor B		Fluence (J/cm²)
	Coded Value	Frequency (Hz)	Coded Value	Time (s)	
1	−1	2	−1	12	1.1
2	−1	2	1	43	3.8
3	1	9	−1	12	4.8
4	1	9	1	43	17.2
5	−1.41	1	0	28	1.2
6	1.41	10	0	28	12.4
7	0	5	−1.41	5	1.1
8	0	5	1.41	50	11.1
9 (C)	0	5	0	28	6.2
10 (C)	0	5	0	28	6.2

* Where (C) is the center point of the design. Additionally, untreated samples were also evaluated for the same responses.

2.4. Temperature Increment Measurement

The emitted light energy was absorbed by the solution as heat. The amount of heat absorbed was calculated for each run. The temperature changes for the gallic acid solution were recorded immediately after each treatment. A temperature measuring RTD (ThermoProbe Inc., Jackson, MS, USA) was used to measure the temperature. The initial solution temperature was recorded to be

21.2 °C. The measurements were duplicated for each observation and the temperature change % and heat gain [13] were calculated using Equations (1) and (2):

$$\text{Temp. increase \%} = (T_f - T_i)/T_i \times 100 \qquad (1)$$

$$\text{Heat gain (J/cm}^2) = [4.19 \times (T_f - T_i) \times V \times \varrho]/A \qquad (2)$$

where, T_f, T_i are final and initial temperature (°C); V = chamber volume (m^3); ϱ = liquid density (kg/m^3), A = area [cm^2]. The constant 4.19 (kJ/kg-K) is taken as the specific heat of the gallic acid solution which is assumed equal to that of water [13].

2.5. Colour

The colour of the treated and untreated samples was measured using a colorimeter (HunterLab, model LabScan™ XE Plus, Hunter Associates Laboratory, Reston, VA, USA). Colour was expressed in CIE L* (whiteness or brightness), a* (redness/greenness) and b* (yellowness/blueness) coordinates. Two replicate measurements were performed, and results were averaged. The total colour difference (ΔE) and browning index (BI) were calculated [14] using the following Equations (3) and (4):

$$\Delta E = ((L - L_o)^2 + (a - a_o)^2 + (b - b_o)^2)^{1/2} \qquad (3)$$

$$BI = 100 \times (x - 0.31)/0.172 \qquad (4)$$

where:

$$x = (a^* + 1.75 \times L^*)/(5.645 \times L^* + a^* - 3.012 \times b^*) \qquad (5)$$

In Equation (3), the L_o, a_o and b_o are the colour values for untreated samples, and the constants in Equations (4) and (5) were taken from the literature [14].

2.6. Total Phenolic Content, Gallic Acid Content and Antioxidant Activity Determination

Total phenolic content (TPC) were estimated using the Folin–Ciocalteau's (FC) method with modifications [15]. Briefly, an aliquot (5 mL) of the gallic acid solution was transferred to a glass tube; reactive 10^{-1} diluted FC reagent (20 mL) is added after 5 min; sodium carbonate (Na$_2$CO$_3$, 5 mL, 7.5% w/v) was added and the mixture shaken. After 30 min of incubation at ambient temperature in the dark, 200 μL samples were placed in 96-well plates. Finally, the absorbances were measured in a spectrophotometer (Infinite Pro M200 series, Tecan™, Männedorf, Switzerland) at 765 nm and compared to a gallic acid calibration curve for TPC (prepared using 0 to 1 mg/mL concentration gallic acid solution). Results were expressed as mg gallic acid equivalent (GAE)/100 mL. All measurements were done in duplicate.

Gallic acid content (GAC) was determined using HPLC (Agilent 1100 system, Agilent Technologies, Santa Clara, CA, USA) equipped with a Zorbax SB-C18 column according to the methodology presented in [16]. This was carried out to measure the changes in gallic acid concentration due to photodegradation. Results were expressed as mg GAC/100 mL solution.

To determine the antioxidant activity (AA) of gallic acid solutions, 2,2-diphenyl-1-picrylhydrazyl (DPPH) free radical scavenging assay was used. A standard curve was constructed using Trolox™ (20 μM) solution. For sample wells, gallic acid (20 μL) was added. In both standard and sample wells of a 96-well microtiter plate, 1 mM DPPH (20 μL) was added. The blank well consisted of HPLC grade methanol (200 μL). The plate was incubated for 10 min at room temperature in the dark. Then the plate absorbances were read at 519 nm by a microtiter plate reader (Tecan™ Infinite M200 Pro). All reagents were dissolved in HPLC grade methanol. Antioxidant capacity reported in mM Trolox™ equivalents (TE) per mL of solution.

2.7. Statistical Analysis

All the data were expressed as mean ± SD after carrying out technical and biological replicate experiments. Tukey's test was used to test for differences at a significance level of $p \leq 0.05$ where appropriate. The Pearson's correlation analysis was employed to assess the relationship between selected parameters and variables. The comprehensive statistical analyses of all obtained results were performed by Hierarchical Cluster Analysis using Ward method. The significance of the impact of pulsed light treatment parameters was evaluated using response surface methodology (RSM) approach. All statistical analyses were performed using Statsoft Inc's Statistica 13 software (Tulsa, OK, Canada).

3. Results

3.1. Impact of PL on the Temperature Increment of Gallic Acid Solutions

This test was carried out to quantify the energy imparted to gallic acid solutions by PL application. The heat energy absorbed by gallic acid solution and thereby temperature increment due to volumetric heating showed a proportionality with the fluence delivered. There was a strong and significant positive correlation between the temperature increment % and fluence of PL treatment as shown in Figure 2 ($r = 0.974$; $p < 0.05$). The temperature increment was the lowest (10.6%) in case of fluence level of 1.07 J/cm^2. Similarly, the increment was highest (65.3%) in case of highest fluence level of 17.2 J/cm^2. The heat absorbed by gallic acid solution after PL application varied from 0.884 to 5.44 J/cm^2.

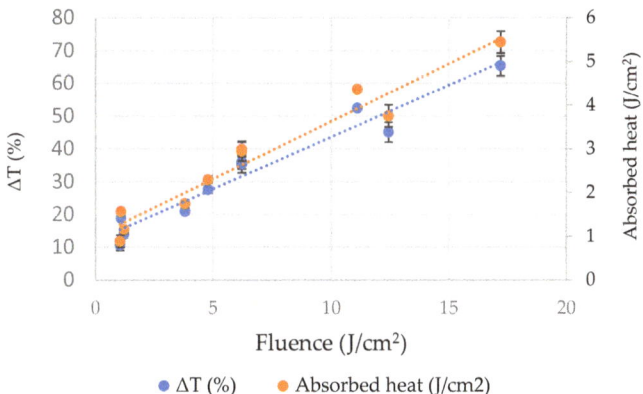

Figure 2. The impact of PL on temperature increment of gallic acid aqueous solution.

The temperature increment during PL treatment can be attributed to energy absorption by samples by virtue of the photothermal effect, which shows an increase in temperature by light absorption. Whenever light interacts with a sample, it decays exponentially as per the Beer-Lamberts law and thus get converted into heat energy in the sample [2]. Similar observations have been made by researchers doing experiments with milk [17], fruit juices [18].

3.2. Colour Measurement of PL Treated Gallic Acid Solution

The L* and b* colour parameters of all PL treated gallic acid solutions were significantly different ($p < 0.05$) from that of untreated sample (Figure 3). In the case of the a* coordinate the vast majority of the PL-treated samples exhibited significantly different values. Only the sample treated by 1.070 J/cm^2 did not differ from the control. More specifically, the L* parameter was equal to 93.49–95.78 and 96.96 in the case of PL treated and untreated solution, respectively. In the case of the a* colour parameter, which represents the share of red and green colour, the changes were smaller. For instance, the untreated gallic acid solution was characterized by a* = 1000 whereas PL application resulted in increment of this coordinate to 1.84 and 1.80 in the case of fluence 11.10 and 17.20 J/cm^2, respectively. The difference between these samples expressed by changes of red/green share was statistically irrelevant ($p > 0.05$). As aforementioned, the b* component of colour of gallic acid solution was significantly affected by PL application. The highest fluence (17.20 J/cm^2) resulted in the biggest change of blue/yellow colour share −7.62. For comparison, the b* of untreated gallic acid solution was equal to 0.54.

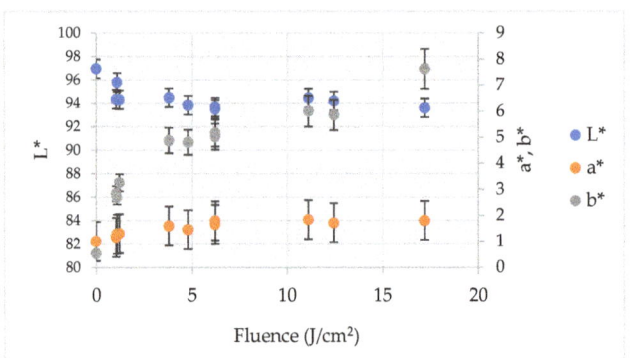

Figure 3. The impact of pulsed light on L*, a* and b* colour parameters of gallic acid aqueous solution.

The Pearson's correlation analysis proved that the relation between a* and fluence had significant and positive character with $r = 0.820$ ($p < 0.05$). In turn, the Pearson's correlation coefficient established for the relation between b* and fluence was even bigger—$r = 0.950$ ($p < 0.05$). To the contrary, no significant linear dependency was found between fluence and L* ($r = −0.456$; $p > 0.05$). The changes of optical parameters clearly demonstrate that PL provoked browning of gallic acid solution. The total colour difference (ΔE) and browning index (BI) depended strongly on fluence as presented in Figure 4. The Pearson's correlation coefficient for the relationship with fluence was equal to $r = 0.967$ and $r = 0.902$, for ΔE and BI, respectively. These changes of colour may be attributed to the photodegradation of gallic acid by the UV light component of PL [19]. The possible mechanism that could be involved in browning may be related to the production of reactive oxygen species (ROS) due to the presence of oxygen in the gallic acid solutions. ROS may oxidize gallic acid and form quinones or semiquinones [20]. These products may undergo further different reactions, i.e., polymerization, and form brown pigments [21]. The temperature increment may also play a role in browning of gallic acid solution, since the correlation between colour changes expressed by BI and ΔE was significant ($p < 0.05$) and positive ($r = 0.926$ and $r = 0.953$, respectively).

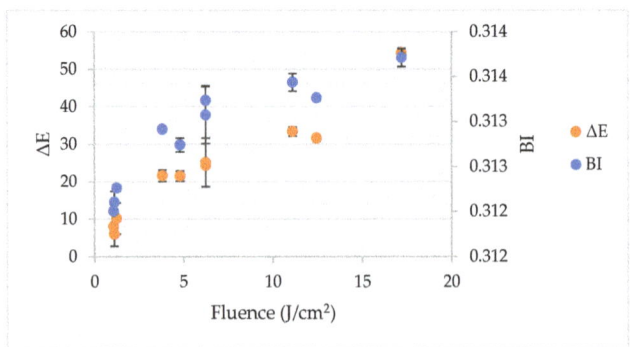

Figure 4. The impact of pulsed light on total colour difference (ΔE) and browning index (BI) of gallic acid aqueous solution.

3.3. Impact of PL on Total Polyphenol Content and Antioxidant Activity Expressed by DPPH Assay

Table 2 presents the gallic acid content (GAC), total phenolics content (TPC) and antioxidant activity of all investigated gallic acid solution samples. Almost all samples treated by PL exhibited smaller concentrations of gallic acid as measured by the HPLC method. However, the difference was significant only for samples that were treated by 3.82 J/cm^2. In this case the concentration of gallic acid was equal to 43.15 mg/100 mL, which means that it was 13.7% lower in comparison to untreated solution. Interestingly, the delivery of higher fluence did not necessarily cause higher degradation of gallic acid as measured by the HPLC method against TPC. It could indicate that there exist some optimal parameters for gallic acid decomposition by light.

Table 2. Total phenolics content (TPC) and antioxidant activity measured by DPPH assay due to PL treatment.

Run	Fluence (J/cm^2)	TPC (mg GAE/100 mL) *	GAC (mg /100 mL) *	Antioxidant Activity (mM TEAC/mL) *
Untreated	0.0	50.6 ± 1.07 [a]	50.00 ± 0.58 [ab]	10.29 ± 0.88 [a]
1	1.07	47.23 ± 0.33 [b]	50.64 ± 2.32 [b]	9.55 ± 1.01 [ac]
2	3.82	46.63 ± 0.57 [b]	43.15 ± 0.78 [c]	9.4 ± 0.5 [abc]
3	4.80	45.99 ± 1.19 [b]	49.67 ± 0.88 [ab]	9.26 ± 0.01 [bc]
4	17.20	41.27 ± 0.37 [c]	46.65 ± 0.74 [ac]	8.22 ± 0.76 [b]
5	1.24	46.8 ± 1.32 [b]	48.02 ± 0.87 [ab]	9.29 ± 0.17 [abc]
6	12.44	44.82 ± 0.37 [b]	47.90 ± 0.85 [ab]	9.31 ± 0.22 [abc]
7	1.11	45.77 ± 0.97 [b]	49.83 ± 0.86 [ab]	9.75 ± 0.35 [a]
8	11.11	45.61 ± 1.11 [b]	47.44 ± 0.86 [ab]	9.27 ± 0.92 [bc]
9 (C)	6.22	45.99 ± 0.59 [b]	48.02 ± 0.85 [ab]	9.96 ± 0.21 [a]
10 (C)	6.22	46.47 ± 0.56 [b]	47.89 ± 0.86 [ab]	9.92 ± 0.09 [a]

* Values expressed as mean ± SD followed by letters a–c wherein, means followed by same letter are not significantly different ($p > 0.05$).

These results are very interesting especially when compared to TPC results since they indicate that even if the gallic acid content was not changed, the TPC was altered. The TPC of control sample was 50.6 mg GAE/100 mL. All samples subjected to PL treatment were characterized by significantly ($p < 0.05$) smaller TPC which ranged between 41.27 and 47.23 mg GAE/100 mL. The biggest decline of TPC, equal to 18.4% in comparison to the reference sample, was found for the trial with the highest delivered fluence (17.2 J/cm^2).

The smallest applied fluence (1.07 J/cm^2) also contributed to a significant decrease of TPC by 6.6% when compared to the control solution. It means that the reactions gallic acid can undergo are very photosensitive. Even though there was no degradation of gallic acid under this condition some compounds that could react with FC reagent were formed. However, within the fluence range of 1.07–12.44 J/cm^2 the TPC remained the same from statistical point of view. It means that despite high photosensitivity of gallic acid, there is a wide spectrum of fluence that cause similar degradation level. Such information is extremely important when designing PL-based technology of processing phenolic-abundant products. It is worth emphasizing that the degradation of gallic acid maintained a negative linear relation with fluence. The Pearson's correlation coefficient for this dependency was $r = -0.856$ ($p < 0.05$). What's more, the relationship between TPC and b*, ΔE and BI was significant as well ($p < 0.05$) and correlation coefficients were equal to −0.771; −0.851 and −0.654, respectively. Indeed, it indicates that products of degradation of gallic acid contribute directly to browning of the investigated solution [4]. The photoinduced decomposition of gallic acid was also reported by Benitez et al. [5].

Further, it must be mentioned that we observed the measured TPC was slightly lower than the measured gallic acid content for all samples. Ideally, as the model gallic acid solution did not had any other component, the TPC should be equal to the gallic acid content. It has been previously reported [22] that radiation by UV, which is also a constituent of pulsed light, may create reactive oxygen species by photooxidation of oxygen which is dissolved in water. These intermediates can form other reactive molecules like hydroxyl radicals and hydrogen peroxide, which included with the products of photodecomposition of gallic acid, have been reported to interfere with the Folin–Ciocalteu reagent. For instance, hydrogen peroxide can lead to lower TPC values if present in the system together with gallic acid, as found by Rangel et al. [23].

In the current study, almost all samples exhibited antioxidant activity similar to control solution. The decrease of free radical scavenging activity was found only for sample treated by the highest fluence – 17.2 J/cm^2. Comparing these results with gallic acid concentration it can be stated that within the fluence range of 1.07 to 12.4 J/cm^2 the products of gallic acid oxidation exhibited some antioxidant potential. Similar results were reported by Oms-Oliu et al. [24] for mushrooms subjected to PL. The authors have found that depending on the fluence the antioxidant activity can be decreased or increased which is related with decomposition of phenolics and polymerization reaction of the quinones. Also, Llano et al. [25] reported that application of PL with fluence 8–16 J/cm^2 helps to maintain the antioxidant activity of fresh-cut apples.

3.4. Response Surface Methodology (RSM) and Cluster Analysis

RSM analysis was applied to estimate the values of investigated variables of gallic acid solution depending on frequency (x) and time (y) of PL treatment. Equations (6–11) depict the response model equation, wherein, a significant model was obtained for all responses, with a high coefficient of regression R^2 (>0.9) for temperature increment, color change and browning index responses. The gallic acid content model showed lowest R^2 (0.74), whereas R^2 of total phenolic content and antioxidant activity were within acceptable range (0.8–0.9). The results of the analysis of variance (ANOVA) for the response surface models are given in Table 3. The model itself was found significant for all responses ($p < 0.05$) and was adequate for navigating the design space. The plots of RSM have been given in Figure 5:

$$\Delta T\ (\%) = -1.208 + 5.443x - 0.449x^2 + 0.143y - 0.0006y^2 + 0.1245xy,\ R^2 = 0.98 \quad (6)$$

$$\Delta E = -1.279 + 1.342x - 0.082x^2 + 0.398y - 0.0034y^2 + 0.089xy,\ R^2 = 0.95 \quad (7)$$

$$BI = 0.311 + 0.0003x - 0.00002x^2 + 0.0001y - 0.000001y^2 + 0.0000003xy,\ R^2 = 0.99 \quad (8)$$

$$TPC\ (mg\ GAE/100\ ml) = 44.474 + 0.466x - 0.240x^2 + 0.145y - 0.0015y^2 - 0.0202\ xy,\ R^2 = 0.82 \quad (9)$$

$$GAC\ (mg/100\ mL) = 52.742 - 0.108x - 0.027x^2 - 0.227y - 0.0004y^2 + 0.018xy,\ R^2 = 0.74 \quad (10)$$

$$AA\ (mM\ TEAC/mL) = 8.096 + 0.382x - 0.031x^2 + 0.095y - 0.0019y^2 - 0.004xy,\ R^2 = 0.88 \quad (11)$$

Based on the ANOVA results in Table 3, it could be concluded that individual interactions of both PL frequency (x) and PL treatment time (y) significantly affected ($p < 0.005$) temperature change, total color change and browning index. Based on the signs of the corresponding coefficient in Equations (6)–(8), it could be concluded that higher PL frequency and higher PL time led to higher values of temperature change, total color change and browning index. Total phenolic content was significantly affected by only PL frequency ($0.005 < p < 0.05$) and not PL treatment time ($p > 0.05$), whereas gallic acid content and antioxidant activity were significantly affected by only PL treatment time ($0.005 < p < 0.05$) and not PL frequency ($p > 0.05$).

The effect of the mutual interaction of PL frequency and PL treatment time (x*y), which describes the PL fluence that governs the microbial lethality level during the PL process, was only significant ($0.005 < p < 0.05$) for temperature change, and did not affect other responses. Based on this, it could be inferred that even a process requiring high fluence could be optimized to minimize quality loss (at same levels of microbial lethality) by understanding the relative importance of PL frequency and time on the response. The effect of the quadratic interaction of PL frequency (x^2) was only significant ($0.005 < p < 0.05$) for temperature change and browning index, while that of PL treatment time (y^2) was only significant ($0.005 < p < 0.05$) for antioxidant activity and browning index. This is often indicative of the curvilinear relationship between the parameters.

Table 3. Analysis of Variance (ANOVA) table for the influence of PL frequency and time on the physicochemical properties of gallic acid solution.

Parameter	Temperature Change (% ΔT)			Total Color Change (ΔE)			Browning Index (BI)		
	Type III SS	F Value	Pr > F	Type III SS	F Value	Pr > F	Type III SS	F Value	Pr > F
Model	2.82×10^3	-	3.23×10^{-5} ***	1.85×10^3	-	1.70×10^{-3} **	3.09×10^{-6}	-	8.70×10^{-15} ***
Frequency (x)	1.38×10^3	2.17×10^2	1.24×10^{-4} ***	7.48×10^2	3.33×10	4.47×10^{-3} **	1.10×10^{-6}	1.51×10^2	2.53×10^{-4} ***
x^2	1.01×10^2	1.59×10	1.63×10^{-2} *	3.35	1.49×10^{-1}	7.19×10^{-1}	1.92×10^{-7}	2.64×10	6.79×10^{-3} *
Time (y)	1.22×10^3	1.92×10^2	1.57×10^{-4} ***	9.53×10^2	4.24×10	2.87×10^{-3} **	1.76×10^{-6}	2.42×10^2	9.96×10^{-5} ***
y^2	1.25×10^{-1}	1.96×10^{-2}	8.96×10^{-1}	3.51	1.56×10^{-1}	7.13×10^{-1}	1.22×10^{-7}	1.67×10	1.50×10^{-2} *
x*y	1.85×10^2	2.89×10	5.77×10^{-3} *	9.43×10	4.20	1.10×10^{-1}	8.75×10^{-10}	1.20×10^{-1}	7.46×10^{-1}

Parameter	Total Phenolic Content (TPC)			Gallic Acid Content (GAC)			Antioxidant Activity (AA)		
	Type III SS	F Value	Pr > F	Type III SS	F Value	Pr > F	Type III SS	F Value	Pr > F
Model	2.56×10	-	4.39×10^{-7} ***	3.87×10	-	1.80×10^{-6} ***	3.31	-	1.51×10^{-6} ***
Frequency (x)	1.13×10	1.00×10	3.39×10^{-2} *	6.55×10^{-1}	2.65×10^{-1}	6.34×10^{-1}	2.75×10^{-1}	2.84	1.67×10^{-1}
x^2	2.90×10^{-1}	2.58×10^{-1}	6.38×10^{-1}	3.64×10^{-1}	1.47×10^{-1}	7.21×10^{-1}	4.68×10^{-1}	4.84	9.26×10^{-2}
Time (y)	4.62	4.11	1.12×10^{-1}	2.16×10	8.72	4.19×10^{-2} *	1.44	1.49×10	1.82×10^{-2} *
y^2	6.66×10^{-1}	5.94×10^{-1}	4.84×10^{-1}	5.54×10^{-2}	2.24×10^{-2}	8.88×10^{-1}	1.02	1.06×10	3.13×10^{-2} *
x*y	4.88	4.35	1.05×10^{-1}	3.77	1.52	2.85×10^{-1}	1.62×10^{-1}	1.67	2.65×10^{-1}

* $p < 0.05$; ** $p < 0.005$; *** $p < 0.0005$.

Based on the total sum of squares in Table 3 for significant interactions, effect of PL treatment time dominated the value of all responses except temperature increment and total phenolic content as compared to PL frequency. Thus, PL treatment time must be minimized preferably over PL treatment time for reducing color change, gallic acid deterioration, browning index and antioxidant activity.

Cluster analysis (Figure 6) allowed the samples to be distinguished into two big aggregates depending on the fluence. Cluster AI was formed by the reference gallic acid solution and samples treated by rather small fluences <3.82 J/cm². In turn, all samples that were treated by higher energies were gathered in Cluster BI. Within the clusters, the samples also exhibit some dissimilarity—in Cluster AI, the reference sample formed separate, one item aggregate while in the Cluster BI, similar behavior was observed for sample treated by the highest fluence. Based on that data it can be estimated that when it comes to treatment of polyphenols rich liquid food, there is a threshold value of fluence which does not lead to relevant quality changes. Results obtained for gallic acid model solution indicate that treatment below 3.82 J/cm² maintains the quality (expressed by the investigated physicochemical properties) almost unchanged from the statistical point of view whereas delivery of higher than 3.82 J/cm² values of fluence lead to relevant modification of the quality. A similar optimization approach of imposing limits on reciprocation time and frequency was proposed in our previous work [26] for improving the quality of reciprocating agitation thermal processing, which suggests the parallel between optimization approaches for both thermal and non-thermal processing technologies. What more, the application of fluence higher than 11.1 J/cm² results in severe modification of the quality giving the product with significantly different properties than solutions treated by other parameters. However, it is worth emphasizing that obtained results are valid for model solution and should be rather considered as leads when it comes to the design and analysis of PL treatment of real food systems.

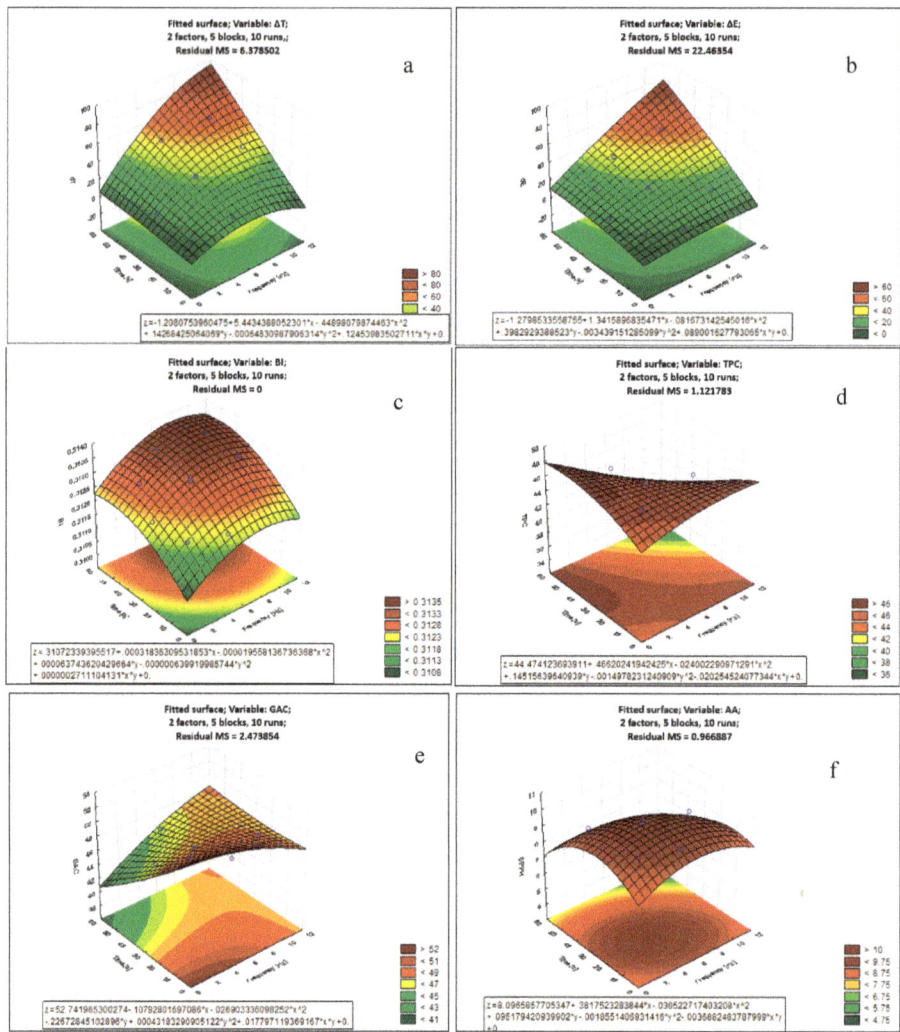

Figure 5. RSM plots showing the impact of PL on the properties of gallic acid aqueous solution. from top to bottom: (**a**) temperature change, ΔT; (**b**) total colour difference, ΔE; (**c**) browning index, BI; (**d**) Total phenolic content, TPC; (**e**) Gallic acid content, GAC; (**f**) antioxidant capacity.

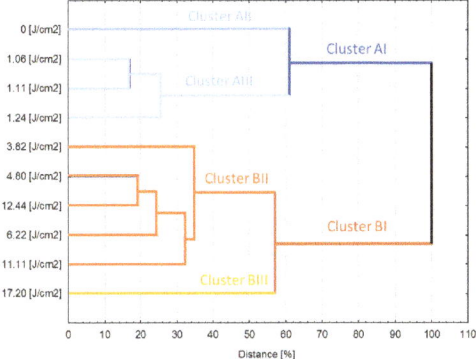

Figure 6. Results of the cluster analysis performed based on of all investigated variables.

4. Conclusions

Pulsed light (PL) processing happens to be a promising non-thermal technology for food preservation and processing. It has been shown to have great potential for decontaminating food products by destroying pathogenic microorganisms. However, the effect of PL on food bioactive and nutritional compounds is unclear. The aim of this study was to understand how the PL treatment affects the gallic acid aqueous solution, which is taken as a model system of phenolic-abundant liquid food matrices. Several parameters of the gallic acid solution were tested by a design of experiments approach for the effect of PL processing-colour, total phenolic concentration, antioxidant activity using DPPH free radical assay. It was found that PL can modify the optical properties of gallic acid and cause reactions and degradation of gallic acid. The absorbed light energy resulted in a proportional temperature increase. The L* and b* color values decreased significantly, with an increase in the browning index on PL treatment. The gallic acid content and the total polyphenol content changed initially, but remained constant after a critical fluence was reached, whereas the antioxidant activity did not vary significantly, except at the highest fluence level tested (17.20 J/cm^2). This suggests that despite the photosensitivity of the gallic acid, the degradation products still had similar antioxidant potential even at relatively high fluences. Based on a cluster analysis, a critical fluence level (<3.82 J/cm^2) was suggested, below which PL shall have minimal effect on the overall quality of model gallic acid solution. With less than 10% degradation of gallic acid and less than 20% degradation of the antioxidant activity, it can be said that the PL technology has excellent ability to process liquid foods rich in polyphenols and antioxidants.

Author Contributions: Conceptualization, A.W. and A.P.S.; methodology, A.W., A.P.S., A.S.; formal analysis, A.W.; investigation, A.W., R.M., A.P.S., A.S.; writing—original draft preparation, R.M.; writing—review and editing, A.W.; visualization, A.W., R.M.; supervision, A.P.S.

Funding: This research was funded by the National Science and Engineering Research Council of Canada (NSERC) Discovery Grant Number RGPIN-2018-04735 and NSERC Collaborative Research and Development Grant Number CRDPJ-522364-17 in collaboration with Solaris Disinfection Inc., Mississauga, ON, Canada.

Acknowledgments: Artur Wiktor would like to acknowledge the financial support of the Dekaban Foundation.

Conflicts of Interest: The authors declare no conflict of interest. The funders had no role in the design of the study; in the collection, analyses, or interpretation of data; in the writing of the manuscript, or in the decision to publish the results.

References

1. Rahman, M.S. *Handbook of food preservation*, 2nd ed.; CRC Press: Boca Raton, FL, USA, 2007. [CrossRef]
2. Palmieri, L.; Cacace, D. High intensity pulsed light technology. In *Emerging Technologies for Food Processing*; Academic Press: Amsterdam, The Netherlands, 2005; pp. 279–306. [CrossRef]
3. Bernal, J.; Mendiola, J.A.; Ibáñez, E.; Cifuentes, A. Advanced analysis of nutraceuticals. *J. Pharm. Biomed. Anal.* **2011**, *55*, 758–774. [CrossRef]
4. Li, H.; Guo, A.; Wang, H. Mechanisms of oxidative browning of wine. *Food Chem.* **2008**, *108*, 1–13. [CrossRef]
5. Benitez, F.J.; Real, F.J.; Acero, J.L.; Leal, A.I.; Garcia, C. Gallic acid degradation in aqueous solutions by UV/H_2O_2 treatment, Fenton's reagent and the photo-Fenton system. *J. Hazard Matter.* **2005**, *126*, 31–39. [CrossRef]
6. Shamsudin, R.; Noranizan, M.A.; Yap, P.Y.; Mansor, A. Effect of repetitive ultraviolet irradiation on the physico-chemical properties and microbial stability of pineapple juice. *Innov. Food Sci. Emerg. Technol.* **2014**, *10*, 166–171. [CrossRef]
7. Shah, N.N.A.K.; Rahman, R.A.; Shamsudin, R.; Adzahan, N.M. Furan development in Dean Vortex UV-CC treated pummelo (*Citrus Grandis* L. Osbeck) fruit juice. In Proceedings of the International Conference on Sustainable Agriculture for Food, Energy and Industry in Regional and Global Context, Serdang, Malaysia, 25–27 August 2015.
8. Caminiti, I.M.; Palgan, I.; Muñoz, A.; Noci, F.; Whyte, P.; Morgan, D.J.; Lyng, J.G. The effect of ultraviolet light on microbial inactivation and quality attributes of apple juice. *Food Bioproc. Technol.* **2012**, *5*, 680–686. [CrossRef]
9. Feng, M.; Ghafoor, K.; Seo, B.; Yang, K.; Park, J. Effects of ultraviolet-C treatment in Teflon coil on microbial populations and physic-chemical characteristics of watermelon juice. *Innov. Food Sci. Emerg. Technol.* **2013**, *9*, 133–139. [CrossRef]
10. Charles, F.; Vidal, V.; Olive, F.; Filgueiras, H.; Sallanon, H. Pulsed light treatment as new method to maintain physical and nutritional quality of fresh-cut mangoes. *Innov. Food Sci. Emerg. Technol.* **2013**, *18*, 190–195. [CrossRef]
11. Pollock, A.M.; Pratap Singh, A.; Ramaswamy, H.S.; Ngadi, M.O. Pulsed light destruction kinetics of *L. monocytogenes*. *LWT* **2017**, *84*, 114–121. [CrossRef]
12. Pataro, G.; Muñoz, A.; Palgan, I.; Noci, F.; Ferrari, G.; Lyng, J.G. Bacterial inactivation in fruit juices using a continuous flow pulsed light (PL) system. *Food Res. Int.* **2011**, *44*, 1642–1648. [CrossRef]
13. Singh, R.P.; Heldman, D.R. *Introduction to Food Engineering*, 5th ed.; Academic Press: San Diego, CA, USA, 2014. [CrossRef]
14. Kasim, R.; Kasin, M.U. Biochemical changes and color properties of fresh-cut green bean (*Phaseolus vulgaris* L. cv. *gina*) treated with calcium chloride during storage. *Food Sci. Technol.* **2015**, *35*, 266–272.
15. Vega-Gálvez, A.; Di Scala, K.; Rodríguez, K.; Lemus-Mondaca, R.; Miranda, M.; López, J.; Perez-Won, M. Effect of air-drying temperature on physico-chemical properties, antioxidant capacity, colour and total phenolic content of red pepper (*Capsicum annuum*, L. var. *Hungarian*). *Food Chem.* **2009**, *117*, 647–653. [CrossRef]
16. Sawant, L.; Prabhakar, B.; Pandita, N. Quantitative HPLC analysis of ascorbic acid and gallic acid in *Phyllanthus emblica*. *J. Anal. Bioanal. Tech.* **2010**, *1*, 2. [CrossRef]
17. Innocente, N.; Segat, A.; Manzocco, L.; Marino, M.; Maifreni, M.; Bortolomeoli, I.; Ignat, A.; Nicoli, M.C. Effect of pulsed light on total microbial count and alkaline phosphatase activity of raw milk. *Int. Dairy J.* **2014**, *39*, 108–112. [CrossRef]
18. Ferrario, M.; Alzamora, S.M.; Guerrero, S. Inactivation kinetics of some microorganisms in apple, melon, orange and strawberry juices by high intensity light pulses. *J. Food. Eng.* **2013**, *118*, 302–311. [CrossRef]
19. Du, Y.; Chen, H.; Zhang, Y.; Chang, Y. Photodegradation of gallic acid under UV irradiation: Insights regarding the pH effect on direct photolysis and the ROS oxidation-sensitized process of DOM. *Chemosphere* **2014**, *99*, 254–260. [CrossRef]
20. Eslami, A.C.; Pasanphan, W.; Wagner, B.A.; Buettner, G.R. Free radicals produced by the oxidation of gallic acid: An electron paramagnetic resonance study. *Chem. Cent. J.* **2010**, *4*, 15. [CrossRef] [PubMed]
21. Altunkaya, A.; Gökmen, V. Effect of various inhibitors on enzymatic browning, antioxidant activity and total phenol content of fresh lettuce (*Lactuca sativa*). *Food Chem.* **2008**, *107*, 1173–1179. [CrossRef]

22. Song, K.; Mohseni, M.; Taghipour, F. Application of ultraviolet light-emitting diodes (UV-LEDs) for water disinfection: A review. *Water Res.* **2016**, *94*, 341–349. [CrossRef] [PubMed]
23. Sánchez-Rangel, J.C.; Benavides, J.; Heredia, J.B.; Cisneros-Zevallos, L.; Jacobo-Velázquez, D.A. The Folin–Ciocalteu assay revisited: improvement of its specificity for total phenolic content determination. *Anal. Methods* **2013**, *5*, 5990–5999.
24. Oms-Oliu, G.; Aguiló-Aguayo, I.; Martín-Belloso, O.; Soliva-Fortuny, R. Effects of pulsed light treatments on quality and antioxidant properties of fresh-cut mushrooms (*Agaricus bisporus*). *Postharvest Biol. Technol.* **2010**, *56*, 216–222. [CrossRef]
25. Llano, K.R.A.; Marsellés-Fontanet, A.R.; Martín-Belloso, O.; Soliva-Fortuny, R. Impact of pulsed light treatments on antioxidant characteristics and quality attributes of fresh-cut apples. *Innov. Food Sci. Emerg. Technol.* **2016**, *33*, 206–215. [CrossRef]
26. Singh, A.; Pratap Singh, A.; Ramaswamy, H.S. A controlled agitation process for improving quality of canned green beans during agitation thermal processing. *J. Food Sci.* **2016**, *81*, E1399–E1411. [CrossRef] [PubMed]

© 2019 by the authors. Licensee MDPI, Basel, Switzerland. This article is an open access article distributed under the terms and conditions of the Creative Commons Attribution (CC BY) license (http://creativecommons.org/licenses/by/4.0/).

Article

Analysis of Apple Candying by Microwave Spectroscopy

Juan A. Tomas-Egea, Pedro J. Fito and Marta Castro-Giraldez *

Instituto Universitario de Ingeniería de Alimentos para el Desarrollo, Universitat Politècnica de València, Camino de Vera s/n, 46022 Valencia, Spain
* Correspondence: marcasgi@upv.es

Received: 28 June 2019; Accepted: 2 August 2019; Published: 4 August 2019

Abstract: Process control in the industry requires fast, safe and easily applicable methods. In this sense, the use of dielectric spectroscopy in the microwave range can be a great opportunity to monitor processes in which the mobility and quantity of water is the main property to produce a quality and safety product. The candying of fruits is an operation in which the samples are first osmotically dehydrated and then exposed to a hot air-drying operation. This process produces changes in both the structure of the tissue and the relationships between water, the solid matrix and the added soluble solids. The aim of this paper is to develop a dielectric tool to predict the water/sucrose states throughout the candying of apple, by considering the complexity of the tissue and describing the different transport phenomena and the different transition processes of the sucrose inside the sample.

Keywords: dielectric spectroscopy; permittivity; dehydration; candying; hot air drying; isotherms; sucrose

1. Introduction

Dehydration is probably one of the most important methods of foods preservation. The main objective in dehydrating agricultural products is the removal of water in the foods up to a certain level of water activity, at which microbial spoilage and deterioration chemical reactions are minimized [1]. Apples are consumed either fresh or in the form of various processed products such as juice, jam, marmalade and dried products [2], being the dried apples in extensive demand. Candying is an industrial operation, which consists on osmotic dehydration (OD) followed by a hot-air drying (HAD) treatment. OD consists of the immersion of foods in hypertonic solutions with the objective of producing a water flow out of the food and a simultaneous flow of solutes inside the tissue. In fruits, these mass transfer phenomena affect the apple structure. Fruits are formed by vegetal cells, conforming the parenchyma tissue, which is a complex structure with intercell connections (plasmodesmata), intracellular and extracellular spaces [3–5]. Intracellular volume is mainly occupied by vacuoles, which are fundamentally a water solution with multiple solutes. The membrane of the cell is named protoplast, it is selectively permeable (active and passive protein channels) and controls turgor and the cell growth [6]. The cell membrane has protein channels named aquaporin and calcium protein channel that are responsible of the water transport and the cell homeostasis [7]. The cell wall provides mechanical resistance to the cell. In the union between cell walls and protoplast are bonds of Na^+, protein microtubules and conduits named plasmodesmata that allow the transport between the adjacent cells (symplastic pathways). Extracellular volume comprehends the cell wall and the spaces between cells [3]. OD produces different phenomena in the cellular tissue, first dehydration with shrinkage and solute intake in extracellular space. Continuous shrinkage produces a breakdown between wall and protoplast called plasmolysis causing mechanical driving forces (swelling/shrinkage) [8–11]. The changes in cellular tissue affect the water mobility and its distribution [12].

HAD influences on fruit quality because it produces chemical, physical and biological changes in food [13]. Moreover, the internal structure undergoes deformation and could be locally damaged. Removal of water adds rigidity to the external layers and simultaneously builds up moisture gradients, which create shrinkage stresses [14]. The understanding of these effects is still limited, and more studies are necessary in this field in order to improve food quality and nutritional characteristics of dehydrated foods.

During the candying process, crystallization of sucrose occurs. Sucrose crystallizes in supersaturated aqueous solutions. The sucrose molecule has eight hydroxyl groups, which can be involved in hydrogen bond formation. In sufficiently diluted aqueous solutions, all the hydroxyl groups form hydrogen bonds with water molecules. If the concentration increases, the molecules start to interact forming an intramolecular bond and then two intermolecular bonds [15]. If sucrose concentration in the solution increases, aggregation phenomena occur between sucrose molecules, leading to stable three-dimensional nucleus [16]. After nucleus formation, the crystal growth consists in the incorporation of sucrose molecules to the crystal lattice, which requires the migration of hydration water from the crystal surface to the solution [17].

The analysis of the electromagnetic field (EMF) properties in range of microwaves could be a good tool to monitor and improve apple candying. The development of sensors to determine the dielectric properties of biological tissues, in range of radiofrequency and microwaves, has been demonstrated to be a useful tool for monitoring the quality of many products: Poultry [18–22], pork [23–27], beef [28], goat [29], meat products [30], cheese [31], mandarin [32], potato [33], wheat [34], agricultural products [35–37], pomegranate [38] and apple [39–41], and for monitoring processes of pork meat HAD [42] and salting [43], orange peel drying [44], cheese salting [45], brewing [46], puffing of amaranth seeds [47], OD of kiwi [48] and apple [49].

The EMF is a flux of photons [50] and the interaction with matter can be modeled by Schrodinger's equation [51] attending to the quantum theory. However, at the macroscopic level, it is possible to apply the Maxwell's equations [52], where the physical property that describes the electric effect is the complex permittivity and for the magnetic effect is the complex permeability [50]. In the microwave range ($1.24 \cdot 10^{-6}$ to $1.24 \cdot 10^{-3}$ eV or 300 MHz to 300 GHz), these interactions are described by γ-dispersion and ionic conductivity. γ-dispersion is caused by the induction and orientation of dipolar molecules, being water the most important in biological systems [53]. These phenomena generate electric energy accumulation caused mainly by water spin reorientation and it is represented by the real part of the permittivity (ε'). On the other hand, a part of the electrical energy of photons is transformed in other energies (mechanical or calorific) due to the collisions or frictions associated to the increase of molecular mobility; this part of the electric energy is called the loss factor (ε''). In the microwave range, the vibration of chemical species with very high ionic strength causes a part of the losses of electrical energy; this is called ionic conductivity (σ) [54].

The aim of this paper is to develop a dielectric tool to predict the water/sucrose states throughout the candying of apple, by considering the complexity of the tissue and describing the different transport phenomena and the different transition processes of the sucrose inside the sample.

2. Materials and Methods

Apples (var. Granny Smith) were bought from a local supermarket and kept refrigerated until use. The apples were cut with a caliper and a cork borer in cylinders (1 cm thickness, 2 cm diameter) from the parenchymatic tissue. There were prepared 126 samples in order to obtain 7 isotherms (Figure 1). Eighteen samples were used to obtain the isotherm of raw apple (I_0): Three samples to characterize the raw material, it is without hot air dehydration, and 15 exposed to hot air dehydration. The remaining 108 samples were dehydrated osmotically to obtain six isotherms (I_1 to I_6); 18 samples were used for each selected time of OD: Three samples to characterize osmotic dehydrated samples, it is without hot air dehydration, and 15 exposed to hot air dehydration. In conclusion, seven isotherms were obtained, considering that the samples of an isotherm have the same solid matrix/sucrose weight relation.

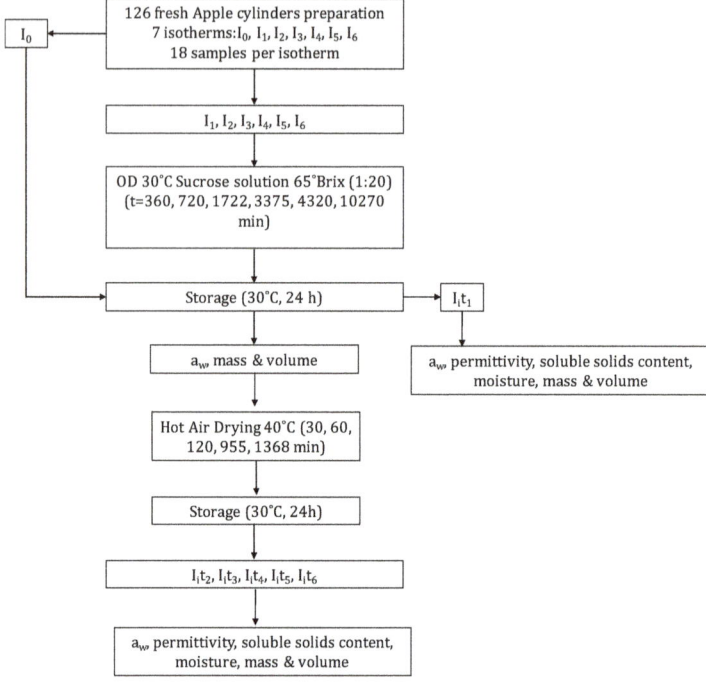

Figure 1. Flow diagram of the experimental procedure, where I represents each isotherm, subindices i from 0 to 6 of each isotherm represent a concrete sucrose/solid matrix weight relation, t_1 to t_6 represent the different times of hot-air drying (HAD).

Sucrose solution (65% w/w, 30 °C), prepared with commercial sugar and distilled water, was used as an osmotic agent. The relation between the fruit and the osmotic solution was of 1:20 (w/w) to avoid changes in the solution during the process. The system was maintained at 30 °C in a constant-temperature chamber. To prevent evaporation the vessel was covered with a sheet of plastic wrap. Preliminary kinetic studies were done at the same working conditions in order to select the OD treatment times. The OD treatment times in the preliminary studies were: 180, 360, 720, 1463, 1577, 1722, 3375, 4320, 7200, 8640, 10270, 14590 and 23230 min. Based on the results [54], OD treatment times were selected for this research: 0, 360, 720, 1722, 3375, 4320 and 10270 min. After the treatment, the samples were removed from the solution and blotted with a paper to remove the superficial osmotic solution. Then, the samples were kept at 30 °C for 24 h, on AquaLab disposable sample containers, closed with parafilm®. The mass, volume and water activity of the 126 samples were measured after the repose. Moreover, permittivity, moisture and soluble solids content (°Brix) of three samples of each OD treatment time were measured to characterize each sucrose/solid matrix weight relation (x_S/x_{SM}), considering as the solid matrix mass is neither water nor solutes. The remaining samples in each isotherm were hot air dried (times of HAD treatment: 30, 60, 120, 955 and 1368 min). Three samples were used in each HAD treatment time. The drying experiments were carried out at 40 °C drying air temperature. The convective dryer was designed and built in the Food Technology Department of Universitat Politècnica de València, has a control unit for setting the velocity and temperature of air, which is heated through electrical resistances. Air velocity was kept at a constant value of 1.5 ± 0.2 m s^{-1} in all experiments.

After the drying treatment, samples were maintained at 30 °C for 24 h, on AquaLab disposable sample containers, closed with parafilm®®. After this repose time, the permittivity, mass, volume, water activity, moisture and soluble solids content were measured.

Volume measurements were analyzed by image analysis and the software Adobe Photoshop®® CS5 (Adobe Systems Inc., San Jose, CA, USA) to get the diameter and the thickness of the samples. The images of the samples were obtained with a digital camera (Canon EOS 550D, with a size of 2592 × 1728 pixels and a resolution of 16 pixel/mm).

Mass was determined by using a Mettler Toledo Balance (±0.0001 g; Mettler-Toledo, Inc., Columbus, OH, USA). Measurements were done in structured samples (not minced), thus the obtained a_w is considered to be the surface a_w [54].

Water activity was measured in the structured samples with a dew point hygrometer Decagon (Aqualab®® series 3TE) with precision ±0.003.

Moisture content in the apple cylinders was determined gravimetrically at 60 °C in a vacuum oven until constant weight was reached [55]. Sugar content was determined in a refractometer (ABBE, ATAGO Model 3-T, Tokyo, Japan).

The system used to measure permittivity consists of an Agilent 85070E open-ended coaxial probe (Agilent, Santa Clara, CA, USA) connected to an Agilent E8362B Vector Network Analyzer (Agilent, Santa Clara, CA, USA). The system was calibrated using three different types of loads: Open (air), short-circuit and 30 °C Milli-Q water. Once the calibration was carried out, 30 °C Milli-Q water was measured again to check calibration suitability. Permittivity was measured from 500 MHz to 20 GHz. The measurements were performed in triplicate. Dielectric constant (ε') was modeled adjusting the experimental data using Traffano-Schiffo model [20] (Equation (1)) in order to obtain information of γ-dispersion:

$$log\varepsilon'(\omega) = log\varepsilon'_\infty + \sum_{n=1}^{3} \frac{\Delta log\varepsilon'_n}{1 + e^{(log\omega^2 - log\tau_n^2) \cdot a_n}}, \quad (1)$$

where n represents α, β or γ dispersion, $log\varepsilon'$ represents the decimal logarithm of the dielectric constant, $log\varepsilon'_\infty$ the logarithm of the dielectric constant at high frequencies, $log\omega$ represents the decimal logarithm of the angular velocity (obtained from the frequency), $\Delta log\varepsilon'_n$ ($\Delta l\varepsilon'_n = log\varepsilon'_n - log\varepsilon'_{n-1}$) the amplitude of the n dispersion, $log\tau_n$ the logarithm of the angular velocity at relaxation time for each n dispersion and a_n are the dispersion slopes. In this work, this model was applied for γ-dispersion only.

3. Results

In order to understand the mechanisms that govern the relationship between water and sucrose in plant tissue matrix during OD and HAD treatments, and thus develop dielectric predictive tools that not only explain the water state but also explain the state of the whole internal liquid phase of the vegetal tissue, a kinetic analysis of the variation of overall mass, water and sucrose was proposed as a first step. Overall mass, water and sucrose mass variations throughout the OD treatment are shown in Figure 2. These parameters were estimated by Equations (2)–(4), respectively.

$$\Delta M = \frac{M_t - M_0}{M_0}, \quad (2)$$

$$\Delta M_w = \frac{M_t x_{wt} - M_0 x_{w0}}{M_0}, \quad (3)$$

$$\Delta M_s = \frac{M_t x_{st} - M_0 x_{s0}}{M_0} \quad (4)$$

where M represents the mass (kg), x_i is the mass fraction of the compound i (kg$_i$/kg$_T$), being the different compounds represented by subscripts: w the water, and s the soluble solids; moreover, the subscripts t represent the treatment time, being 0 the initial value.

In Figure 2, it can be observed that the greatest loss of mass occurs during the first 1722 min of the OD treatment. From that moment, there was no variation in the total mass of the sample or in the mass of water. In contrast, the increase in soluble solids occurred during the first 720 min, being this content stabilized from this point.

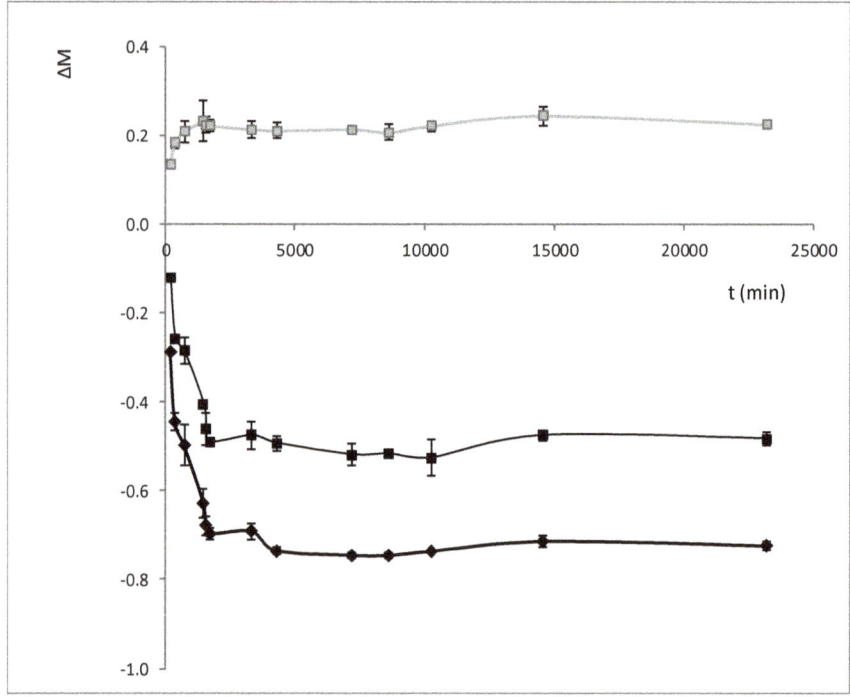

Figure 2. Evolution of overall mass (■), water mass (♦) and sucrose mass (■) through the osmotic treatment.

The mass variation evolution does not allow us to observe correctly the chemical equilibrium that happens when the plasmolysis occurs [54]. The plasmolysis or breakdown the bonds between the protoplast and the wall precedes tissue shrinkage and swelling [56], and these phenomena change the amount of liquid phase but not its composition. For this reason, the relationship between the sucrose content and the solid matrix with respect to the treatment time is shown in Figure 3.

In Figure 3, it is possible to observe how the sucrose/solid matrix weight relation reaches a maximum (cellular plasmolysis) at 1722 min. Seven different osmodehydration treatment times were selected which correspond to seven different sucrose/solid matrix weight relations: 0 min of OD (1.33 kg$_{sucrose}$/kg$_{solid\ matrix}$), 360 min of OD (3.58 kg$_{sucrose}$/kg$_{solid\ matrix}$), 720 min of OD (4.27 kg$_{sucrose}$/kg$_{solid\ matrix}$), 1722 min of OD (7.28 kg$_{sucrose}$/kg$_{solid\ matrix}$), 3375 min of OD (4.31 kg$_{sucrose}$/kg$_{solid\ matrix}$), 4320 min of OD (3.20 kg$_{sucrose}$/kg$_{solid\ matrix}$) and 10,270 min of OD (6.82 kg$_{sucrose}$/kg$_{solid\ matrix}$). Two OD treatment times were selected before the plasmolysis in order to study the samples before the chemical equilibrium. After the plasmolysis, the samples suffered mechanical phenomena (swelling and shrinkage), which were responsible for the osmotic solution intake or outflow; from this time there were no longer diffusional driving forces. After 1722 min of OD, the samples suffered a shrinkage causing the outflow of liquid phase, and therefore the decrease in the sucrose content (decrease in the weight relation sucrose/solid matrix). Two OD treatment times were selected in this period (3375 and 4320 min of OD). After the shrinkage, the swelling phenomenon occurred, which could be observed by the increase of sucrose/solid matrix weight relation. One OD treatment time was selected in this period: 10,270 min.

Figure 4 shows the relationship between the water activity of samples in equilibrium (24 h after HAD treatment) and moisture expressed in dry matter ($x_w/1 - x_w$), or isotherm at 30 °C. Moreover, Figure 4 shows the isotherm of pure water/sucrose solution at 30 °C from [57], where it is possible to observe that the apple isotherm data are around the isotherm of pure solution. The range of water activity measured covers some state transitions associated with water and sucrose. Thus, it is necessary to analyze the possible transitions that occur during the HAD treatment.

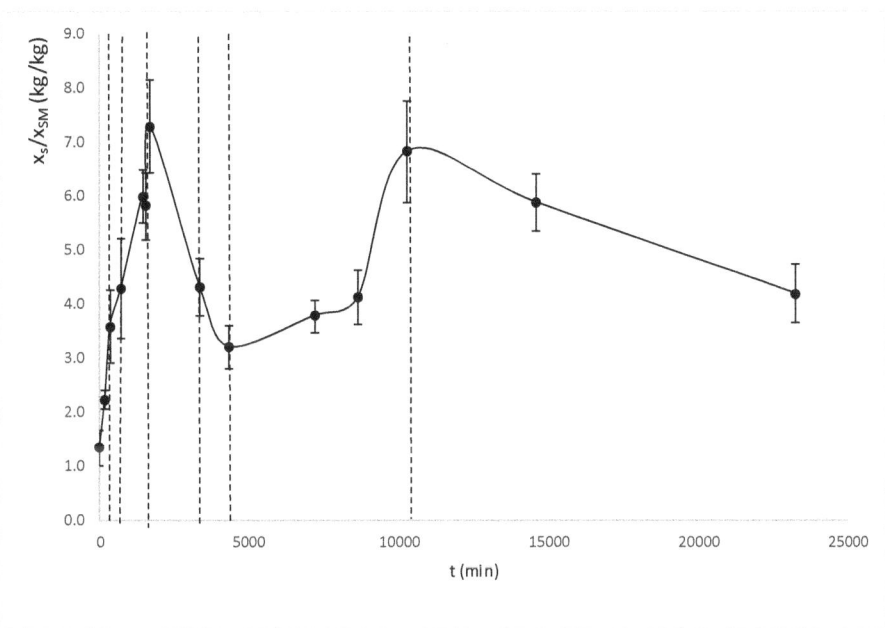

Figure 3. Sucrose/solid matrix relation with regard to osmotic dehydration (OD) time. Dotted lines represent the OD samples times chosen for the samples that will be dehydrated later by HAD.

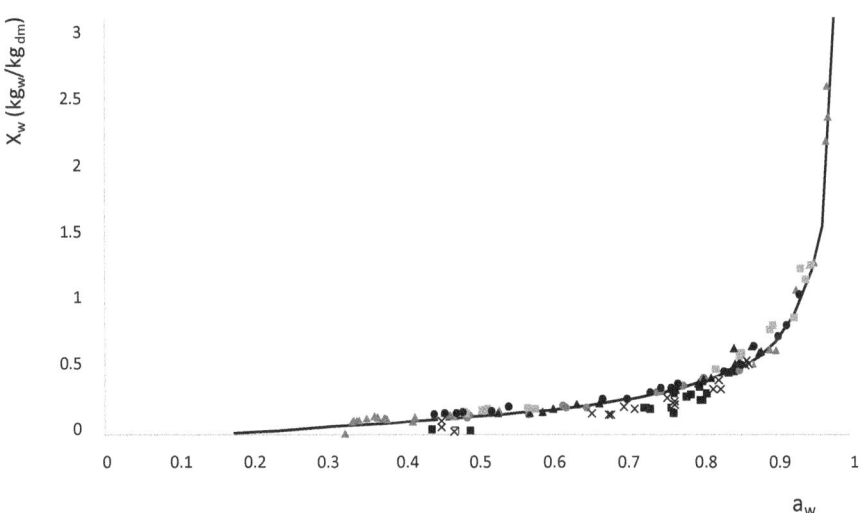

Figure 4. Moisture (kg water/kg dry matter) with regard to the water activity, where: (—) represents a water-sucrose solution [57], (▲) 1.33 kg$_{sucrose}$/kg$_{solid\ matrix}$, (■) 3.58 kg$_{sucrose}$/kg$_{solid\ matrix}$, (●) 4.27 kg$_{sucrose}$/kg$_{solid\ matrix}$, (▲) 7.28 kg$_{sucrose}$/kg$_{solid\ matrix}$, (×) 4.31 kg$_{sucrose}$/kg$_{solid\ matrix}$, (●) 3.20 kg$_{sucrose}$/kg$_{solid\ matrix}$ and (■) 6.82 kg$_{sucrose}$/kg$_{solid\ matrix}$.

Figure 5 shows the diagram state of the sucrose–water solution represented as water activity versus temperature. In this figure, it is possible to observe how at 30 °C the samples crossed the saturation and supersaturation curve as the viscosity of the liquid phase increased. This represents that the liquid phase analyzed could be in the stable, metastable or supersaturated region. Therefore, when developing a predictive tool for the water state during an OD or HAD treatment, it is important to predict not only the composition but also to predict the water mobility or the state of the liquid phase of the tissue being treated.

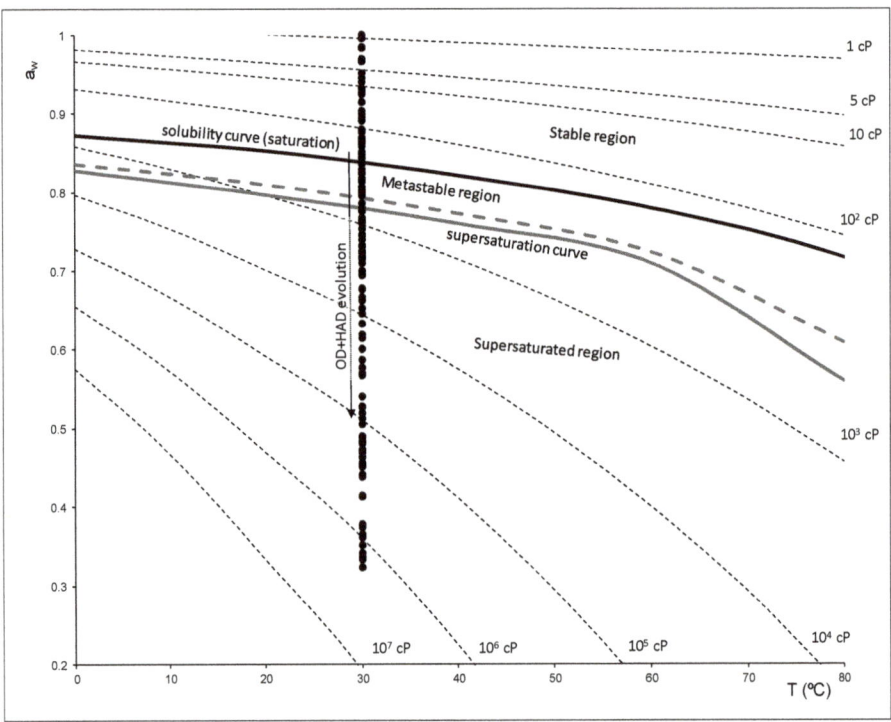

Figure 5. Diagram state of sucrose solution represented as water activity vs. temperature, where (▬) and (▬) lines represent the saturation and supersaturation curves obtained from [16], (---) lines represent different values of viscosity obtained from Lewis (1990) [58], and (●) represent the different experimental values of samples with OD and HAD treatments.

In order to develop a tool to predict the state of the liquid phase throughout the treatment of OD or HAD, the dielectric properties were analyzed in the range of microwaves, where the γ (dipolar effect) affects mainly the water molecules, generating an orientation and induction effect that was greater when the mobility of the molecules was greater.

Figure 6 shows the dielectric constant and loss factor spectra of samples treated 360 min with OD and different HAD treated times, as an example of spectra variation throughout the HAD treatment. The dielectric phenomena by orientation and induction occurred in the range of radiofrequency and microwaves. The phenomena were three, α, β and γ and occurred in a large frequency range. It is considered that each phenomenon has a maximum effect when the losses are maximum and the frequency at which the maximum effect occurs are called the relaxation frequency. In order to understand the relationship of each phenomenon with the molecules affected, it is necessary to adjust the spectrum to a model that obtains the relaxation values of each phenomenon. The Traffano-Schiffo model [20] has been applied to obtain the relaxation values in the γ-dispersion. In Figure 6, it is possible to observe some relaxation frequencies decreasing with the HAD treatment time.

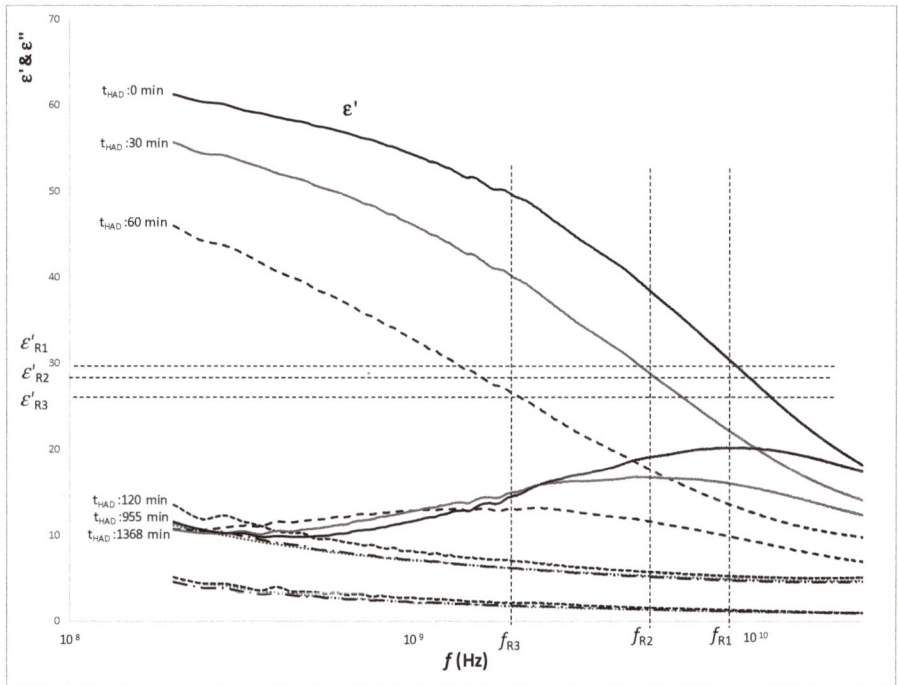

Figure 6. Dielectric constant and loss factor spectra of samples treated 360 min with OD and different HAD treatment times (t$_{HAD}$), where: f_R is the relaxation frequency, and sub-indices R1, R2 and R3 refer to the relaxation of γ-dispersion of samples dehydrated 0, 30 and 60 min, respectively.

The dielectric constant is the part of the permittivity that describes the orientation (electrical storage) of the water molecule. Dielectric constant is higher when the movement capacity of the water molecules is greater, as well as higher the number of water molecules is.

Figure 7 shows the relation between the dielectric constant at relaxation frequency and the water activity, where it is possible to observe a linear relation between the dielectric constant at the relaxation frequency and the water activity, regardless of the sucrose/solid matrix weight relation, the structure changes or the rheologic transitions. Therefore, only the water mobility affected the water orientation, and thus, dielectric constant in relaxation frequency represents an excellent tool to predict the water activity but not the transitions of water/sucrose.

The linear regression obtained was ($R^2 = 0.944$):

$$a_w = 0.0346 \cdot \varepsilon'_{relaxation} + 0.2363. \tag{5}$$

The constant of the equation (0.2363), which represents the activity of water for a null dielectric constant or null water capacity for orientation, coincides with the water activity at which sucrose crystallizes (0.23) [59].

Figure 8 shows the relationship between the moisture in dry basis and the relaxation frequency in the different regions described in Figure 5. As it is possible to observe in this figure, only the samples that were in the stable region showed a linear relation, in this region the liquid phase had great mobility and the viscosity was low. Samples located in a metastable or supersaturated region moved away from linearity. This meant that it was possible to discriminate the regions of the state diagram of sucrose/water by analyzing the frequency of relaxation.

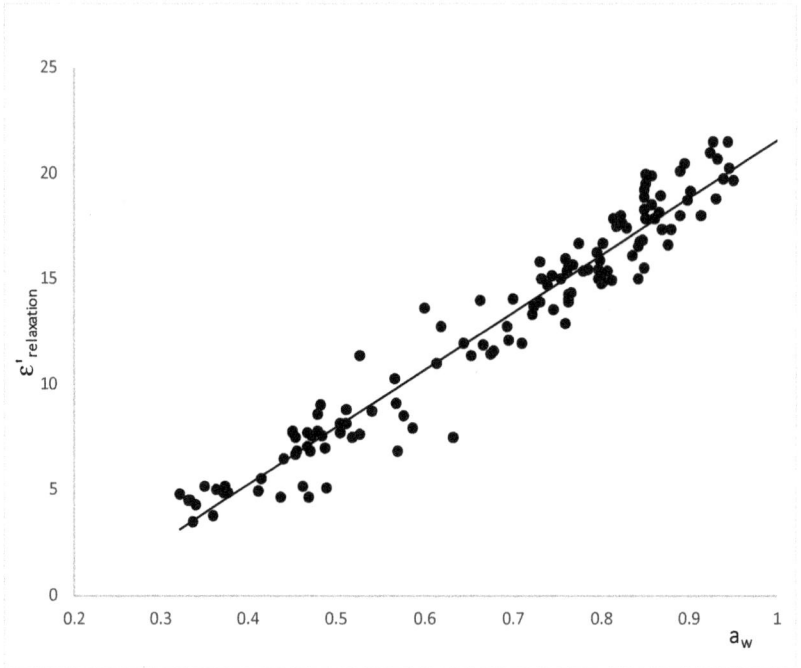

Figure 7. Relation between the dielectric constant at the relaxation frequency and the water activity.

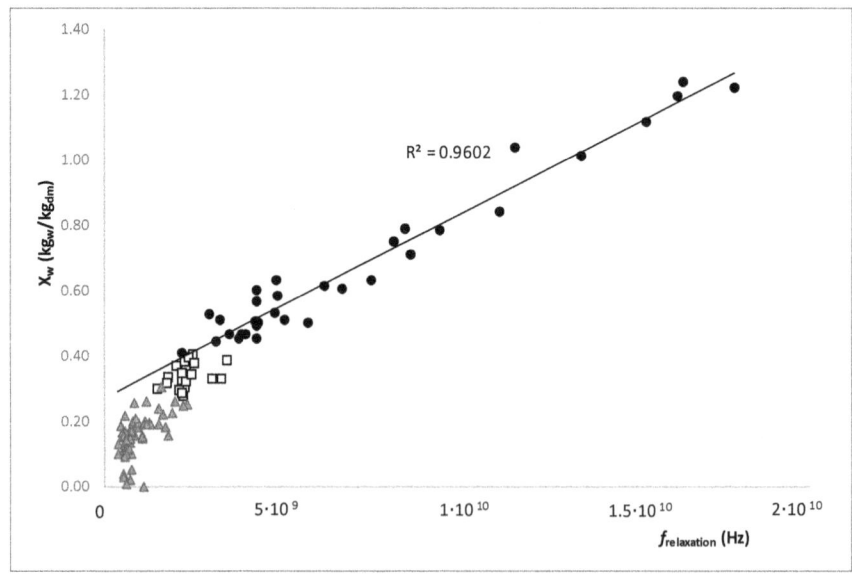

Figure 8. Relation between the moisture in dry basis (kg$_{water}$/kg$_{dry\ matter}$) with the relaxation frequency, where (●) represents the samples in stable region, (□) represents the samples in metastable region and (▲) the samples in the supersaturated region.

With the purpose of being able to use dielectric measurements in the microwave range to predict the sorption isotherm of apples in the process of OD and HAD, besides being able to correctly describe the water/sucrose state, the frequency of relaxation was compared with the loss factor.

Figure 9 shows the semilogarithmic relationship of the relaxation frequency and the loss factor at relaxation frequency, segregated according to the stability region. As it is possible to observe the relation between both was linear in all the regions, and did not allow the separation. Nevertheless, at the very low mobility zone in the supersaturated region, new behavior appeared because the relaxation frequency remained constant while the loss factor was reduced, this might be due to the vitrification process (glass transition).

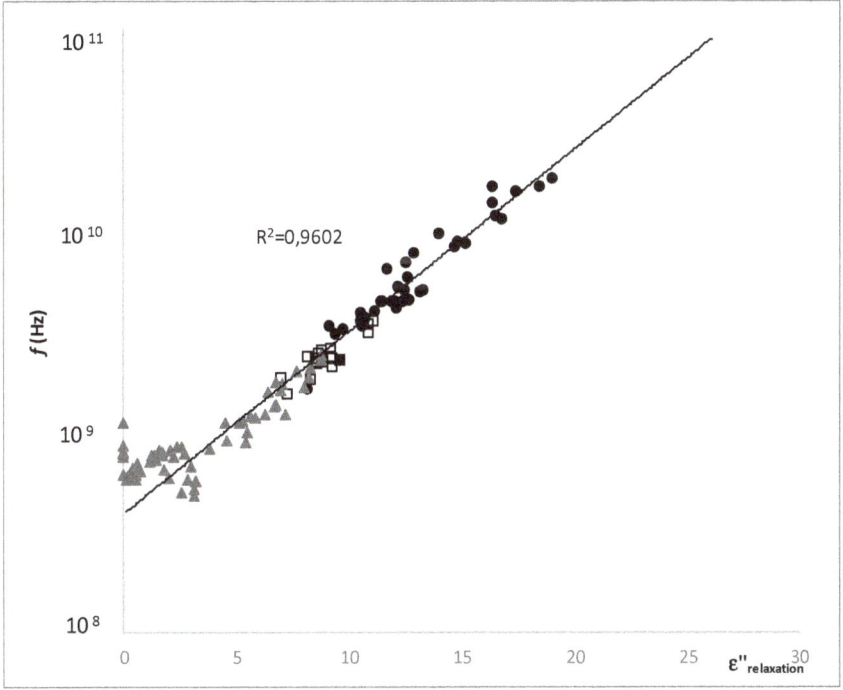

Figure 9. Semi-logarithmic relation between the relaxation frequency and the loss factor at relaxation frequency, where (●) represents the samples in stable region, (□) represents the samples in metastable region, (▲) the samples in the supersaturated region and (—) represents the linear regression of all data together.

4. Discussion

The treatment of OD produces compositional and structural transformations within the tissue and the HAD treatments produce water losses, structural transformations and transitions in the liquid phase. In order to analyze the two treatments it is necessary to include several critical points of both processes such as plasmolysis, compression/relaxation phenomena or saturation and supersaturation of the liquid phase. In Figure 3, it is possible to observe how the sucrose/solid matrix weight relation reaches a maximum (plasmolysis) at 1722 min.

Samples treated with OD and HAD, plasmolyzed or not, cross the saturation and supersaturation curve as the viscosity of the liquid phase increases. This represents that the liquid phase of samples analyzed could be in the stable, metastable or supersaturated region. Therefore, when developing a predictive tool for describing the water state during an OD or HAD treatment, it is important to predict not only the composition but also to predict the water mobility or the state of the liquid phase of the tissue being treated.

The Traffano-Schiffo model allows obtaining the relaxation values in the microwave range, it allows us to represent the dielectric constant at relaxation frequency versus the water activity, where it is possible to observe a linear relation between both variables. Therefore, only the water mobility affects to the water orientation, and thus, dielectric constant in relaxation frequency represents an excellent tool to predict the water activity but not the transitions of water/sucrose. Moreover, the good fit of the data is shown in the good values of the correlation coefficient and also in the coincidence between the constant of the equation and the water activity at which sucrose crystallizes (0.23).

The relationship between the moisture in the dry basis and the relaxation frequency in the different regions described shows that only the samples that are in the stable region have a linear relation. In this region, the liquid phase has great mobility and the viscosity is low. Samples located in a metastable or supersaturated region move away from linearity. This means that it is possible to discriminate the regions of the state diagram of sucrose/water by analyzing the frequency of relaxation.

The relaxation frequency and the loss factor at the relaxation frequency had a semilogarithmic, relationship in all the regions, and did not allow the separation. Nevertheless, at the very low mobility zone in the supersaturated region, new behavior appeared because the relaxation frequency remained constant while the loss factor was reduced, this might be due to the vitrification process (glass transition).

In the dielectric analysis in γ-dispersion, samples before and after the OD plasmolysis showed no differences, therefore, in this case of parenchymatic tissue, the effect of membrane plasmolysis had no effect in the water orientation or induction. Therefore, in similar fruit tissue, with high quantity of parenchymatic tissue and low vascular tissue, the behavior could be expected to be similar.

5. Conclusions

The use of specific frequencies of the dielectric properties in γ-dispersion did not allow us to correctly analyze the mobility of water and it makes it necessary to determine the dielectric properties at the relaxation frequency. It was demonstrated that the dielectric constant (at relaxation frequency) was linearly related to the water activity and not to the moisture, it means that it was affected by water mobility and therefore by the structure. In addition, it was possible to determine sucrose supersaturation processes by analyzing the relaxation frequency, which depends on the deformation of the water molecule. Therefore, it was demonstrated that the use of dielectric properties in γ-dispersion at relaxation frequency allowed us not only to monitor the OD and HAD processes of the apple candying, but also to predict the supersaturation state of the liquid phase until vitrification.

Author Contributions: All the authors contributed to the same extent in the experimental part, in the analysis of the data and in the subsequent writing of the manuscript.

Funding: This research was funded by THE SPANISH MINISTERIO DE ECONOMÍA, INDUSTRIA Y COMPETITIVIDAD, Programa Estatal de I+D+i orientada a los Retos de la Sociedad AGL2016-80643-R, Agencia Estatal de Investigación (AEI) and Fondo Europeo de Desarrollo Regional (FEDER).

Acknowledgments: The author Juan Ángel Tomás-Egea wants to thank the FPI Predoctoral Program of the Universidad Politécnica de Valencia for its support.

Conflicts of Interest: The authors declare no conflict of interest. The funders had no role in the design of the study; in the collection, analyses, or interpretation of data; in the writing of the manuscript, or in the decision to publish the results.

References

1. Krokida, M.K.; Marinos-Kouris, D. Rehydration kinetics of dehydrated products. *J. Food Eng.* **2003**, *57*, 1–7. [CrossRef]
2. Sacilik, K.; Elicin, A.K. The thin layer drying characteristics of organic apple slices. *J. Food Eng.* **2006**, *73*, 281–289. [CrossRef]

3. Ahmed, I.; Qazi, I.M.; Jamal, S. Developments in osmotic dehydration technique for the preservation of fruits and vegetables. *Innov. Food Sci. Emerg. Technol.* **2016**, *34*, 29–43. [CrossRef]
4. Castro-Giraldez, M.; Tylewicz, U.; Fito, P.J.; Dalla Rosa, M.; Fito, P. Analysis of chemical and structural changes in kiwifruit (Actinidia deliciosa cv Hayward) through the osmotic dehydration. *J. Food Eng.* **2011**, *105*, 599–608. [CrossRef]
5. Fito, P.; LeMaguer, M.; Betoret, N.; Fito, P.J. Advanced food process engineering to model real foods and processes: The "SAFES" methodology. *J. Food Eng.* **2007**, *83*, 173–185. [CrossRef]
6. Salisbury, F.; Ross, C. Plant Physiology. In *Plant Physiology and Plant Cells*, 4th ed.; Salisbury, F., Ross, C., Eds.; Wadsworth Pub. Co.: Belmont, CA, USA, 1991; pp. 3–26.
7. Agre, P.; Bonhivers, M.; Borgnia, M.J. The aquaporins, blueprints for cellular plumbing systems. *J. Biol. Chem.* **1998**, *273*, 14659–14662. [CrossRef] [PubMed]
8. Nieto, A.B.; Vicente, S.; Hodara, K.; Castro, M.A.; Alzamora, S.M. Osmotic dehydration of apple: Influence of sugar and water activity on tissue structure, rheological properties and water mobility. *J. Food Eng.* **2013**, *119*, 104–114. [CrossRef]
9. Vicente, S.; Nieto, A.B.; Hodara, K.; Castro, M.A.; Alzamora, S.M. Changes in structure, rheology, and water mobility of apple tissue induced by osmotic dehydration with glucose or trehalose. *Food Bioprocess. Technol.* **2012**, *5*, 3075–3089. [CrossRef]
10. Mavroudis, N.E.; Dejmek, P.; Sjöholm, I. Osmotic-treatment-induced cell death and osmotic processing kinetics of apples with characterised raw material properties. *J. Food Eng.* **2004**, *63*, 47–56. [CrossRef]
11. Alzamora, S.M.; Castro, M.A.; Nieto, A.B.; Vidales, S.L.; Salvatori, D.M. The role of tissue microstructure in the textural characteristics of minimally processed fruits. In *Minimally Processed Fruits and Vegetables*; Alzamora, S.M., Tapia, S., Lopez-Malo, A., Eds.; Aspen Publishers Inc.: Frederick, MD, USA, 2000; pp. 153–171.
12. Mauro, M.A.; Dellarosa, N.; Tylewicz, U.; Tappi, S.; Laghi, L.; Rocculi, P.; Dalla Rosa, M. Calcium and ascorbic acid affect cellular structure and water mobility in apple tissue during osmotic dehydration in sucrose solutions. *Food Chem.* **2016**, *195*, 19–28. [CrossRef]
13. Lewicki, P.P. Design of hot air drying for better foods. *Trends Food Sci. Technol.* **2006**, *17*, 153–163. [CrossRef]
14. Lewicki, P.P.; Jakubczyk, E. Effect of hot air temperature on mechanical properties of dried apples. *J. Food Eng.* **2004**, *64*, 307–314. [CrossRef]
15. Mathlouthi, M. X-ray diffraction study of the molecular association in aqueous solutions of d-fructose, d-glucose, and sucrose. *Carbohydr Res.* **1981**, *91*, 113–123. [CrossRef]
16. Mathlouthi, M.; Genotelle, J. Role of water in sucrose crystallization. *Carbohydr. Polym.* **1998**, *37*, 335–342. [CrossRef]
17. Mathlouthi, M. Amorphous sugar. In *Sucrose*; Mathlouthi, M., Reiser, P., Eds.; Springer: Boston, MA, USA, 1995; pp. 75–100.
18. Traffano-Schiffo, M.V.; Castro-Giraldez, M.; Colom, R.J.; Fito, P.J. Innovative photonic system in radiofrequency and microwave range to determine chicken meat quality. *J. Food Eng.* **2018**, *239*, 1–7. [CrossRef]
19. Traffano-Schiffo, M.V.; Castro-Giraldez, M.; Herrero, V.; Colom, R.J.; Fito, P.J. Development of a non-destructive detection system of Deep Pectoral Myopathy in poultry by dielectric spectroscopy. *J. Food Eng.* **2018**, *237*, 137–145. [CrossRef]
20. Traffano-Schiffo, M.V.; Castro-Giraldez, M.; Colom, R.J.; Fito, P.J. Development of a spectrophotometric system to detect white striping physiopathy in whole chicken carcasses. *Sensors* **2017**, *17*, 1024. [CrossRef]
21. Trabelsi, S.; Roelvink, J. Investigating the influence of aging on radiofrequency dielectric properties of chicken meat. *J. Microw. Power Electromagn. Energy* **2014**, *48*, 215–220. [CrossRef]
22. Samuel, D.; Trabelsi, S.; Karnuah, A.B.; Anthony, N.B.; Aggrey, S.E. The use of dielectric spectroscopy as a tool for predicting meat quality in poultry. *Int. J. Poult. Sci.* **2012**, *11*, 551. [CrossRef]
23. Castro-Giraldez, M.; Toldrá, F.; Fito, P. Development of a dielectric spectroscopy technique for the determination of key bichemical markers of meat quality. *Food Chem.* **2011**, *127*, 228–233. [CrossRef]
24. Castro-Giraldez, M.; Toldrá, F.; Fito, P. Low frequency dielectric measurements to assess post-mortem ageing of pork meat. *LWT Food Sci. Technol.* **2011**, *44*, 1465–1472. [CrossRef]
25. Castro-Giraldez, M.; Toldrá, F.; Fito, P. Low-frequency dielectric spectrum to determine pork meat quality. *Innov. Food Sci. Emerg. Technol.* **2010**, *2*, 376–386. [CrossRef]
26. Castro-Giráldez, M.; Aristoy, M.C.; Toldrá, F.; Fito, P. Microwave dielectric spectroscopy for the determination of pork meat quality. *Food Res. Int.* **2010**, *1*, 2369–2377. [CrossRef]

27. Kent, M.; Peymann, A.; Gabriel, C.; Knight, A. Determination of added water in pork products using microwave dielectric spectroscopy. *Food Control.* **2002**, *13*, 143–149. [CrossRef]
28. Brunton, N.P.; Lyng, J.; Zhang, L.; Jacquier, J.C. The use of dielectric properties and other physical analyses for assessing protein denaturation in beef biceps femoris muscle during cooking from 5 to 85 °C. *Meat Sci.* **2006**, *72*, 236–244. [CrossRef]
29. Mohiri, A.; Burhanudin, Z.A.; Ismail, I. Dielectric properties of slaughtered and non-slaughtered goat meat. In Proceedings of the RF and Microwave Conference (RFM), IEEE International, Seremban, Negeri Sembilan, Malaysia, 12–14 December 2011; pp. 393–397.
30. Lyng, J.G.; Zhang, L.; Brunton, N.P. A survey of the dielectric properties of meats and ingredients used in meat product manufacture. *Meat Sci.* **2005**, *69*, 589–602. [CrossRef]
31. Everard, C.D.; Fagan, C.C.; O'Donell, C.P.; O'Callaghan, D.J.; Lyng, J. Dielectric properties of process cheese from 0.3 to 3GHz. *J. Food Eng.* **2006**, *75*, 415–422. [CrossRef]
32. Traffano-Schiffo, M.V.; Castro-Giraldez, M.; Colom, R.J.; Fito, P.J. New Spectrophotometric System to Segregate Tissues in Mandarin Fruit. *Food Bioprocess. Technol.* **2018**, *2*, 399–406. [CrossRef]
33. Cuibus, L.; Castro-Giraldez, M.; Fito, P.J.; Fabbri, A. Application of infrared thermography and dielectric spectroscopy for controlling freezing process of raw potato. *Innov. Food Sci. Emerg. Technol.* **2014**, *1*, 80–87. [CrossRef]
34. Nelson, S.O.; Trabelsi, S. Dielectric spectroscopy of wheat from 10 MHz to 1.8 GHz. *Meas. Sci. Technol.* **2006**, *17*, 2294–2298. [CrossRef]
35. Venkatesh, M.S.; Raghavan, G.S.V. An overview of microwave processing and dielectric properties of agri-food materials. *Biosyst. Eng.* **2004**, *88*, 1–18. [CrossRef]
36. Nelson, S.O. Dielectric spectroscopy in agriculture. *J. Non Cryst. Solids* **2005**, *351*, 2940–2944. [CrossRef]
37. Nelson, S.O. Dielectric properties of agricultural products and some applications. *Res. Agric. Eng.* **2008**, *54*, 104–112. [CrossRef]
38. Castro-Giraldez, M.; Fito, P.J.; Ortola, M.D.; Balaguer, N. Study of pomegranate ripening by dielectric spectroscopy. *Postharvest Biol. Technol.* **2013**, *86*, 346–353. [CrossRef]
39. Shang, L.; Guo, W.; Nelson, S.O. Apple variety identification based on dielectric spectra and chemometric methods. *Food Anal. Methods* **2015**, *8*, 1042–1052. [CrossRef]
40. Castro-Giraldez, M.; Fito, P.J.; Chenoll, C.; Fito, P. Development of a dielectric spectroscopy technique for determining key chemical components of apple maturity. *J. Agric. Food Chem.* **2010**, *6*, 3761–3766. [CrossRef]
41. Castro-Giraldez, M.; Fito, P.J.; Chenoll, C.; Fito, P. Development of a dielectric spectroscopy technique for the determination of apple (Granny Smith) maturity. *J. Food Eng.* **2010**, *4*, 749–754. [CrossRef]
42. Traffano-Schiffo, M.V.; Castro-Giraldez, M.; Colom, R.J.; Fito, P.J. Study of the application of dielectric spectroscopy to predict the water activity of meat during drying process. *J. Food Eng.* **2015**, *166*, 285–290. [CrossRef]
43. Castro-Giraldez, M.; Fito, P.J.; Fito, P. Application of microwaves dielectric spectroscopy for controlling pork meat (*Longissimus dorsi*) salting process. *J. Food Eng.* **2010**, *4*, 484–490. [CrossRef]
44. Talens, C.; Castro-Giraldez, M.; Fito, P.J. Study of the effect of microwave power coupled with hot air drying on orange peel by dielectric spectroscopy. *LWT Food Sci. Technol.* **2016**, *66*, 622–628. [CrossRef]
45. Velazquez-Varela, J.; Castro-Giraldez, M.; Cuibus, L.; Tomas-Egea, J.A.; Socaciu, C.; Fito, P.J. Study of the cheese salting process by dielectric properties at microwave frequencies. *J. Food Eng.* **2018**, *224*, 121–128. [CrossRef]
46. Velazquez-Varela, J.; Castro-Giraldez, M.; Fito, P.J. Control of the brewing process by using microwaves dielectric spectroscopy. *J. Food Eng.* **2013**, *119*, 633–639. [CrossRef]
47. Castro-Giraldez, M.; Fito, P.J.; Andrés, A.M.; Fito, P. Study of the puffing process of amaranth seeds by dielectric spectroscopy. *J. Food Eng.* **2012**, *11*, 298–304. [CrossRef]
48. Castro-Giraldez, M.; Fito, P.J.; Fito, P. Application of microwaves dielectric spectroscopy for controlling osmotic dehydration of kiwifruit (*Actinidia deliciosa* cv Hayward). *Innov. Food Sci. Emerg. Technol.* **2011**, *12*, 623–627. [CrossRef]
49. Castro-Giraldez, M.; Fito, P.J.; Fito, P. Application of microwaves dielectric spectroscopy for controlling long time osmotic dehydration of parenchymatic apple tissue. *J. Food Eng.* **2011**, *2*, 227–233. [CrossRef]
50. Baker-Jarvis, J.; Kim, S. The Interaction of Radio-Frequency Fields with Dielectric Materials at Macroscopic to Mesoscopic Scales. *J. Res. Natl. Inst. Stand. Technol.* **2012**, *117*, 1–60. [CrossRef]

51. Roychoudhuri, C.; Kracklauer, A.F.; Creath, K. *The Nature of Light: What Is a Photon*; CRC Press: New York, NY, USA, 2008. [CrossRef]
52. Horie, K.; Ushiki, H.; Winnik, F.M. *Molecular Photonics*; Wiley: New York, NY, USA, 2000. [CrossRef]
53. Crane, C.A.; Pantoya, M.L.; Weeks, B.L. Spatial observation and quantification of microwave heating in materials. *Rev. Sci. Instrum.* **2013**, *84*, 084705. [CrossRef]
54. Castro-Giraldez, M.; Fito, P.J.; Fito, P. Nonlinear thermodynamic approach to analyze long time osmotic dehydration of parenchymatic apple tissue. *J. Food Eng.* **2011**, *102*, 34–42. [CrossRef]
55. AOAC. *Official Method 934.06 Moisture in Dried Fruits*; AOAC: Rockville, MD, USA, 2000.
56. Tylewicz, U.; Fito, P.J.; Castro-Giraldez, M.; Fito, P.; Dalla Rosa, M. Analysis of kiwifruit osmodehydration process by systematic approach systems. *J. Food Eng.* **2011**, *104*, 438–444. [CrossRef]
57. Starzak, M.; Mathlouthi, M. Temperature dependence of water activity in aqueous solutions of sucrose. *Food Chem.* **2006**, *96*, 346–370. [CrossRef]
58. Lewis, D.F. Structure of sugar confectionery. In *Sugar Confectionery Manufacture*; Jackson, E.B., Ed.; Blackie: Glasgow, UK, 1990; pp. 331–350.
59. Roos, Y.; Karel, M. Water and molecular weight effects on glass transitions in amorphous carbohydrates and carbohydrate solutions. *J. Food Sci.* **1991**, *56*, 1676–1681. [CrossRef]

© 2019 by the authors. Licensee MDPI, Basel, Switzerland. This article is an open access article distributed under the terms and conditions of the Creative Commons Attribution (CC BY) license (http://creativecommons.org/licenses/by/4.0/).

Article

Effect of Thermo-Sonication and Ultra-High Pressure on the Quality and Phenolic Profile of Mango Juice

Abdul Ghani Dars [1,†], Kai Hu [1,†], Qiudou Liu [1], Aqleem Abbas [2], Bijun Xie [1] and Zhida Sun [1,*]

[1] College of Food Science and Technology, Huazhong Agricultural University, Wuhan 430070, China
[2] Department of Plant Pathology, College of Plant Science and Technology, Huazhong Agricultural University, Wuhan 430070, China
* Correspondence: sunzhida@mail.hzau.edu.cn; Tel.: +86-27-8728-3201; Fax: +86-27-8728-2966
† These authors contributed equally to this work.

Received: 15 June 2019; Accepted: 26 July 2019; Published: 29 July 2019

Abstract: Consumer demand for safe and nutritious fruit juices has led to the development of a number of food processing techniques. To compare the effect of two processing technologies, thermo-sonication (TS) and ultra-high pressure (UHP), on the quality of mango juice, fresh mango juice was treated with TS at 25, 45, 65 and 95 °C for 10 min and UHP at 400 MPa for 10 min. The phenolic profile of mango was also analyzed using the newly developed ultra-performance liquid chromatography-electrospray ionization-quadrupole time of flight-mass spectrometry (UPLC-Q-TOF-HRMS^n) and, based on this result, the effect of TS and UHP on the phenolics of mango juice was evaluated. Both treatments had minimal effects on the °Brix, pH, and titratable acidity of mango juice. The residual activities of three enzymes (polyphenol oxidase, peroxidase, and pectin methylesterase), antioxidant compounds (vitamin C, Total phenolics, mangiferin derivatives, gallotannins, and quercetin derivatives) and antioxidant activity sharply decreased with the increase in the temperature of the TS treatment. Nevertheless, the UHP treatment retained antioxidants and antioxidant activity at a high level. The UHP process is likely superior to TS in bioactive compounds and antioxidant activity preservation. Therefore, the mango juice products obtained by ultra-high-pressure processing might be more beneficial to health.

Keywords: mango juice; thermo-sonication; ultra-high pressure; physicochemical property; phenolic compounds

1. Introduction

Mango *(Mangifera indica* L.) is one of the most economically important and popular tropical fruits and is widely recognized worldwide for its admirable sensorial characteristics (sweet taste, bright color, and delicious flavor) and nutritional composition (carbohydrates such as glucose, fructose and sucrose, vitamins, minerals, fiber, and phytochemical), as well as its popularity and high production. Mango is considered as a 'king of fruits' due to extensive appeal across Africa, the Americas, Australia, Europe and Asia [1,2].

Mango is consumed as fresh fruit as well as processed products, like juice, pulp, powder, mash, pickles and syrup [2,3] and, among these products, juice has the higher consumption and higher economic value [2,4]. Due to a higher consumption, the mango juice market has increased considerably [5,6]. As for fruit juice processing, a number of studies have focused on the effects of processing technologies on juice quality [2,3,7,8]. Among them, thermal processing was widely applied in the food industries to preserve juice. However, since fruits are usually susceptible to thermal processing, this can lead to considerable damage to bioactive constituents and sensory characteristics of fruit products. Studies on non-thermal processing technologies such as ultraviolet, pulsed electric field, ultrasound and ultra-high pressure (UHP) are being carried out because they cater to the consumer demand of fresh-like, minimally processed foods. When high power ultrasound at low frequencies (20–100 kHz) propagates in liquid, cavitation (formation and collapse of bubbles) occurs. As a result, elevation of localized pressure and temperature causes the alteration of the properties of food products either physically or chemically [7,8]. Individual ultrasound processing effects on juice products were widely reported in different fruit cultivars [8–10]. However, few papers about the combinative effects of ultrasound and heat on fruit juice quality are available. UHP is a promising processing technology to meet the needs of fresh juice. At present, the studies are mainly focused on the influence of the technology on macromolecular substances such as protein, pectin and fiber. However, the effect on bioactive components (small molecule) is rarely reported [11].

Mango is rich in antioxidants such as polyphenols, flavonoids and other phytochemicals [12]. Previous reports have shown that polyphenols can modulate immune response activities [13,14]. In addition, polyphenols prevent genetic toxicity by reducing the exposure to oxidative and carcinogenic factors [15,16]. Regular consumption of mango fruits supplies a considerable number of polyphenols which have beneficial physiological effects [17–19]. Mango polyphenols exhibit anti-inflammatory and cancer cytotoxic properties in multiple cancer types, including malignancies of the colon and breast [13,15,16]. A previous study showed that the content of polyphenols in mango peel is higher than that in pulp [14]. Mango polyphenolics are mainly rich in gallic acid, gallotannins, galloyl glycosides, and flavonoids [12,20]. It has been proved that these phenolic compounds in mango are the main bioactive ingredients beneficial to health [19]. However, no information about the effect of TS and UHP treatments on the phenolic profile of mango juice is available. Current consumers look for healthy food products, maintaining mango juice's inherent physical and chemical properties of interest (e.g., texture, pH, titratable acidity, total soluble solids, total phenolics, total carotenoids and vitamin C) and nutritional quality during its processing. Hence, the objective of the present study was to evaluate the effects of TS and UHP treatments on the quality and phenolic profile of mango juice.

2. Materials and Methods

2.1. Preparation of Juice Samples

Chinese mangoes cv. Kensington Pride were purchased from a producer in Hainan province and brought to full ripeness by maintaining at 20–23 °C and 90% relative humidity (RH). Mangoes were washed with tap water, then dried and cut into small pieces. Kernel and bruised portions were discarded. Peel and a small part of pulp were collected and stored at −20 °C for phenolic identification within two weeks. The remaining pulp was used to obtain juice by a domestic juice extractor (AUX-PB953, Foshan Haixun Electric Appliances Co., Ltd., Foshan, China). After filtration using a sterilized double-layered muslin cloth, the juice was vortex mixed and stored in 50-mL pre-sterilized PET bottles at 4 °C for further treatment within 2 h.

2.2. Thermo-Sonication (TS) and Ultra-High Pressure (UHP) Treatment

TS treatment was performed using a 250-W ultrasonic processor (Ningbo Xingzhi Biotechnoligy Co., Ltd., Ningbo, China) at four different temperatures 25, 45, 65, and 95 °C for 10 min. UHP treatment was carried out using an ultra-high pressure processor (Bao Tou KeFa High Pressure Technology Co., Ltd., Baotou, China) at 400 MPa for 10 min (based on previous optimization). The juice without treatment was considered as control. All treatments were conducted in triplicate. Brix, pH, acidity, polyphenol oxidase (PPO), peroxidase (POD), and pectin methylesterase (PME) were measured immediately after treatment. The remaining juice was stored at −20 °C until further analysis for vitamin C, total phenolic content, antioxidant activity and quantification of phenolic compounds within two weeks.

2.3. Determination of °Brix, pH, and Acidity

A hand refractometer WYT-80 (Quanzhou Wander Experimental Instrument Co., Ltd., Quanzhou, China) was used to measure °Brix. A digital pH meter (Delta 320 pH meter, Metller Toledo Instruments Co., Ltd., Shanghai, China) was used to measure pH. The titratable acidity was measured according to the method suggested by the "Association of Official Analytical Chemists" (AOAC, 2000) with a 0.1 M NaOH solution as the titration solution.

2.4. Determination of PPO, POD and PME Residual Activities

The samples of mango juice were centrifuged at 8000 rpm for 15 min at 4 °C for the enzyme activity assay. Polyphenol oxidase (PPO) and peroxidase (POD) activity were measured according to the protocol of Macdonald and Schanchke [21]. For the determination of PPO, 1.5 mL supernatant was mixed with 0.5 mL catechol (0.5 mol/L) and 3.0 mL potassium phosphate buffer (0.2 mol/L, pH 6.8). The absorbance was recorded at 410 nm within 3 min. With respect to POD, 0.32 mL potassium phosphate buffer (0.2 mol/L, pH 6.8), 0.32 mL pyrogallol (5 g/100 mL) and 0.6 mL H_2O_2 (0.147 mol/L) were mixed and variation in absorbance at 420 nm within 3 min was noted. Pectinmethylesterase (PME) assay was conducted according to the method used by Saeeduddin et al. [7]. Briefly, 10 mL supernatant was mixed with 40 mL pectin solution (1 g/100 mL) containing 0.15 mol/L NaCl. The pH was adjusted to 7.7 by the addition of 0.05 mol/L NaOH and the time taken was recorded. The reaction system was incubated at 50 ± 2 °C. Residual activities of PPO, POD and PME were calculated using the following equation:

$$\text{Enzyme activity (\%)} = 100 A_t / A_0 \tag{1}$$

where A_0 and A_t are the enzyme activities of the control and treatment samples respectively.

2.5. Determination of Total Phenolic Content

The total phenolic content was determined according to the method carried out by Tong et al. [22]. Approximately 100 µL mango juice from each sample was mixed with 0.4 mL distilled water and 0.5 mL diluted Folin–Ciocalteu reagent (1:10, v:v). The mixtures were incubated for 5 min at room temperature and 1 mL 7.5% sodium carbonate (w/v) was added. The absorbance was measured at 765 nm after maintaining at 30 min in dark. A standard curve was obtained with gallic acid and the result was expressed as mg of gallic acid equivalents (GAE)/mL juice.

2.6. Determination of Vitamin C

Vitamin C content was determined using a simplified method reported by Sulaiman and Ooi [23]. Approximately 25 mL of diluted solution was titrated against 0.1‰ 2, 6-dichlorophenolindophenol sodium (DCIPS) until the solution became a light pink color and persisted for 15 s. The calibration of 0.1‰ DCIPS solution was performed with 1 mg/mL ascorbic acid. The results were calculated and expressed as mgL^{-1} juice.

2.7. Determination of Total Antioxidant Activity

The total antioxidant activity assay was tested using the method reported by Li et al. [24] The juice (0.4 mL) was centrifuged and mixed with 4 mL reagent mixture (sulfuric acid (0.6 mol/L), sodium phosphate (28 mmol/L) and ammonium molybdate (4 mmol/L)). The mixture was kept at 95 °C for 90 min and the absorbance was measured at 695 nm. Ascorbic acid was used as standard and the result was presented as mg ascorbic acid equivalent/mL juice.

2.8. Extraction of Polyphenolic Compounds for UPLC/UPLC-Q-TOF-HRMS^n

Mango pulp was thawed and homogenized in an appropriate ratio of 10 g of pulp to 30 mL of a solvent mixture (ethanol/methanol/acetone, 1/1/1). Similarly, mango peel was thawed and homogenized in a ratio of 5 g of peel to 15 mL of a solvent mixture (ethanol/methanol/acetone, 1/1/1). For the quantification analysis of polyphenolics in mango juice before and after different treatments, the extraction was conducted using the same method used for mango pulp. Afterwards, the resultant solution of pulp, peel and juice was filtered through cheese cloth. The solvents were removed at 40 °C by rotary evaporation under reduced pressure, and the aqueous residue was centrifuged to remove insoluble precipitates. Polyphenolics were partitioned in a 20-mL Waters C18 cartridge. Compounds not adsorbed to the cartridge were partitioned into ethyl acetate using a separatory funnel. The ethyl acetate phase was combined with the methanol elute from the C18 cartridges, and the solvent were removed under reduced pressure. The residual was dissolved in chromatographic acetonitrile and used for UPLC and UPLC-Q-TOF-HRMS^n.

2.9. The Quantification and Identification of Polyphenolic Compounds in Mango by UPLC/UPLC-Q-TOF-HRMS^n

For the quantification and identification of polyphenolic compounds, LC analysis was conducted using Acquity Ultra Performance Liquid Chromatography system (Waters, Milford, MA, USA), equipped with a C18 column (2.1 × 100 mm, 1.7 µm, Waters, Milford, MA, USA). The column, constant at 40 °C, was eluted with a linear gradient mobile phase at 0–28 min: 2–50% B, 28–28.5 min: 50–100% B, 28.5–30.5 min: 100% B, 30.5–32 min: 100–2% B, 32–34 min: 2% B, where A = water with 0.1% acetic acid and B = acetonitrile. The flow rate was 0.3 mL/min, and the injected volume was 1 µL.

The mass spectrometric data of the full scan mode was collected using a G2-XS QT of MS (Waters, Milford, MA, USA). The scan range was from m/z 100 to 2000 with a scan time of 0.3 s. The source temperature was set at 120 °C with a cone gas flow of 50 L/h. The gas flow was set to 800 L/h at a temperature of 400 °C. The capillary was set at 1 kV for ESI$^-$ mode with the cone voltage at 40 V. The MS^n analysis was carried out on a Waters-Micro mass Quattro Premier triple quadrupole mass spectrometer. The collision energy was optimized according to the specific precursor ions.

2.10. Statistical Analysis

All tests were performed in triplicate and data were expressed as the means ± the standard deviation. One-way analysis of variance (ANOVA) followed by LSD multiple comparison were conducted using the SPSS 20 (IBM, Armonk, NY, USA). Differences with a p value < 0.05 were considered significant.

3. Results and Discussion

3.1. Effects of TS and UHP on °Brix, pH and Titratable Acidity

The effects of TS and UHP on °Brix, pH and titratable acidity are shown in Table 1. All treatments were not significantly different from each other when addressing the °Brix, pH and titratable acidity of mango juice. These results are in consistent with the previous reports [9,10,25]. These previous studies showed no significant variations in the °Brix, pH and titratable acidity of other fruit juices as a result of various non-thermal processing food technology. This indicated that TS and UHP are promising tools since they improved the quality of fruit juice without causing the significant change of basic physicochemical indexes.

3.2. Effects of TS and UHP on Inactivation of PPO, POD and PME

The effects of different TS treatments on PPO, POD and PME in mango juice are shown in Table 2. The highest enzyme inactivation was exhibited in the sample treated with TS at 95 °C, which showed the residual activities of PPO, POD and PME as 3.47, 1.61 and 2.24% respectively. The increase in temperature interval significantly deactivates PPO, POD and PME as described in previous report [26]. The time duration of the ultrasonic treatment causes the formation of free radicals, which are then involved in inactivation of enzyme activities [27]. The formation of cavities due to bubbles development and disappearance is related to enzyme inactivation [28]. This can induce sharp increase in temperature and pressure in a localized ultrasound generating area, which may be a major factor in enzyme deactivation. As a result, the temperature and other mechanical forces during ultrasonic pasteurization have a combined role in the enzyme inactivation. The mango juice was treated with UHP treatment at 400 MPa for 10 min. However, just a slight effect on the enzyme inactivation was recorded. These three enzymes were likely pressure resistant, and the corresponding mechanism needs to be further studied.

3.3. Effects of TS and UHP on Antioxidant Compounds and Antioxidant Capacity

The effects of thermo-sonication on ascorbic acids and total phenolics are mentioned in Table 3. The amount of ascorbic acid in fresh mango juice (control) was 117.47 ± 1.12 mg/L. It sharply decreased from 116.26 ± 0.89 to 33.12 ± 1.35 mg/mL with the temperature increasing from 25 to 95 °C, corresponding to 98.97, 64.22, 47.16 and 28.19% of residual quantity when TS treatments were conducted at 25, 45, 65 and 95 °C, respectively. There is no significant difference between the vitamin C contents of control and TS at 25 °C. This means TS at normal temperature has no effect on the vitamin C contents of mango juice. However, it exerts a significant effect of decomposition or oxidation of vitamin C when the temperature is high (>45 °C) and, with the increase in temperature, the loss of vitamin C becomes more prominent. Similar results were reported in apple juice and watermelon juice when treated with TS at different temperatures [29,30]. From Table 3, vitamin C content was 114.16 ± 1.02 mg/mL when treated with UHP at 400 MPa, showing a slight reduction (2.82%) compared with control. It is likely that the density of the reactive system increased during UHP treatment, which promoted the decomposition of vitamin C. The total phenolic content of the control sample was 1.76 ± 0.08 mg/mL. It decreased from 1.73 ± 0.05 to 0.592 ± 0.005 mg GAE/mL, with the temperature increasing from 25 to 95 °C, reducing by 1.70, 51.36, 58.86 and 66.36% respectively. This suggests that TS at relatively low temperature has no significant effect on total phenolic content of mango juice, but significantly reduce it when temperature is high. A similar result was obtained in pear juice [7]. Mango juice treated with UHP at 400 MPa showed a higher total phenolic content (1.82 ± 0.003) than the control, though this is not significant. The increase in the content of phenolic compounds during UHP treatment might be due to the secretion of the bound forms of these compounds in juice [4,31]. At 25 °C, vitamin C and polyphenol content hardly changed after ultrasonic treatment as compared to control. This indicated sonication only exerted a minor effect on antioxidants. Nevertheless, the high reduction in the content of ascorbic acid and total phenols at 45–95 °C indicates that both entities were highly heat sensitive. UHP is a promising non-thermal technology for food processing and can effectively protect bioactive components and avoid losses caused by heat treatment.

The effects of TS on the antioxidant capacity of mango juice are shown in Table 3. There was no significant difference in total antioxidant activity between control (0.867 ± 0.006 mg AAE/mL) and TS25 (0.862 ± 0.008 mg AAE/mL) samples. However, with the increase in temperature from 45 °C to 95 °C, the total antioxidant activity decreased from 0.792 ± 0.006 to 0.572 ± 0.005 mg AAE/mL, leading to 8.65–34.03% loss. The antioxidant capacity of fruit juice is attributed to the presence of antioxidant compounds such as ascorbic acid and total phenol [9]. The decreasing trend of the antioxidant capacity of samples is consistent with the decrease in vitamin C and polyphenol content by TS treatments at high temperature. However, ultra-high-pressure treatment was found to retain the antioxidant activity (0.831 ± 0.003 mg AAE/mL) of mango juice at a high level of 95.85%, which indicates that UHP is a promising technique for the protection of antioxidant and free radical scavenging capacity.

3.4. Identification of Polyphenolic Compounds

The polyphenols in mango pulp and peel were identified by UPLC-Q-TOF-HRMS^n (Supporting Information, Figures S1 and S2), and the results are shown in Tables 4 and 5 respectively. Approximately 22 compounds from the peel and 14 compounds from the pulp were identified using the UPLC-Q-TOF-HRMS^n technique. All of the phenolic compounds identified in mango pulp, except for iriflophenone di-O-galloyl-glucoside, were included in those of peel. Therefore, the MSs of phenolic compounds in mango peel are discussed here (Table 4). Among these compounds, five compounds, i.e., **2**, **4**, **9**, **10** and **11**, were identified as benzophenone derivatives (Table 4). The molecule ions at m/z 575.1039, 727.1071 and 879.1256 were identified as maclurin mono-O-galloyl-glucoside (compound **2**), maclurin-di-O-galloyl-glucoside (Compounds **4**, **10** and **11**) and maclurin tri-O-galloyl-glucoside (Compound **9**). The fragment ions were obtained by the successive loss of galloyl or H_2O. The MS^2 results showed that the fragment ions of compound **2** were found at *m/z* 423.09, 303.05, 285.04, 261.04 and 193.02, corresponding to the loss of galloy moiety, 272.0469 Da ion, H_2O, $2H_2O$ and 110.0352 Da ion, respectively. Similar fragment loss was found in compounds **4**, **9**, **10** and **11**. The above results show that benzophenone derivatives in mango were all maclurin-gallic glucosides with different substitution degrees. Compounds **3**, **6** and **7** were identified as mangiferin and their derivatives (Table 4). Compound **3** was identified and detected as mangiferin at *m/z* 421.0771 ($[M - H]^-$). Mangiferin is a glycosylated xanthine found in several varieties of mango. It is reported to be not only a typical biomarker for resistance against *Fm* infection but also have pharmacological activities in different organs and tissues, such as protecting the heart, neurons, liver, and kidneys and preventing or delaying the onset of diseases [32]. Compounds **6** and **7** were identified as mangiferin gallate and iso-mangiferin gallate as a result of the loss of galloyl moiety and H_2O. In this study, five tannins, compounds **1**, **5**, **8**, **16** and **20**, were identified as gallic acid, tetra-O-galloyl-glucoside, iso-tetra-O-galloyl-glucoside, penta-O-galloyl-glucose and hexa-O-galloyl-glucose, respectively. Gallic acid is a widespread tannin in mangoes and has been recognized in other cultivars of mangoes [32]. In common, for most ions, neutral losses of the galloyl fraction (152 Da) and of gallic acid (170 Da) were shown in Table 4. The fragmentation profile created for these polyphenolic compounds were related to gallotannins and benzophenone derivatives [12]. Compounds **12**, **13**, **14**, **15**, **17**, **18**, **19**, **21** and **22** were found to be flavonoids and identified as quercetin 3-O-galactoside, quercetin 3-O-glucoside, iso-quercetin 3-O-glucoside, quercetin 3-O-xyloside, iso-quercetin 3-O-glucoside, iso-quercetin 3-O-xyloside, kaempferol 3-O-glucoside, quercetin 3-O-rhamnoside and iso-quercetin 3-O-rhamnoside, respectiviely. These flavonoids were mainly of different forms of glycosides. Among them, the quercetin compound was the major aglycone. The phenolic compounds identified in pulp are quite similar to the compounds of peel. However, one compound in pulp was different from peel. This compound was identified as iriflophenone di-O-galloyl-glucoside at *m/z* 711.1124 ($[M - H]^-$). Specific fragmentation patterns of all the identified compounds are shown in the Supplementary Materials.

3.5. Effects of TS and UHP on Phenolic Groups

Based on the analysis of UPLC/UPLC-Q-TOF-HRMS^n, as well as the distribution and structure characteristics of phenolic compounds in the chromatogram, the phenolic compounds in mango juice were divided into three main groups, (i) mangiferin/derivatives, (ii) gallotannins, and (iii) quercetin derivatives. Each group was relatively quantified by peak area to study its variation during TS and UHP treatments. As shown in Figure 1, all three phenolic groups exhibited a generally decreasing trend with the development of temperature, which was in accordance with the changes of total phenolics. When treated with TS at 95 °C, the content of phenolic groups in mango juice was lowest. However, quercetin derivatives showed an ultrasonic-resistant ability at low treated temperature (25 °C), but other two kinds of phenolic compounds were unstable under ultrasound treatment, likely due to oxidation and degradation induced by the ultrasonic cavitation effect. In addition, all three phenolic groups were temperature-sensitive. A previous study showed that thermal processing significantly ($p < 0.05$) affected individual phenolic acids, anthocyanins, flavan-3-ols, and flavonols, significantly ($p < 0.05$) reduced total phenolic acid contents in both pinto and black beans and total flavonol contents in pinto beans, and dramatically reduced anthocyanin contents in black beans [33]. The flavonols rutin and quercetin also degraded under thermal processing in an aqueous model system [34]. The combined mangiferin/derivatives may be released by UHP. This shows that ultra-high pressure could be an excellent process technology of mango juice, with a high retention rate of phenolic groups (Figure 1).

Table 1. Effect of different treatments on Brix, pH, titratable acidity in mango juice [a].

Treatments [b]	°Brix	pH	Titratable Acidity (%)
Control	11.80 ± 0.02a	4.75 ± 0.03a	0.16 ± 0.01a
TS25	11.80 ± 0.01a	4.74 ± 0.03a	0.16 ± 0.01a
TS45	11.72 ± 0.02a	4.73 ± 0.01ab	0.17 ± 0.01a
TS65	11.60 ± 0.03a	4.69 ± 0.02b	0.17 ± 0.01a
TS95	11.54 ± 0.06a	4.67 ± 0.02bc	0.17 ± 0.01a
UHP400	11.62 ± 0.04a	4.70 ± 0.02c	0.17 ± 0.01a

[a] Data are presented as means ± the standard deviation. Means with different letters within a column are significantly different (LSD test) at $p < 0.05$. [b] TS25, TS45, TS65 and TS95 represent thermo-sonication treatment at 25, 45, 65 and 95 °C. UHP400 represents ultra-high pressure at 400 MPa for 10 min.

Table 2. Effect of different treatments on the residual activity percentage of POD, PPO and PME in mango juice [a].

Treatment [b]	POD (Residual Activity %)	PPO (Residual Activity %)	PME (Residual Activity %)
Control	100.00 ± 00a	100.00 ± 00a	100.00 ± 00a
TS25	92.57 ± 0.94c	87.73 ± 1.30c	90.76 ± 1.82c
TS45	51.42 ± 1.22d	45.44 ± 2.11d	48.36 ± 1.10d
TS65	37.45 ± 1.15e	31.39 ± 1.71e	34.52 ± 0.77e
TS95	3.47 ± 0.68f	1.61 ± 0.57f	2.24 ± 0.57f
UHP400	98.18 ± 0.80b	93.26 ± 0.82b	96.46 ± 1.76b

[a] Data are presented as means ± the standard deviation. Means with different letters within a column are significantly different (LSD test) at $p < 0.05$. [b] TS25, TS45, TS65 and TS95 represent thermo-sonication treatment at 25, 45, 65 and 95 °C. UHP400 represents ultra-high pressure at 400 MPa for 10 min.

Table 3. Effect of different treatments on antioxidants compounds and antioxidant activity [a].

Treatment [b]	Vitamin C (mg/L)	Total Phenolic Content (mg GAE/mL) [c]	Total Antioxidant Capacity (mg AAE/mL) [d]
Control	117.47 ± 1.12a	1.76 ± 0.08ab	0.867 ± 0.006a
TS25	116.26 ± 0.89ab	1.73 ± 0.05b	0.862 ± 0.008a
TS45	75.45 ± 1.04c	0.856 ± 0.006c	0.792 ± 0.004c
TS65	55.40 ± 0.71d	0.724 ± 0.008d	0.716 ± 0.008d
TS95	33.12 ± 1.35e	0.592 ± 0.005e	0.572 ± 0.005e
UHP400	114.16 ± 1.02b	1.82 ± 0.003a	0.831 ± 0.003b

[a] Data are presented as means ± the standard deviation. Means with different letters within a column are significantly different at (LSD test) $p < 0.05$. [b] TS25, TS45, TS65 and TS95 represent thermo-sonication treatment at 25, 45, 65 and 95 °C. UHP400 represents ultra-high pressure at 400 MPa for 10 min. [c] Total phenolic content expressed as mg gallic acid equivalent per mL juice. [d] Total antioxidant capacity expressed as mg ascorbic acid equivalent per mL juice.

Table 4. Identification of polyphenolic compounds in Kensington Pride mango peel by UPLC-Q-TOF-HRMS[n].

Compound	Retention Time (min)	Identity	Formula	Calculated [M − H] m/z	Observed [M − H] m/z	Error (ppm)	Ion Fragment
1	1.82	Gallic acid	$C_7H_6O_5$	169.0137	169.0218	47.93	125.0325, 97.0374
2	6.822	maclurin mono-O-galloyl-glucoside	$C_{26}H_{24}O_{15}$	575.1037	575.1039	0.35	303.057, 285.0470, 261.0472, 423.0966, 193.0218
3	8.22	mangiferin	$C_{19}H_{18}O_{11}$	421.0771	421.0814	10.21	
4	8.283	maclurin di-O-galloyl-glucoside	$C_{33}H_{28}O_{19}$	727.1147	727.1071	−10.45	
5	8.956	tetra-O-galloyl-glucose	$C_{34}H_{28}O_{22}$	787.0994	787.0882	−14.23	635.0850, 617.0764
6	9.567	mangiferin gallate	$C_{26}H_{22}O_{15}$	573.088	573.0867	−2.27	403.0731
7	10.009	iso-mangiferin gallate	$C_{26}H_{22}O_{15}$	573.088	573.0974	16.40	421.0774
8	10.177	iso-tetra-O-galloyl-glucose	$C_{34}H_{28}O_{22}$	787.0994	787.0882	−14.23	635.0854, 617.0757
9	10.33	maclurin tri-O-galloyl-glucoside	$C_{40}H_{32}O_{23}$	879.1256	879.1214	−4.78	727.1166
10	10.477	maclurin di-O-galloyl-glucoside iso-maclurin	$C_{33}H_{28}O_{19}$	727.1147	727.1071	−10.45	421.0774, 403.0706
11	10.682	di-O-galloyl-glucoside	$C_{33}H_{28}O_{19}$	727.1147	727.1071	−10.45	421.0774, 403.0706
12	10.78	quercetin 3-O-galactoside	$C_{21}H_{20}O_{12}$	463.0877	463.0908	6.69	301.0396, 300.0330,
13	11.03	quercetin 3-O-glucoside	$C_{21}H_{20}O_{12}$	463.0877	463.0909	6.91	301.0396, 300.0330
14	11.439	iso-quercetin 3-O-glucoside	$C_{21}H_{20}O_{12}$	463.0877	463.0909	6.91	301.0396, 300.0330
15	11.624	quercetin 3-O-xyloside	$C_{20}H_{18}O_{11}$	433.0771	433.0811	9.24	301.0396, 300.0330
16	11.676	penta-O-galloyl-glucose	$C_{41}H_{32}O_{26}$	939.1104	939.101	−10.01	787.0900, 769.0813, 617.0764
17	11.906	iso-quercetin 3-O-glucoside	$C_{21}H_{20}O_{12}$	463.0877	463.0909	6.91	301.0397, 300.0333
18	12.081	iso-quercetin 3-O-xyloside	$C_{20}H_{18}O_{11}$	433.0771	433.0811	9.24	301.0396, 300.0330
19	12.196	kaempferol 3-O-glucoside	$C_{21}H_{20}O_{11}$	447.0927	447.0963	8.05	285.0455, 284.0391, 255.0366
20	12.206	hexa-O-galloyl-glucose	$C_{48}H_{36}O_{30}$	1091.1213	1091.1174	−3.57	939.0916
21	12.217	quercetin 3-O-rhamnoside	$C_{21}H_{20}O_{11}$	447.0927	447.096	7.38	300.0331
22	12.318	iso-quercetin 3-O-rhamnoside	$C_{21}H_{20}O_{11}$	447.0927	447.096	7.38	300.034

Table 5. Identification of polyphenolic compounds in Kensington Pride mango pulp by UPLC-Q-TOF-HRMS[n].

Compound	Retention Time (min)	Identity	Formula	Calculated [M−H] m/z	Observed [M−H] m/z	Error (ppm)	Ion Fragment
1	1.811	Gallic acid	$C_7H_6O_5$	169.0137	169.0221	49.70	125.0325
2	6.791	maclurin mono-O-galloyl-glucoside	$C_{26}H_{24}O_{15}$	575.1037	575.103	−1.22	303.062
3	8.321	maclurin di-O-galloyl-glucoside	$C_{33}H_{28}O_{19}$	727.1147	727.1057	−12.38	
4	9.587	mangiferin gallate	$C_{26}H_{22}O_{15}$	573.088	573.088	0.00	301.0057
5	9.783	iriflophenone di-O-galloyl-glucoside	$C_{33}H_{28}O_{18}$	711.1197	711.1124	−10.27	
6	10.187	tetra-O-galloyl-glucose	$C_{34}H_{28}O_{22}$	727.1147	787.0897	−12.32	635.0834, 617.0766
7	10.445	tetra-O-galloyl-glucose	$C_{34}H_{28}O_{22}$	727.1147	787.0897	−12.32	635.0834, 617.0766
8	10.787	quercetin 3-O-galactoside	$C_{21}H_{20}O_{12}$	463.0877	463.0895	3.89	300.032, 301.0393
9	11.055	quercetin 3-O-glucoside	$C_{21}H_{20}O_{12}$	463.0877	463.0902	5.40	300.032, 301.0393
10	11.439	quercetin 3-O-arabinopyranoside	$C_{20}H_{18}O_{11}$	433.0771	433.0796	5.77	300.0334
11	11.76	quercetin 3-O-rhamnoside	$C_{21}H_{20}O_{11}$	447.0927	447.0951	5.37	301.0034
12	12.217	kaempferol 3-O-glucoside	$C_{21}H_{20}O_{11}$	447.0927	447.0954	6.04	284.0297, 255.0339
13	14.215	penta-O-galloyl-glucose	$C_{41}H_{32}O_{26}$	939.1104	939.1005	−10.54	769.0818
14	16.858	maclurin tri-O-galloyl-glucoside	$C_{40}H_{32}O_{23}$	879.1256	879.1292	4.09	

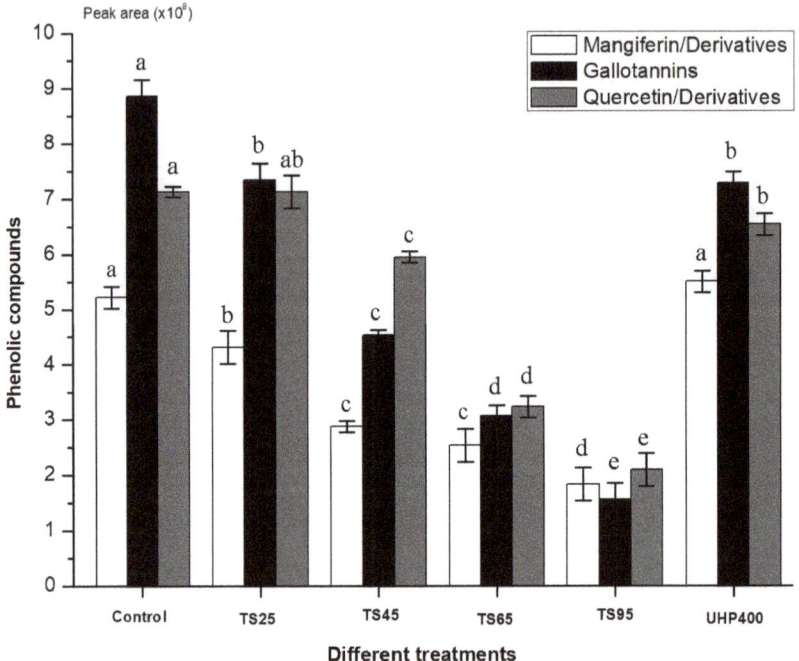

Figure 1. Effects of thermo-sonication (TS) and ultra-high pressure (UHP) on phenolic groups, mangiferin and its derivatives, quercetin derivatives and gallotannins. TS25, TS45, TS65 and TS95 represent thermo-sonication treatment at 25, 45, 65 and 95 °C. UHP400 represents ultra-high pressure at 400 MPa for 10 min. Each column represents a mean and the vertical bars indicate the standard deviation. Means with different letters within the same color columns are significantly different (LSD test, $p < 0.05$).

4. Conclusions

TS and UHP had a minor effect on the basic physicochemical properties of mango juice. TS at high temperature significantly reduced the enzyme activities of PPO, POD and PME. The effects of TS treatment on vitamin C, total phenolics and antioxidant activity were significant, and mainly degradative or oxidative action. However, UHP treatment gave a high level of antioxidants and antioxidant activity of mango juice. The Kensington Pride mango researched in this study was cultivated in Hainan, the southernmost province of China. Abundant phenolic compounds from mango were identified and the effects of TS and UHP on the phenolic profile were analyzed. We find that TS induced the significant degradation of phenolic groups, mangiferin/derivatives, gallotannins, and quercetin derivatives. The UHP treatment was likely superior to TS in bioactive compounds and antioxidant activity preservation except for browning-related enzymes, which needs further study. The mango juice products obtained by ultra-high-pressure processing are more beneficial to health.

Supplementary Materials: The following are available online at http://www.mdpi.com/2304-8158/8/8/298/s1, Figure S1: Identification of polyphenolic compounds in Kensington Pride mango peel by UPLC-Q-TOF-HRMS^n, Figure S2: Identification of polyphenolic compounds in Kensington Pride mango pulp by UPLC-Q-TOF-HRMS^n.

Author Contributions: A.G.D. and K.H. designed the study and interpreted the results. Q.L. collected test data. B.X. and A.A. drafted the manuscript. Z.S., as a corresponding author, supervised the experiment.

Funding: This research was funded by the Hubei province program of agricultural products processing and comprehensive utilization.

Acknowledgments: This work was supported by the Hubei province program of agricultural products processing and comprehensive utilization. The author Dars is also thankful to the Chinese Govt Scholarship (CSC) for providing funds for PhD studies.

Conflicts of Interest: The authors declare no conflict of interest.

References

1. Barbosa Gamez, I.; Caballero-Montoya, K.P.; Ledesma, N.; Sayago-Ayerdi, S.G.; Garcia-Magana, M.L.; Bishop-von-Wettberg, E.J.; Montalvo-Gonzalez, E. Changes in the nutritional quality of five *Mangifera* species harvested at two maturity stages. *J. Sci. Food Agric.* **2017**, *97*, 4987–4994. [CrossRef] [PubMed]
2. Liu, F.; Li, R.; Wang, Y.; Bi, X.; Liao, X. Effects of high hydrostatic pressure and high-temperature short-time on mango nectars: Changes in microorganisms, acid invertase, 5-hydroxymethylfurfural, sugars, viscosity, and cloud. *Innov. Food Sci. Emerg. Technol.* **2014**, *22*, 22–30. [CrossRef]
3. Hoang-Hai, N.; Shpigelman, A.; Van Buggenhout, S.; Moelants, K.; Haest, H.; Buysschaert, O.; Van-Loey, A. The evolution of quality characteristics of mango piece after pasteurization and during shelf life in a mango juice drink. *Eur. Food Res. Technol.* **2016**, *242*, 703–712.
4. Zapata Londono, M.B.; Chaparro, D.; Alberto Rojano, B.; Alzate Arbelaez, A.F.; Restrepo Betancur, L.F.; Maldonado Celis, M.E. Effect of storage time on physicochemical, sensorial, and antioxidant characteristics, and composition of mango (cv. Azucar) juice. *Emir. J. Food Agric.* **2017**, *29*, 367–377.
5. Kaur, C.; Kapoor, H.C. Antioxidants in fruits and vegetables: The millennium's health. *Int. J. Food Sci. Technol.* **2001**, *36*, 703–725. [CrossRef]
6. Ulla, A.; Rahman, M.T.; Habib, Z.F.; Rahman, M.M.; Subhan, N.; Sikder, B.; Alam, M.A. Mango peel powder supplementation prevents oxidative stress, inflammation, and fibrosis in carbon tetrachloride induced hepatic dysfunction in rats. *J. Food Biochem.* **2017**, *41*, e12344. [CrossRef]
7. Saeeduddin, M.; Abid, M.; Jabbar, S.; Wu, T.; Hashim, M.M.; Awad, F.N.; Zeng, X. Quality assessment of pear juice under ultrasound and commercial pasteurization processing conditions. *LWT Food Sci. Technol.* **2015**, *64*, 452–458. [CrossRef]
8. Santhirasegaram, V.; Razali, Z.; Somasundram, C. Effects of thermal treatment and sonication on quality attributes of Chokanan mango (*Mangifera indica* L.) juice. *Ultrason. Sonochem.* **2013**, *20*, 1276–1282. [CrossRef] [PubMed]
9. Abid, M.; Jabbar, S.; Wu, T.; Hashim, M.M.; Hu, B.; Lei, S.; Zeng, X. Effect of ultrasound on different quality parameters of apple juice. *Ultrason. Sonochem.* **2013**, *20*, 1182–1187. [CrossRef] [PubMed]
10. Tiwari, B.K.; O'Donnell, C.P.; Cullen, P.J. Effect of non thermal processing technologies on the anthocyanin content of fruit juices. *Trends Food Sci. Technol.* **2009**, *20*, 137–145. [CrossRef]
11. Wang, C.Y.; Huang, H.W.; Hsu, C.P.; Yang, B.B. Recent Advances in Food Processing Using High Hydrostatic Pressure Technology. *Crit. Rev. Food Sci. Nutr.* **2016**, *56*, 527–540. [CrossRef] [PubMed]
12. Berardini, N.; Carle, R.; Schieber, A. Characterization of gallotannins and benzophenone derivatives from mango (*Mangifera indica* L. cv. 'Tommy Atkins') peels, pulp and kernels by high-performance liquid chromatography/electrospray ionization mass spectrometry. *Rapid Commun. Mass Spectrom.* **2004**, *18*, 2208–2216. [CrossRef] [PubMed]
13. Middleton, E. Effect of plant flavonoids on immune and inflammatory cell function. *Adv. Exp. Med. Biol.* **1998**, *439*, 175–182. [PubMed]
14. Wang, H.; Cao, G.; Prior, R.L. Oxygen Radical Absorbing Capacity of Anthocyanins. *J. Agric. Food Chem.* **1997**, *45*, 304–309. [CrossRef]
15. Cai, Q.; Rahn, R.O.; Zhang, R. Dietary flavonoids, quercetin, luteolin and genistein, reduce oxidative DNA damage and lipid peroxidation and quench free radicals. *Cancer Lett.* **1997**, *119*, 99–107. [CrossRef]
16. Pool-Zobel, B.L.; Bub, A.; Schröder, N.; Rechkemmer, G. Anthocyanins are potent antioxidants in model systems but do not reduce endogenous oxidative DNA damage in human colon cells. *Eur. J. Nutr.* **1999**, *38*, 227–234. [CrossRef] [PubMed]
17. Boyle, S.P.; Dobson, V.L.; Duthie, S.J.; Kyle, J.A.; Collins, A.R. Absorption and DNA protective effects of flavonoid glycosides from an onion meal. *Eur. J. Nutr.* **2000**, *39*, 213–223. [CrossRef]

18. Pedersen, C.B.; Kyle, J.; Jenkinson, A.M.; Gardner, P.T.; Mcphail, D.B.; Duthie, G.G. Effects of blueberry and cranberry juice consumption on the plasma antioxidant capacity of healthy female volunteers. *Eur. J. Clin. Nutr.* **2000**, *54*, 405–408. [CrossRef]
19. Masibo, M.; He, Q. Major mango polyphenols and their potential significance to human health. *Compr. Rev. Food Sci. Food Saf.* **2008**, *7*, 309–319. [CrossRef]
20. Nemec, M.J.; Kim, H.; Marciante, A.B.; Barnes, R.C.; Hendrick, E.D.; Bisson, W.H.; Mertens-Talcott, S.U. Polyphenolics from mango (*Mangifera indica* L.) suppress breast cancer ductal carcinoma in situ proliferation through activation of AMPK pathway and suppression of mTOR in athymic nude mice. *J. Nutr. Biochem.* **2017**, *41*, 12–19. [CrossRef]
21. Macdonald, L.; Schaschke, C.J. Combined effect of high pressure, temperature and holding time on polyphenoloxidase and peroxidase activity in banana (*Musa acuminata*). *J. Sci. Food Agric.* **2000**, *80*, 719–724. [CrossRef]
22. Tong, L.-T.; Liu, L.-Y.; Zhong, K.; Wang, Y.; Guo, L.-N.; Zhou, S.-M. Effects of Cultivar on Phenolic Content and Antioxidant Activity of Naked Oat in China. *J. Integr. Agric.* **2014**, *13*, 1809–1816. [CrossRef]
23. Sulaiman, S.F.; Ooi, K.L. Polyphenolic and Vitamin C Contents and Antioxidant Activities of Aqueous Extracts from Mature-Green and Ripe Fruit Fleshes of *Mangifera* sp. *J. Agric. Food Chem.* **2012**, *60*, 11832–11838. [CrossRef]
24. Li, F.h.; Yuan, Y.; Yang, X.l.; Tao, S.y.; Ming, J. Phenolic Profiles and Antioxidant Activity of Buckwheat (*Fagopyrum esculentum* Moench and *Fagopyrum tartaricum* L. Gaerth) Hulls, Brans and Flours. *J. Integr. Agric.* **2013**, *12*, 1684–1693. [CrossRef]
25. Walkling-Ribeiro, M.; Noci, F.; Cronin, D.A.; Lyng, J.G.; Morgan, D.J. Shelf life and sensory evaluation of orange juice after exposure to thermosonication and pulsed electric fields. *Food Bioprod. Process.* **2009**, *87*, 102–107. [CrossRef]
26. Villamiel, M.; Jong, P.D. Influence of high-intensity ultrasound and heat treatment in continuous flow on fat, proteins, and native enzymes of milk. *J. Agric. Food Chem.* **2000**, *48*, 472–478. [CrossRef]
27. Netsanetshiferaw, T.; Mala, G.; Kamaljit, V.; Lloyd, S.; Raymond, M.; Cornelis, V. The kinetics of inactivation of pectin methylesterase and polygalacturonase in tomato juice by thermosonication. *Food Chem.* **2009**, *117*, 20–27.
28. Ercan, S.S.; Soysal, C. Effect of ultrasound and temperature on tomato peroxidase. *Ultrason. Sonochem.* **2011**, *18*, 689–695. [CrossRef]
29. Abid, M.; Jabbar, S.; Wu, T.; Hashim, M.M.; Hu, B.; Lei, S.; Zeng, X. Sonication enhances polyphenolic compounds, sugars, carotenoids and mineral elements of apple juice. *Ultrason. Sonochem.* **2014**, *21*, 93–97. [CrossRef]
30. Rawson, A.; Tiwari, B.K.; Patras, A.; Brunton, N.; Brennan, C.; Cullen, P.J.; O'Donnell, C. Effect of thermosonication on bioactive compounds in watermelon juice. *Food Res. Int.* **2011**, *44*, 1168–1173. [CrossRef]
31. Liu, F.; Wang, Y.; Li, R.; Bi, X.; Liao, X. Effects of high hydrostatic pressure and high temperature short time on antioxidant activity, antioxidant compounds and color of mango nectars. *Innov. Food Sci. Emerg. Technol.* **2014**, *21*, 35–43. [CrossRef]
32. Oliveira, B.G.; Costa, H.B.; Ventura, J.A.; Kondratyuk, T.P.; Barroso, M.E.; Correia, R.M.; Romao, W. Chemical profile of mango (*Mangifera indica* L.) using electrospray ionisation mass spectrometry (ESI-MS). *Food Chem.* **2016**, *204*, 37–45. [CrossRef]
33. Xu, B.; Chang, S.K.C. Total Phenolic, Phenolic Acid, Anthocyanin, Flavan-3-ol, and Flavonol Profiles and Antioxidant Properties of Pinto and Black Beans (*Phaseolus vulgaris* L.) as Affected by Thermal Processing. *J. Agric. Food Chem.* **2009**, *57*, 4754–4764. [CrossRef]
34. Buchner, N.; Krumbein, A.; Rohn, S.; Kroh, L.W. Effect of thermal processing on the flavonols rutin and quercetin. *Rapid Commun. Mass Spectrom.* **2006**, *20*, 3229–3235. [CrossRef]

 © 2019 by the authors. Licensee MDPI, Basel, Switzerland. This article is an open access article distributed under the terms and conditions of the Creative Commons Attribution (CC BY) license (http://creativecommons.org/licenses/by/4.0/).

Article

The Application of Combined Pre-Treatment with Utilization of Sonication and Reduced Pressure to Accelerate the Osmotic Dehydration Process and Modify the Selected Properties of Cranberries

Malgorzata Nowacka *, Artur Wiktor, Magdalena Dadan, Katarzyna Rybak, Aleksandra Anuszewska, Lukasz Materek and Dorota Witrowa-Rajchert

Department of Food Engineering and Process Management, Faculty of Food Sciences, Warsaw University of Life Sciences, Nowoursynowska 159c, 02-776 Warsaw, Poland
* Correspondence: malgorzata_nowacka@sggw.pl; Tel.: +48-22-593-75-79

Received: 30 June 2019; Accepted: 22 July 2019; Published: 24 July 2019

Abstract: The aim of this study was to investigate the effect of a pretreatment, performed by a combined method based on blanching, ultrasound, and vacuum application, on the kinetics of osmotic dehydration and selected quality properties such as water activity, color, and bioactive compound (polyphenols, flavonoids, and anthocyanins) content. The pretreatment was carried out using blanching, reduced pressure, and ultrasound (20 min, 21 kHz) in various combinations: Blanching at reduced pressure treatment conducted three times for 10 min in osmotic solution; blanching with reduced pressure for 10 min and sonicated for 20 min in osmotic solution; and blanching with 20 min of sonication and 10 min of reduced pressure. The osmotic dehydration was performed in different solutions (61.5% sucrose and 30% sucrose with the addition of 0.1% of steviol glycosides) to ensure the acceptable taste of the final product. The changes caused by the pretreatment affected the osmotic dehydration process by improving the efficiency of the process. The use of combined pretreatment led to an increase of dry matter from 9.3% to 28.4%, and soluble solids content from 21.2% to 41.5%, lightness around 17.3% to 56.9%, as well as to the reduction of bioactive compounds concentration until even 39.2% in comparison to the blanched sample not subjected to combined treatment. The osmotic dehydration caused further changes in all investigated properties.

Keywords: cranberries; reduced pressure; sonication; color; bioactive compounds

1. Introduction

Cranberries are known as fruits rich in bioactive compounds, which provide positive effects and health benefits to the human body. The abundance of bioactive compounds contained in cranberries are also used in a medicine to treat urinary tract infections, gastrointestinal diseases, and support the neurological and cardiovascular conditions. Moreover, cranberries exhibit antivirus, anti-inflammation, and anticancer effects [1]. However, fresh cranberry is very sour, which is why its direct consumption and application is limited. Therefore, a sugar addition is required during processing in order to improve the taste of the fruit and to achieve consumers' acceptance. On the one hand, osmotic dehydration (OD) is a process in which sugars are used to partially remove water from the material, and as an introduction of ingredients of the osmotic solution, usually sucrose, which has a beneficial effect on changing the flavor profile of the product. On the other hand, one of the main disadvantages of OD is time consumption [2].

Non-thermal methods such as pulsed electric fields, sonication, or high hydrostatic pressure have the potential to accelerate the OD kinetics. Among these methods, sonication is one of the technologies which can be used in OD intensification. It could be introduced as a pretreatment step or directly during the process [3–6]. Ultrasounds are series of vibrations at frequency varying between 20 kHz and 100 MHz, which spread in a given medium, i.e., in liquids, solids, or gases. The use of high power ultrasound leads to emergence of acoustic waves, which produce a direct (so-called "sponge effect") and an indirect effect (cavitation) on solid food matrices [4,7,8]. Ultrasonic waves have an impact on nutritional and physico-chemical parameters of food, which are related to good extraction and the recovery of bioactive compounds [9–13]. It has been reported that high power ultrasound has been used to accelerate processes of mass exchange such as OD [5,14–16]. Moreover, ultrasound can be used alone or combined with different techniques [10].

Another promising method of OD facilitation is vacuum impregnation/infusion technology, which can also be described as a reduced pressure treatment. This method allows the air present in the plant pores to be replaced by acting solution in an instant. Due to the use of reduced pressure, mass transfer occurs faster, which is related to a hydrodynamic mechanism [17] and deformation-relaxation phenomena [18]. Isotonic or hypertonic solutions and other solutions containing valuable compounds beneficial for human health can be used for vacuum impregnation [19]. Similarly, OD might be performed under low-pressure conditions, which have a positive effect on shortening the process time in comparison to normal conditions. The use of low pressure during the treatment degasses the tissue, which increases the contact surface of the osmotic substance with the product, and the process of dehydration takes place faster [20].

In recent years, an increase in the number of obese people has been observed, therefore it is advisable to design production processes that enable reduction or replacement of sugar in food products [21]. Consumers poorly perceive the use of synthetic sweeteners, which is why food manufacturers are increasingly willing to use natural sweeteners, for example, steviol glycosides extracted from stevia leaves (*Stevia rebaudiana*). These leaves also contain nutritional and antioxidant compounds such as vitamin C, polyphenols, carotenoids, chlorophylls, and other macro and micronutrients [22–24]. Thanks to the application of steviol glycosides, it is possible to reduce sugar content while maintaining proper sweetness of the product [25–27]. However, it is usually used as an addition to sucrose due to the bitter aftertaste of steviol glycosides [28] and high molecular weight.

The aim of this study was to investigate the effect of pretreatment performed by a combined method based on blanching, ultrasound, and vacuum application on the kinetics of osmotic dehydration in two solutions (61.5% sucrose solution and 30% sucrose solution with the addition of 0.1% of steviol glycosides). Moreover, selected quality properties of osmodehydrated cranberries as water activity, color and polyphenols, flavonoids, and anthocyanins content were evaluated.

2. Materials and Methods

2.1. Material

Swamp cranberry (Vaccinium oxycoccus) bought at the local market (Bronisze, Poland) was used as targeted matrix. Due to seasonality of the matrix, and to ensure reproducible results, cranberries from one batch, harvested in October 2017, have been frozen in a blast cabinet Irinox Shock Freezer HCM 51.20 (Irinox, Treviso, Italy) at a temperature of −25 °C for 5 h. Frozen fruits were stored at −18 °C no longer than 3 months. Only intact fruits of dark red coloring were used in the research.

2.2. Processing

2.2.1. Blanching

Cranberries are a difficult material to process due to their hard peel. In order to ensure proper mass transfer during OD, the material was blanched until the peel broke. Blanching was performed at a temperature of 90 °C for 5 min [29]. The frozen material, weighed on a laboratory scale (RADWAG, Radom, Poland) with an accuracy of ± 0.01 g, and a distilled water were prepared in a ratio of 2:1 (w/w).

After bringing the water to boiling, frozen cranberry fruits were placed in the water, which allowed them to reach the process temperature. During blanching, temperature was controlled by measuring the temperature. After 5 min of blanching, cranberry fruits were drained, dried on a filter paper, and re-weighed.

2.2.2. Combined Pretreatment

Prior to the unconventional processing, cranberry fruits were weighed on a laboratory scale (RADWAG, Radom, Poland) with an accuracy of ±0.01 g and placed in beakers. Blanched cranberries were flooded with osmotic solution in 1:4 ratio. Afterwards, fruits followed unconventional pretreatment with the use of: Sonication—ultrasounds at 21 kHz frequency and 180 W power (ultrasound intensity equal to 3.6 W/g) generated by ultrasonic bath (MKD-3 ULTRASONIC) were applied [30]. Time of sample treatment amounted to 20 min [5,14], and during the ultrasound application changes in temperature of surrounding solution were below 1 °C. Reduced pressure—a specifically prepared desiccator was used in pretreatment to lower the pressure by 300 mmHg (~40 kPa) by vacuum pump. After reducing the pressure (lowering the pressure down to abovementioned value lasted for 20 s), the pretreatment took 10 min and then the vacuum was cut, which took around 5 s.

The unconventional treatment was carried out in duplicate and applied in three variants:

1. A blanched material was placed in osmotic solution and the pressure was lowered three times in 10 min intervals (BL_3 × 10 v),
2. A blanched material was placed in osmotic solution and subjected to lowered pressure for 10 min, then sonicated for 20 min (BL_10 v 20 us),
3. A blanched material was placed in osmotic solution and sonicated for 20 min, afterwards it was subjected to lowered pressure for 10 min (BL_20 us 10 v).

All analyzed samples were summarized in Table 1 with their related abbreviations.

Table 1. Abbreviations of all researched samples of cranberries.

Treatment	Abbreviations	
After treatment, before osmotic dehydration		
Blanched material	BL	
	Osmotic solution	
	Sucrose 61.5% (s)	Sucrose 30% with 0.1% steviol glycosides addition (s + g)
Blanched material in osmotic solution under the pressure lowered three times in 10 min intervals (BL_3 × 10v)	BL_3 × 10 v_s	BL_3 × 10 v_s + g
Blanched material in osmotic solution subjected to lowered pressure for 10 min and sonicated for 20 min (BL_10 v 20us)	BL_10 v 20 us_s	BL_10 v 20 us_s + g
Blanched material in osmotic solution subjected to sonication for 20 min and to lowered pressure for 10 min (BL_20 us 10 v)	BL_20 us 10 v_s	BL_20 us 10 v_s + g
After osmotic dehydration		
Blanched material subjected to osmotic dehydration for 72 h (BL_72 h)	BL_s_72 h	BL_s + g_72 h
Blanched material in osmotic solution under the pressure lowered three times in 10 min intervals, and then subjected to osmotic dehydration for 72 h (BL_3 × 10 v_72 h)	BL_3 × 10 v_s_72 h	BL_3 × 10 v_s + g_72 h
Blanched material in osmotic solution subjected to lowered pressure for 10 min and sonicated for 20 min, and then osmotic dehydration for 72 h (BL_10 v 20 us_72 h)	BL_10 v 20 us_s_72 h	BL_10 v 20 us_s + g_72 h
Blanched material in osmotic solution subjected to sonication for 20 min and to lowered pressure for 10 min, and then osmotic dehydration for 72 h (BL_20 us 10 v_72 h)	BL_20 us 10 v_s_72 h	BL_20 us 10 v_s + g_72 h

BL - blanched material.

2.2.3. Osmotic Dehydration (OD)

OD was conducted in two solutions: 61.5% sucrose solution and 30% sucrose solution with addition of 0.1% steviol glycosides. A sucrose solution of 61.5% concentration was chosen since it is usually used as a standard solution [2,5,31]. However, a solution with the addition of steviol glycosides was used to reduce sugar content in dehydrated fruits. Due to the fact that steviol glycosides are up to 300 times sweeter than sucrose, the addition of little amount (0.1%) of this sweetener is enough to maintain its sweetness comparable with fruits dehydrated in a standard solution [25,32].

The pretreated fruits were placed in beakers and poured with osmotic solution in 1:4 ratio. OD was applied at a temperature of 40 °C for 72 h in a water bath with a stirrer at a rotation rate of 100 rotations per minute and at amplitude of 4. During OD, the process kinetics were determined after 1, 3, 6, 24, 48, and 72 h, measuring the mass changes, dry matter content, and water soluble solids. OD was performed in duplicate.

The analysis of cranberry fruits' osmotic dehydration kinetics were determined based on water loss WL (kg H_2O/kg of cranberry) and dry matter mass gain SG (kg d.m./kg of cranberry) [5]:

$$WL = (m_o X_o^w - m_t X_o^w)/m_o \tag{1}$$

$$SG = (m_t X_t^{ST} - m_o X_o^{ST})/m_o \tag{2}$$

where m_o—initial mass of cranberry prior to dehydration [kg], m_t—final mass of cranberry after dehydration [kg], X_o^w, X_o^{ST}—water and dry matter content prior to dehydration [kg/kg], X_t^w, X_t^{ST}—water and dry matter content after dehydration [kg/kg].

2.3. Cranberries Analysis

2.3.1. Dry Matter Content

Dry matter content was determined by weighing differently pretreated cranberries at 70 °C for 20 h [33]. The analysis was performed in triplicate.

2.3.2. Water Soluble Solids (Brix Index)

Water soluble solids were determined by squeezing the juice out of fruits with linen material. Squeezed extract was placed in refractometer glass ATAGO PAL-3, and results were in Brix degrees (°Bx) [2]. The measurement was done in a triplicate.

2.3.3. Water Activity

Water activity of cranberries were determined after a given time of OD. The measures were held with hygrometer Aqua Lab CX-2 (Decagon Devices Inc., United States) at 25 °C in three repetitions for a few randomly selected fruits [34].

2.3.4. Color

The color test of pretreated and dehydrated samples were performed with the use of chromameter of CM-5 type by Konica Minolta (Japan), with a reflection method in CIE L*a*b* system of the following parameters: Light source D65, angle 8 °C, standard observer CIE 2°. In CIE system L*, parameter refers to sample lightness, while a* coordinate denotes share of green (-) and red color (+), and b* coordinate describes share of blue (-) and yellow color (+). Prior to measurement the apparatus was black and white calibrated [35]. The measurement was repeated five times for each sample.

The absolute color difference was calculated based on the following equation:

$$\Delta E = \sqrt{(\Delta L^*)^2 + (\Delta a^*)^2 + (\Delta b^*)^2} \tag{3}$$

where ΔL*, Δa*, Δb* — indicator of color difference in comparison to blanched cranberry fruits.

2.3.5. Bioactive Compounds

Bioactive compounds as polyphenols, flavonoids, and anthocyanins content were measured using spectrophotometric methods according to Nowacka et al. [30]. The analyses were performed in triplicate. Polyphenolic content was expressed in mg of gallic acid in 100 g dry matter, flavonoid content in mg of quercetin in 1 g of dry matter, and anthocyanin content in mg of cyanidin-3-glucoside per 100 g of fresh mass.

2.4. Statistical Analysis

One-way analysis of variance (ANOVA) was performed in order to determine the impact of pretreatment and osmotic solution on the physical and chemical properties of blanched cranberry fruits. A detailed comparison allowed for a split of samples into homogenic groups. For this purpose, Duncan test was used with confidence interval of 95% (significance level $\alpha = 0.05$). Moreover, all investigated variables were used to perform cluster analysis using Ward method as a criterion for agglomeration and expressing the results using Euclidean distance. The analyses were carried out in Statistica 2013 software.

3. Results and Discussion

3.1. Kinetics of Osmotic Dehydration

OD causes two different types of mass changes. The first one is linked with water removal from dehydrated material towards solution with a higher osmotic potential. The second phenomenon occurring simultaneously is impregnation of the osmotic substance into the processed material [36,37]. Along with the progress of the OD, the mass transfer is hindered, due to the equalization of chemical potentials between material and solution. The intensity of this phenomenon is described by OD kinetics, by the means of water loss (WL) and soluble solid gain (SG) in osmodehydrated material during the process [5].

During OD, an increase of WL was observed for both osmotic solutions—sucrose and sucrose with the addition of steviol glycosides (Figures 1 and 2). In 61.5% of sucrose solution during the whole dehydration process a decrease of the water content in cranberries was observed (Figure 1). The WL was in the range of 0.35 kg H2O/kg (when sonication and reduced pressure treatment were used—BL_20 us 10 v_s) to 0.38 kg H2O/kg (when vacuum treatment was repeated three times—BL_3 × 10 v_s). The utilization of combined pretreatment significantly influenced the course of OD. The highest WL after 72 h of OD was noted after combined pretreatment with blanching and triplicate lowering of the pressure (BL_3 × 10 v_s). In turn, Nowacka et al. [32] reported that US applied as a pretreatment before OD increases the WL during OD, irrespective of the osmotic solution.

The use of 30% of sucrose with the addition of steviol glycosides caused a loss of water, which after 72 h of the process ranged from 0.19 to 0.25 kg H2O/kg, depending on the combined treatment used (Figure 2). The highest WL in the final stage of the OD was noted in the case of cranberry fruits subjected to triple vacuum treatment before the OD process (BL_3 × 10 v_s + g) and to only blanching (BL_s + g). However, in the case of combined methods in which US was used (BL_10 v 20 us_s + g; BL_20 us 10 v_s + g) a statistically lower water loss after 72 h of the OD was noted. Also, Sulistyawati et al. [38] noted statistically unchanged WL after reduced pressure treatment, compared with classical OD. A reverse dependence was noted by Feng et al. [39] for garlic but for a different osmotic solution—25% NaCl. The authors reported a significant increase in water loss as a result of both single vacuum (VOD) and ultrasound pretreatments (UOD), as well as for combined method—vacuum treatment with ultrasonic-assisted osmotic dehydration (VUOD) of garlic. As they have shown, the impact of vacuum treatment was more noticeable than the effect of sonication. However, when the combined method VUOD was used, the highest water loss was noted. Koubaa et al. [13] stated that US pretreatment accelerate the mass and heat transfer, which is linked with a cavitation effect provoked by sonication. Goula et al. [40] also observed a higher WL after US treatment before OD and US-assisted OD of potatoes in NaCl solution. Their findings and the results obtained herein suggest that the impact of the treatment was connected with the utilized osmotic solution.

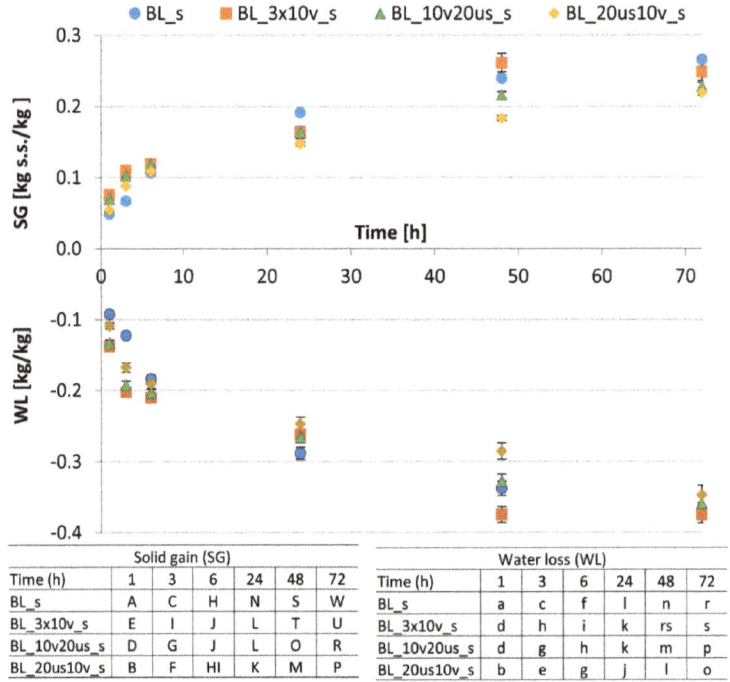

Solid gain (SG)							Water loss (WL)						
Time (h)	1	3	6	24	48	72	Time (h)	1	3	6	24	48	72
BL_s	A	C	H	N	S	W	BL_s	a	c	f	l	n	r
BL_3x10v_s	E	I	J	L	T	U	BL_3x10v_s	d	h	i	k	rs	s
BL_10v20us_s	D	G	J	L	O	R	BL_10v20us_s	d	g	h	k	m	p
BL_20us10v_s	B	F	HI	K	M	P	BL_20us10v_s	b	e	g	j	l	o

Figure 1. Water loss (WL) and solid gain (SG) as a function of the osmotic dehydration time of different treated cranberries in 61.5% sucrose solution. Different letter means significant difference by the Duncan test ($p < 0.05$).

With the progressing of the OD process, the content of dry matter increases, which is related to the penetration of substances from osmotic solution to material [36]. A solid gain (SG) after 72 h of the process performed in 61.5% sucrose was in the range of 0.22 kg d.m./kg of cranberry for BL_20 us 10 v_s to 0.27 kg d.m./kg of cranberry for BL_s (Figure 1). In turn, in the case of the osmotic solution with a lower sugar content (s + g) a lower solid gain was noted (Figure 2), which was connected with a lower driving force of the OD [38]. Similarly, as in the case of 61.5% sucrose, when s + g solution was used (Figure 2), the highest SG was observed for sample blanched before OD (BL_s + g), which reached the value of 0.17 kg d.m./kg of cranberry. Simultaneously, the lowest solid gain after OD, amounted to 0.13 kg d.m./kg of cranberry, and was characterized for material treated with US followed by reduced pressure treatment (BL_20 us 10 v_s + g). Statistical analysis revealed significant changes in SG as a result of both the time of OD and the type of different pretreatments. Contrary, Feng et al. [39] noted an irrelevant increase of SG in garlic dehydrated in NaCl after vacuum and the majority of US treatments [32]. Similar results were presented also in our study for cranberry pretreated by US and dehydrated in different osmotic solutions. Allahdad et al. [41] concluded that the impact of US-assisted OD of pomegranate arils on the SG was dependent on the applied US frequency. When the 40 kHz was used the SG was significantly higher than in the case of both 21 kHz and material without US treatment. This suggests that the treatment type and the treatment parameters should be adjusted for each material individually.

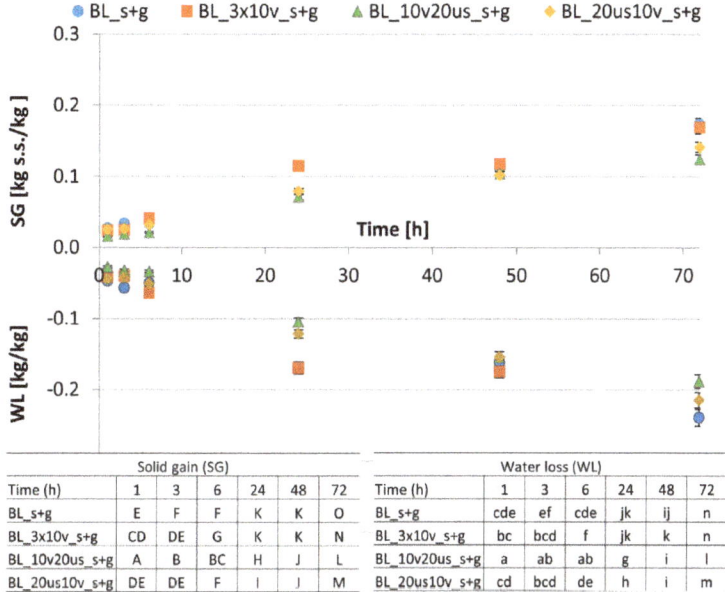

Figure 2. Water loss (WL) and solid gain (SG) as a function of the osmotic dehydration time of different treated cranberries in 30% sucrose solution with 0.1% addition of steviol glycosides. Different letter means significant difference by the Duncan test ($p < 0.05$).

3.2. Physical Properties—Dry Matter Content, Water Activity, Water Soluble Compounds, and Color

Depending on the variety, fresh cranberry fruits are characterized by a dry substance ranging from 12.1% to 14.5%, and water-soluble compounds are from 7 to 9.4 °Bx [42]. In the case of blanched fruits, similar results of water-soluble compounds (9.3 ± 0.2 °Bx) and dry matter content (14.6 ± 0.1%) were observed (Table 2). However, the combined pretreatment, consisting of the application of reduced pressure or combining reduced pressure with ultrasound conducted in an osmotic solution, caused a significant increase in the dry matter and Brix index in all cases. The increase in these parameters was related to the penetration of the components in the osmotic solution used during the pretreatment [36]. Pretreatment in a sucrose solution resulted in an increase in the dry matter content by 32.2% to 39.7%, depending on the combination used, whereas in the solution with the addition of steviol glycosides by 10.3% to 13.7%, compared to blanched cranberries. For water soluble compounds, a greater increase of brix index in cranberry fruits was observed while a higher concentration of sucrose solution was used. Blanched fruits treated with ultrasound before or after the application of reduced pressure treatment in sucrose solution (BL_20 us 10 v and BL_10 v 20 us_s) were characterized by the highest water-soluble compounds of 15.9 ± 0.4 and 15.9 ± 0.5° Bx, respectively. The higher increase in the brix index in the ultrasound-treated tissue could be related to the intensification of the OD process enhanced with ultrasound waves [42] and with the changes of the plant tissue subjected to sonication [43]. However, such changes were not noticed during the OD process in the solution with a lower sucrose content with the addition of steviol glycosides.

After 72 h of OD process a significant increase in the dry matter content and Brix index was observed, in the range from 28.9% to 45.9% and from 22.4 to 43.1 °Bx, respectively. The highest content of these parameters after the OD process was found in the blanched material (BL_s_72 h), and then in the tissue subjected to reduced pressure treatment (BL_3 × 10 v_s_72 h). In the case of combined pretreatment, the dry matter content was slightly lower and it was 45.2% for BL_10 v 20 us_s_72 h and 41.9% for BL_20 u s10 v_s_72 h. The similar effect, with lower values, was observed for cranberries subjected to the OD process in sucrose solution with the addition of steviol glycosides, which was linked to the lower driving force of the process [36].

Water activity is one of the key parameters determining the durability of food, affecting the survival and growth of microorganisms, enzymatic reactions, and oxidation. Water activity for most food products of plant origin is close to 1 [44]. For blanched and differently treated cranberries, water activity was in the range from 0.934 to 0.966 (Table 1). After the OD process a significant decrease in water activity for both solutions was noticed. In a solution with a higher concentration of sugar (61.5%) the water activity was in the range of 0.862 (BL_3 × 10 v_s_72 h) to 0.896 (BL_20 us 10 v_s_72 h). OD allowed reducing the water activity which limits only the growth of bacteria [44]. However, in the case of OD in a 30% sucrose solution with the addition of steviol glycosides, a smaller decrease was observed and the water activity reached a value from 0.914 (BL_3 × 10 v_s + g_72 h) to 0.919 (BL_20 us 10 v_s + g_72 h).

Table 2. Dry matter content, water activity, water soluble compounds, and color parameters in different treated cranberries, different letters in columns means significant difference by the Duncan test ($p < 0.05$).

Material	Dry Matter Content [%]	Water Activity [-]	Water Soluble Compounds [°Bx]	L*	a*	b*	Total Color Difference ΔE
BL	14.6 ± 0.1a	0.952 ± 0.009 fg	9.3 ± 0.2 a	9.1 ± 0.7 a	51.8 ± 0.6 i	15.7 ± 0.7 bc	–
After treatment s solution							
BL_3 × 10 v_s	19.3 ± 0.1 c	0.937 ± 0.006 de	14.5 ± 0.2 c	16.1 ± 0.8 h	40.7 ± 0.7 f	17.2 ± 0.6 efg	13.2 ± 0.8 c
BL_10 v 20 us_s	20.4 ± 0.1 c	0.950 ± 0.004 f	15.9 ± 0.5 d	21.1 ± 0.7 i	35.0 ± 0.6 cd	13.8 ± 0.7 a	20.8 ± 0.5 h
BL_20 us 10 v_s	19.3 ± 0.1 c	0.934 ± 0.005 e	15.9 ± 0.4 d	13.8 ± 1.3 d	35.5 ± 0.6 d	14.5 ± 0.8 a	17.1 ± 0.7 ef
s + g solution							
BL_3 × 10 v_s + g	16.6 ± 0.1 b	0.952 ± 0.007 fg	11.8 ± 0.2 b	14.7 ± 0.4 e	43.0 ± 0.9 g	14.1 ± 0.4 a	10.7 ± 0.6 b
BL_10 v 20 us_s + g	16.3 ± 0.3 b	0.963 ± 0.003 g	11.9 ± 0.1 b	12.9 ± 0.8 c	46.5 ± 0.8 h	16.4 ± 0.7 bcd	6.6 ± 0.8 a
BL_20 us 10 v_s + g	16.1 ± 0.1 b	0.966 ± 0.002 fg	11.9 ± 0.1 b	11.0 ± 0.5 b	47.3 ± 0.7 h	18.6 ± 0.6 h	5.8 ± 0.6 a
After osmotic dehydration s solution							
BL_s_72 h	45.9 ± 0.6 h	0.898 ± 0.018 b	43.1 ± 0.4 g	14.8 ± 0.7 e	34.2 ± 0.9 bc	16.7 ± 0.8 cde	18.6 ± 0.8 g
BL_3 × 10 v_s_72 h	45.2 ± 0.8 h	0.862 ± 0.014 a	42.5 ± 0.5 g	16.2 ± 0.5 h	34.0 ± 0.8 b	17.7 ± 0.7 g	19.3 ± 0.7 g
BL_10 v 20 us_s_72 h	43.1 ± 0.7 g	0.895 ± 0.002 b	36.1 ± 0.8 h	15.6 ± 0.7 fgh	31.9 ± 1.1 a	16.4 ± 0.5 bcd	21.0 ± 1.0 h
BL_20 u s10 v_s_72 h	41.9 ± 0.7 g	0.896 ± 0.009 b	36.1 ± 0.4 h	15.4 ± 1.1 ef	31.8 ± 0.8 a	15.9 ± 1.0 bc	21.0 ± 0.7 h
s + g solution							
BL_s + g_72 h	34.2 ± 1.1 f	0.916 ± 0.009 c	22.4 ± 0.2 e	12.9 ± 0.7 c	37.5 ± 1.0 e	15.6 ± 0.9 b	14.9 ± 0.9 d
BL_3 × 10 v_s + g_72 h	32.9 ± 0.9 f	0.914 ± 0.004 c	26.5 ± 0.8 g	16.1 ± 0.1 gh	37.3 ± 2.9 e	15.6 ± 2.3 b	16.3 ± 2.6 e
BL_10 v 20 us_s + g_72 h	28.9 ± 1.0 d	0.928 ± 0.005 d	23.0 ± 1.5 e	15.9 ± 0.3 fgh	35.7 ± 0.5 d	17.5 ± 0.3 fg	17.6 ± 0.8 f
BL_20 us 10 v_s + g_72 h	31.0 ± 0.9 e	0.919 ± 0.005 c	25.6 ± 0.9 f	15.4 ± 0.2 efg	34.0 ± 0.9 b	16.6 ± 1.6 cde	19.0 ± 0.5 g

Color is one of the most important features of the product, which consumers pay special attention to [35]. Cranberry is characterized by a red color, which changes during processing [33]. The blanched fruits were characterized by L*, a*, and b* parameters equal to 9.1, 51.8, and 15.7, respectively. The color changes were not unambiguous (Table 1). The use of reduced pressure treatment and combined treatment with US, and lowered pressure resulted in a significant increase in lightness (L*) and a decrease in the share of red color (a*), compared to the only blanched sample. OD processing caused further lightening and lower value of a* parameter, which is related to mass transfer form tissue to the surrounding solution. What is worth mentioning is that higher changes were observed for combined treatment, which could be linked to the acceleration of OD processes by the application of reduced pressure treatment and US [13,35]. Moreover, a significant increase in the total color difference ΔE was found with the value higher than five, which confirms that the treatment performed had a significant effect on the overall change in the color of the product. However, higher changes were noted for the samples processed in 61.5% sucrose solution, probably due to the higher driving force of OD process. The changes of color of OD fruits pretreated by blanching and blanching combined with reduced pressure in comparison to intact, blanched material are also related to higher sugar concentration. The presence of sugar affects the light reflection and thus changes the lightness, redness, and the yellowness of the material. Moreover, as described and discussed further, dehydrated samples contained less of the pigment compounds such as anthocyanins or flavonoids, which explains the smaller values of a* parameter when compared to the reference (BL) material. The sequence of application of reduced pressure treatment and ultrasound did not play a significant role in shaping the color of fruits after OD.

3.3. Bioactive Compounds

Cranberries contain a lot of different bioactive compounds [21]. Figure 3 presents the total phenolic content of cranberry fruits subjected to different pretreatment before and after OD. The highest total phenolic content (5241 ± 179 mg/100g d.m.) was stated for blanched samples. Combination of blanching with other pretreatment methods like reduced pressure treatment or sonication led to a decrease of phenolic compounds in comparison to the blanched material. In general, processing performed with 61.5% sucrose solution lead to smaller phenolic content of cranberries in comparison to fruits processed in a ternary solution composed of sucrose and steviol glycosides. Such situations are associated with higher leakage of polar phenolics to the surroundings during treatment in a more concentrated solution. The difference in phenolic content between samples processed with concentrated sucrose and ternary solution varied from 22.7% to 39.2% for material coded as BL_3 × 10 v_s and BL_20 us 10 v_s, respectively. The introduction of additional processing steps like reduced pressure treatment and sonication contributed to a significant reduction of phenolics in investigated fruits in comparison to less complicated operations. One of the possible explanations of such behavior is related to a sonoporation phenomenon which increases cell membrane permeability [45]. On the one hand, the sonoporation leads to degradation of the cellular structure and thus it may enhance the extractability of bioactive compounds. On the other hand, it implies improved mass transfer between fruits and osmotic medium which can positively impact on water loss but negatively influence to the phenolic phase promoting its leaching. Another explanation of smaller phenolic content of samples subjected to treatment with sonication is associated with the free radical formation during US application which can lead to degradation of antioxidants [46,47]. The impact of ultrasound on enzymatic activity is ambiguous and the effect of US on enzymes seems to depend on the matrix, treatment parameters, and type of the enzyme alike [48,49]. For instance, it has been reported that the combination of ascorbic acid with ultrasound was successful in the inactivation of polyphenoloxidase and peroxidase, whereas individual application of ascorbic acid and sonication did not give the desired inactivation of abovementioned enzymes in apples [50]. In turn, ultrasound treatment of diluted avocado puree increased activity of polyphenols oxidase by 180% depending on treatment [51]. Based on the results presented in the current study it can be stated that ruptured cellular structure and liberated chemicals were more susceptible for enzymatic degradation as the activity of enzymes was modified by sonication. This explanation is also supported by the fact that the phenolics content of fruits subjected only to blanching and reduced pressure treatment before OD was lower when compared to samples treated by a combination of blanching, vacuum treatment, and sonication. It is, furthermore, reported that ultrasounds can provoke conformation changes of enzymes, lead to dissociation of enzyme aggregates, promote collisions between substrate-enzyme, or activate latent enzymes of the tissue [49]. However, when ultrasound was used as treatment for 30 min for blanched cranberries it resulted in higher total phenolic content [52] than when combined methods with reduced pressure were used. OD for 72 h led to a further decrease of phenolics. Combined treatment in most cases did not cause significant differences in comparison to blanched osmodehydrated samples. Only fruits subjected to reduced pressure combined with sonication and dehydrated in 61.5% sucrose solution exhibited significantly higher phenolic content (1802 mg GAE/100 g d.m.) in comparison to the blanched material processed in the same osmotic medium (1465 mg GAE/100 g d.m.).

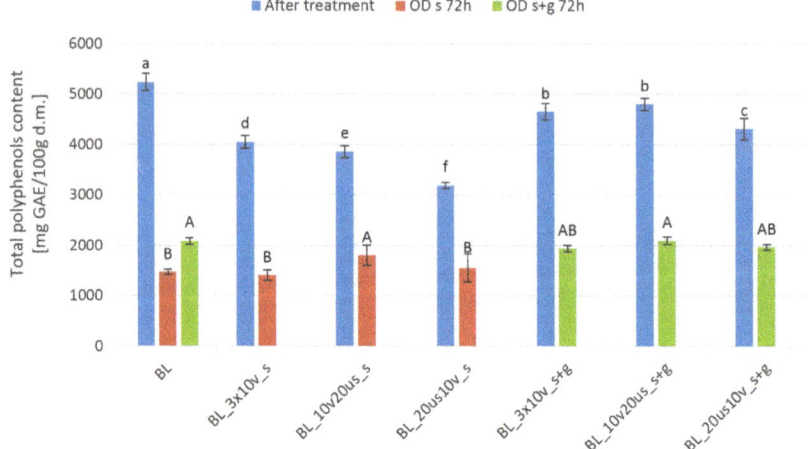

Figure 3. Total phenolic content in different treated cranberries: Blanched (BL), reduced pressure (BL_3 × 10 v), reduced pressure with sonication treatment (BL_10 v 20 us), sonication treatment with reduced pressure (BL_20 us 10 v), and after osmotic dehydration (OD) in 61.5% sucrose solution (s) and in 30% sucrose solution with 0.1% addition of steviol glycosides (s + g). Different lowercase letters means significant difference by the Duncan test ($p < 0.05$) for samples after treatment, and different capital letter means significant difference by the Duncan test ($p < 0.05$) for samples after 72 h osmotic dehydration.

Flavonoids content of investigated cranberries ranged from 30.0 to 45.8 mg/g d.m. and from 10.5 to 16.7 mg/g d.m. before and after 72 h of OD, respectively (Figure 4). The highest flavonoids content was stated for blanched samples before OD. The use of combined treatment decreased flavonoids content from 16% to 35% in the case of samples treated by sonication combined with reduced pressure in 61.5% sucrose solution (BL_20 us_10 v_s) and by reduced pressure treatment combined with sonication in 30% sucrose solution with addition of steviol glycosides (BL_10 v_20_us_s + g). As in the case of phenolic compounds, OD for 72 h significantly decreased flavonoids content regardless of the treatment protocol. In general, samples osmodehydrated in ternary solution (s + g) were characterized by higher flavonoids content than the samples processed in 61.5% sucrose solution (s). For instance, flavonoids content of fruits treated by reduced pressure followed by sonication in 30% sucrose solution with the addition of steviol glycosides was 26.2% higher in comparison to the material processed by the same method but in 61.5% sucrose solution. Such difference was most probably linked to the worse mass transfer of less concentrated solution [53]. It is worth emphasizing that samples subjected to sonication exhibited a higher concentration of flavonoids in comparison to the fruits treated only by reduced pressure treatment. What is more, when sonication was preceded by reduced pressure application, the retention of flavonoids was higher. Similar results were obtained for blanched cranberries treated with ultrasound for 30 min in sucrose solution [52]. These results point out that the effect of ultrasound on improvement of extractability is more pronounced when air in the extracellular space is replaced by liquid. This explanation fits to the data reported by Liu et al. [54] who found out that cavitation is enhanced in degassed water.

Figure 5 presents the anthocyanins content of fruits subjected to different treatments. Samples before OD were characterized by higher anthocyanins content, which ranged from 638 to 985 mg c3g/100 g f.m. The use of combined treatment decreased anthocyanins content in comparison to blanched material. Samples subjected to triple pressure reduction were characterized by significantly higher concentration of anthocyanins in comparison to samples treated by reduced pressure and sonication, regardless of the type of osmotic solution. However, longer sonication (30 min) of blanched cranberries resulted in higher content of anthocyanins, as mentioned in our previous study [52]. What is interesting is that after OD the trend was reversed, which indirectly confirms the ambiguous character of sonication either to improve extractability of bioactive compounds or to cause their degradation. The utilization of 61.5% sucrose solution resulted in better anthocyanins preservation than the use of low concentrated (30%) sucrose solution with the addition of steviol glycosides. These findings fall into place with a protective role of sugar on anthocyanins stability due to the decreased water mobility as demonstrated by Tsai et al. [55] and Watanabe et al. [56].

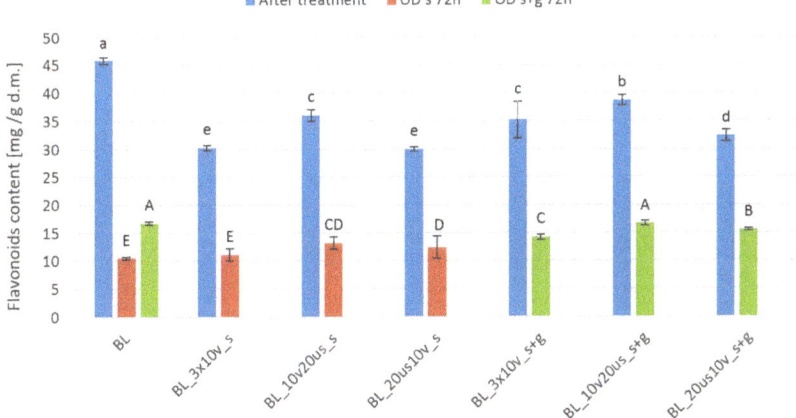

Figure 4. Flavonoids content in different treated cranberries: Blanched (BL), reduced pressure (BL_3 × 10 v), reduced pressure with sonication treatment (BL_10 v 20 us), sonication treatment with reduced pressure (BL_20 us 10 v), and after osmotic dehydration (OD) in 61.5% sucrose solution (s) and in 30% sucrose solution with 0.1% addition of steviol glycosides (s + g). Different lowercase letters means significant difference by the Duncan test ($p < 0.05$) for samples after treatment, and different capital letters means significant difference by the Duncan test ($p < 0.05$) for samples after 72 h osmotic dehydration.

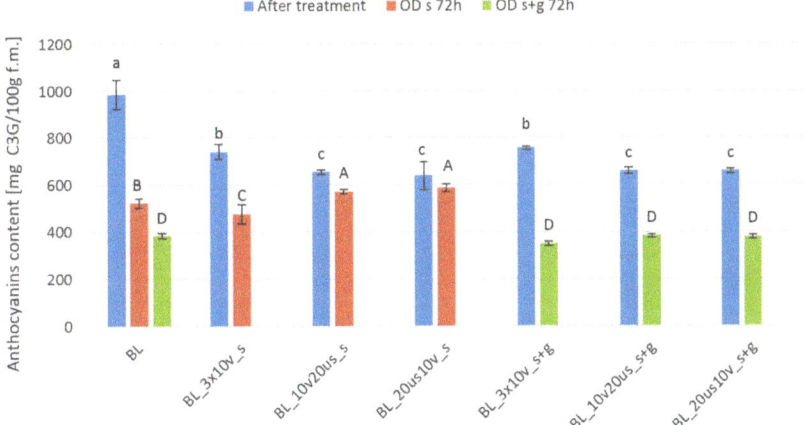

Figure 5. Anthocyanins content in different treated cranberries: Blanched (BL), reduced pressure (BL_3 × 10 v), reduced pressure with sonication treatment (BL_10 v 20 us), sonication treatment with reduced pressure (BL_20 us 10 v), and after osmotic dehydration (OD) in 61.5% sucrose solution (s) and in 30% sucrose solution with 0.1% addition of steviol glycosides (s + g). Different lowercase letters means significant difference by the Duncan test ($p < 0.05$) for samples after treatment, and different capital letters means significant difference by the Duncan test ($p < 0.05$) for samples after 72 h osmotic dehydration.

3.4. Cluster Analysis

Figures 6 and 7 present the results of cluster analysis performed on the basis of all investigated variables for samples not subjected and subjected to OD, respectively. Samples before OD were divided into three groups. The first cluster was composed of only blanched material, whereas second and third group was built by samples subjected to treatment performed in ternary (30% of sucrose with the addition of steviol glycosides) and 61.5% sucrose solution, respectively. Based on the cluster analysis results it can be stated that samples subjected to dehydration in less concentrated solutions were more similar to blanched material than samples processed in a concentrated sucrose solution. For materials after dehydration, clusters were formed upon the type of osmotic medium as well. It means that all fruits dehydrated in 61.5% sucrose solution formed one group while processing in 30% sucrose solution with the addition of steviol glycosides resulted in formation of a separate group.

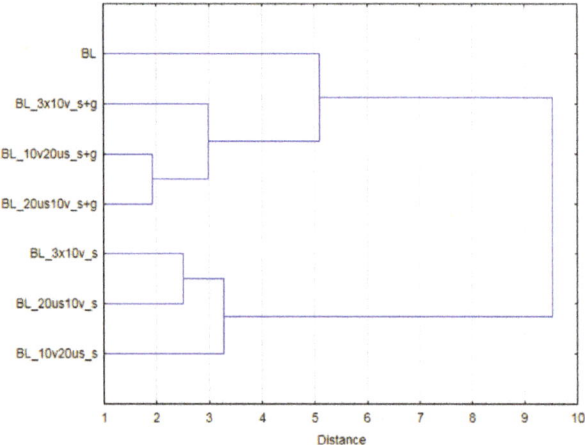

Figure 6. The results of cluster analysis (considering all analyzed variables) for samples not subjected to osmotic dehydration (BL - blanched material).

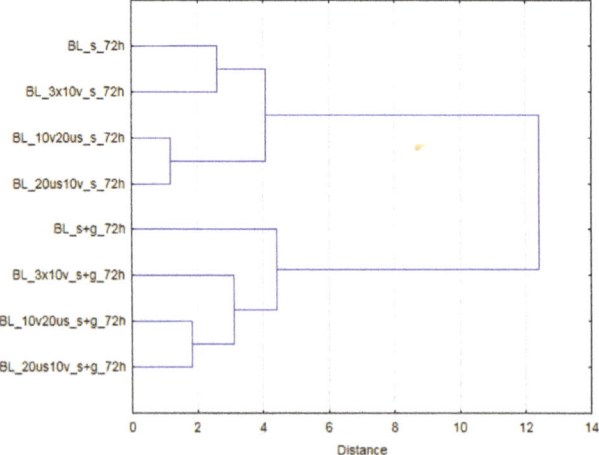

Figure 7. The results of cluster analysis (considering all analyzed variables) for samples subjected to osmotic dehydration.

4. Conclusions

Unconventional pretreatment of cranberries caused a significant increase of osmotic dehydration effectiveness, particularly in the case of a material dehydrated in 61.5% sucrose solution, previously blanched, submitted to 20 min sonication, followed by low pressure application.

Cranberries subjected to combined treatment, in particular to ultrasounds, had comparable or higher polyphenolic, anthocyanin and flavonoids content than a blanched tissue subjected to osmotic dehydration. Taking into account evaluated physical and chemical properties of dehydrated cranberries and the osmotic dehydration process it has been concluded that the best combined pretreatment method was a 20 min sonication followed by a 10 min lowered pressure treatment. However, further optimization studies are required.

Author Contributions: Conceptualization, M.N. and D.W.-R.; Methodology, A.W., M.N., K.R., A.A., and M.D.; Validation, K.R., A.A., and L.M.; Formal analysis, M.N. and A.W.; Investigation, M.N., K.R., L.M., and A.A.; Writing—original draft preparation, M.N., A.W., M.D., and D.W.-R.; Writing—review and editing, M.N., A.W., M.D., and D.W.-R.; Visualization, M.N. and A.W.; Supervision, M.N., A.W. and D.W.-R.

Funding: This research was funded by MINISTRY OF SCIENCE AND HIGHER EDUCATION within a framework of Iuventus Plus programme in the years 2015–2018, grant number IP2014 033173. The work was also co-financed by a statutory activity subsidy from the Polish Ministry of Science and Higher Education for the Faculty of Food Sciences of Warsaw University of Life Sciences.

Conflicts of Interest: The authors declare no conflict of interest.

References

1. Jeszka-Skowron, M.; Zgoła-Grześkowiak, A.; Stanisz, E.; Waśkiewicz, A. Potential health benefits and quality of dried fruits: Goji fruits, cranberries and raisins. *Food Chem.* **2017**, *221*, 228–236. [CrossRef] [PubMed]
2. Rząca, M.; Witrowa-Rajchert, D.; Tylewicz, U.; Dalla, R.M. Mass exchange in osmotic dehydration process of Kiwi fruits. *ŻYWNOŚĆ Nauk Technol Jakość.* **2009**, *6*, 140–149. (in Polish).
3. Bromberger, S.M.; Schmaltz, S.; Wesz, R.F.; Salvalaggio, R.; Lisiane, d.M.T. Effects of pretreatment ultrasound bath and ultrasonic probe, in osmotic dehydration, in the kinetics of oven drying and the physicochemical properties of beet snacks. *J. Food Process. Preserv.* **2018**, *42*, 13393. [CrossRef]
4. Witrowa-Rajchert, D.; Wiktor, A.; Sledz, M.; Nowacka, M. Selected emerging technologies to enhance the drying process: A review. *Dry Technol.* **2014**, *32*, 1386–1396. [CrossRef]
5. Nowacka, M.; Tylewicz, U.; Laghi, L.; Dalla, R.M.; Witrowa-Rajchert, D. Effect of ultrasound treatment on the water state in kiwifruit during osmotic dehydration. *Food Chem.* **2014**, *144*, 18–25. [CrossRef] [PubMed]
6. Misra, N.N.; Koubaa, M.; Roohinejad, S.; Juliano, P.; Alpas, H.; Inácio, R.S.; Saraiva, J.A.; Barba, F.J. Landmarks in the historical development of twenty first century food processing technologies. *Food Res. Int.* **2017**, *97*, 318–339. [CrossRef] [PubMed]
7. Cárcel, J.A.; García-Pérez, J.V.; Benedito, J.; Mulet, A. Food process innovation through new technologies: Use of ultrasound. *J. Food Eng.* **2012**, *110*, 200–207. [CrossRef]
8. Awad, T.S.; Moharram, H.A.; Shaltout, O.E.; Asker, D.; Youssef, M.M. Applications of ultrasound in analysis, processing and quality control of food: A review. *Food Res. Int.* **2012**, *48*, 410–427. [CrossRef]
9. Roselló-soto, E.; Galanakis, C.M.; Brnčić, M.; Orlien, V.; Francisco, J.; Mawson, R.; Knoerzer, K.; Tiwari, B.K.; Barba, F.J. Clean recovery of antioxidant compounds from plant foods, by-products and algae assisted by ultrasounds processing. Modeling approaches to optimize processing conditions. *Trends. Food Sci. Technol.* **2015**, *42*, 134–149. [CrossRef]
10. Zinoviadou, K.G.; Galanakis, C.M.; Brn, M.; Grimi, N.; Boussetta, N.; Mota, M.J.; Saraiva, J.A.; Patras, A.; Tiwari, B.; Barba, F.J. Fruit juice sonication: Implications on food safety and physicochemical and nutritional properties. *Food Res. Int.* **2015**, *77*, 743–752. [CrossRef]
11. Barba, F.J.; Brianceau, S. Effect of alternative physical treatments (ultrasounds, pulsed electric fields, and high-voltage electrical discharges) on selective recovery of bio-compounds from fermented grape pomace. *Food Bioprocess. Technol.* **2015**, *8*, 1139–1148. [CrossRef]

12. Zhu, Z.; Wu, Q.; Di, X.; Li, S.; Barba, F.J.; Koubaa, M.; Roohinejad, S.; Xiong, X.; He, J. Multistage recovery process of seaweed pigments: investigation of ultrasound assisted extraction and ultra-filtration performances. *Food Bioprod. Process.* **2017**, *104*, 40–47. [CrossRef]
13. Koubaa, M.; Barba, F.J.; Grimi, N.; Mhemdi, H.; Koubaa, W.; Boussetta, N.; Vorobiev, E. Recovery of colorants from red prickly pear peels and pulps enhanced by pulsed electric field and ultrasound. *Innov. Food Sci. Emerg. Technol.* **2016**, *37*, 336–344. [CrossRef]
14. Nowacka, M.; Tylewicz, U.; Romani, S.; Dalla Rosa, M.; Witrowa-Rajchert, D. Influence of ultrasound-assisted osmotic dehydration on the main quality parameters of kiwifruit. *Innov. Food Sci. Emerg. Technol.* **2017**, *41*, 71–78. [CrossRef]
15. Amami, E.; Khezami, W.; Mezrigui, S.; Badwaik, L.S.; Bejar, A.K.; Perez, C.T.; Kechaou, N. Effect of ultrasound-assisted osmotic dehydration pretreatment on the convective drying of strawberry. *Ultrason. Sonochem.* **2017**, *36*, 286–300. [CrossRef] [PubMed]
16. Corrêa, J.L.G.; Rasia, M.C.; Mulet, A.; Cárcel, J.A. Influence of ultrasound application on both the osmotic pretreatment and subsequent convective drying of pineapple (Ananas comosus). *Innov. Food Sci. Emerg. Technol.* **2017**, *41*, 284–291. [CrossRef]
17. Fito, P.; Andrés, A.; Chiralt, A.; Pardo, P. Coupling of hydrodynamic mechanism and deformation-relaxation phenomena during vacuum treatments in solid porous food-liquid systems. *J. Food Eng.* **1996**, *27*, 229–240. [CrossRef]
18. Castelló, M.L.; Igual, M.; Fito, P.J.; Chiralt, A. Influence of osmotic dehydration on texture, respiration and microbial stability of apple slices (var. Granny Smith). *J. Food Eng.* **2009**, *91*, 1–9. [CrossRef]
19. Tappi, S.; Tylewicz, U.; Romani, S.; Siroli, L.; Patrignani, F.; Dalla, R.M.; Rocculi, P. Optimization of vacuum impregnation with calcium lactate of minimally processed melon and shelf-life study in real storage conditions. *J. Food Sci.* **2016**, *81*, E2734–E2742. [CrossRef] [PubMed]
20. Deng, Y.; Zhao, Y. Effect of pulsed vacuum and ultrasound osmopretreatments on glass transition temperature, texture, microstructure and calcium penetration of dried apples (Fuji). *LWT - Food Sci. Technol.* **2008**, *41*, 1575–1585. [CrossRef]
21. Kowalska, K.; Olejnik, A. Beneficial effects of cranberry in the prevention of obesity and related complications: Metabolic syndrome and diabetes—A review. *J. Funct. Foods.* **2016**, *20*, 171–181. [CrossRef]
22. Kovačević, D.B.; Maras, M.; Barba, F.J.; Granato, D.; Roohinejad, S.; Mallikarjunan, K.; Montesano, D.; Lorenzo, J.M.; Putnik, P. Innovative technologies for the recovery of phytochemicals from Stevia rebaudiana Bertoni leaves: A review. *Food Chem.* **2018**, *268*, 513–521. [CrossRef] [PubMed]
23. Koubaa, M.; Area, F.S. Current and new insights in the sustainable and green recovery of nutritionally valuable compounds from *Stevia rebaudiana* Bertoni. *J. Agric. Food Chem.* **2015**, *63*, 6835–6846. [CrossRef] [PubMed]
24. José, F.; Nieves, M.; Belda-galbis, C.M.; José, M.; Rodrigo, D. Stevia rebaudiana Bertoni as a natural antioxidant/antimicrobial for high pressure processed fruit extract: Processing parameter optimization. *Food Chem.* **2014**, *148*, 261–267.
25. Nowacka, M.; Fijalkowska, A.; Wiktor, A.; Rybak, K.; Dadan, M.; Witrowa-Rajchert, D. Changes of mechanical and thermal properties of cranberries subjected to ultrasound treatment. *Int. J. Food Eng.* **2017**, *13*, 1–12. [CrossRef]
26. Oliveira, F.I.P.; Rodrigues, S.; Fernandes, F.A.N. Production of low calorie Malay apples by dual stage sugar substitution with Stevia-based sweetener. *Food Bioprod. Process.* **2012**, *90*, 713–718. [CrossRef]
27. Carbonell-Capella, J.M.; Barba, F.J.; Esteve, M.J.; Frígola, A. High pressure processing of fruit juice mixture sweetened with Stevia rebaudiana Bertoni: Optimal retention of physical and nutritional quality. *Innov. Food Sci. Emerg. Technol.* **2013**, *18*, 48–56. [CrossRef]
28. Chranioti, C.; Chanioti, S.; Tzia, C. Comparison of spray, freeze and oven drying as a means of reducing bitter aftertaste of steviol glycosides (derived from *Stevia rebaudiana* Bertoni plant) — Evaluation of the final products. *Food Chem.* **2016**, *190*, 1151–1158. [CrossRef]
29. Wiktor, A.; Sledz, M.; Nowacka, M.; Rybak, K.; Chudoba, T.; Lojkowski, W.; Witrowa-Rajchert, D. The impact of pulsed electric field treatment on selected bioactive compound content and color of plant tissue. *Innov. Food Sci. Emerg. Technol.* **2015**, *30*, 69–78. [CrossRef]

30. Nowacka, M.; Wiktor, A.; Anuszewska, A.; Dadan, M.; Rybak, K.; Witrowa-Rajchert, D. The application of innovative technologies as pulsed electric field, ultrasound and microwave-vacuum drying in the production of dried cranberry snacks. *Ultras. Sonochem.* **2019**, *56*, 1–13. [CrossRef]
31. Ciurzyńska, A.; Kowalska, H.; Czajkowska, K.; Lenart, A. Osmotic dehydration in production of sustainable and healthy food. *Trends. Food Sci. Technol.* **2016**, *50*, 186–192. [CrossRef]
32. Nowacka, M.; Tylewicz, U.; Tappi, S.; Siroli, L.; Lanciotti, R.; Romani, S.; Witrowa-Rajchert, D. Ultrasound assisted osmotic dehydration of organic cranberries (*Vaccinium oxycoccus*): Study on quality parameters evolution during storage. *Food Control* **2018**, *93*, 40–47. [CrossRef]
33. AOAC International. Official Methods of Analysis of AOAC International. Available online: https://www.aoac.org/aoac_prod_imis/AOAC/Publications/Official_Methods_of_Analysis/AOAC_Member/Pubs/OMA/AOAC_Official_Methods_of_Analysis.aspx (accessed on 10 June 2015).
34. Nowacka, M.; Fijalkowska, A.; Wiktor, A.; Dadan, M.; Tylewicz, U.; Dalla, R.M.; Witrowa-Rajchert, D. Influence of power ultrasound on the main quality properties and cell viability of osmotic dehydrated cranberries. *Ultrasonics* **2018**, *83*, 33–41. [CrossRef] [PubMed]
35. Fijalkowska, A.; Nowacka, M.; Witrowa-Rajchert, D. The physical, optical and reconstitution properties of apples subjected to ultrasound before drying. *Ital. J. Food Sci.* **2017**, *29*, 343–356.
36. Kowalska, H.; Lenart, A. Mass exchange during osmotic pretreatment of vegetables. *J. Food Eng.* **2001**, *49*, 137–140. [CrossRef]
37. Waliszewski, K.; Delgado, J.; Garcia, M. Equilibrium concentration and water and sucrose diffusivity in osmotic dehydration of pineapple slabs. *Dry Technol.* **2002**, *20*, 527–553. [CrossRef]
38. Sulistyawati, I.; Dekker, M.; Fogliano, V.; Verkerk, R. Osmotic dehydration of mango: Effect of vacuum impregnation, high pressure, pectin methylesterase and ripeness on quality. *LWT - Food Sci. Technol.* **2018**, *98*, 179–186. [CrossRef]
39. Feng, Y.; Yu, X.; Yagoub, A.E.G.A.; Xu, B.; Wu, B.; Zhang, L.; Zhou, C. Vacuum pretreatment coupled to ultrasound assisted osmotic dehydration as a novel method for garlic slices dehydration. *Ultrason Sonochem.* **2018**, *50*, 363–372. [CrossRef]
40. Goula, A.M.; Kokolaki, M.; Daftsiou, E. Use of ultrasound for osmotic dehydration. The case of potatoes. *Food Bioprod. Process.* **2017**, *105*, 157–170. [CrossRef]
41. Teleszko, M. American cranberry (*Vaccinium macrocarpon* L.)—possibility of using it to produce bio-food. *ŻYWNOŚĆ Nauk. Technol. Jakość.* **2011**, *6*, 132–141. (in Polish).
42. Fernandes, F.A.N.; Oliveira, F.I.P.; Rodrigues, S. Use of ultrasound for dehydration of papayas. *Food Bioprocess. Technol.* **2008**, *1*, 339–345. [CrossRef]
43. Fijalkowska, A.; Nowacka, M.; Wiktor, A.; Sledz, M.; Witrowa-Rajchert, D. Ultrasound as a pretreatment method to improve drying kinetics and sensory properties of dried apple. *J. Food Process. Eng.* **2016**, *39*, 256–265. [CrossRef]
44. Şen, F.; Karaçali, İ.; Eroğul, D. Effects of storage conditions and packaging on moisture content, water activity and tissue hardness of dried apricots. *Meyve. Bilim. Sci.* **2015**, *2*, 45–49.
45. Wang, M.; Zhang, Y.; Cai, C.; Tu, J.; Guo, X.; Zhang, D. Sonoporation-induced cell membrane permeabilization and cytoskeleton disassembly at varied acoustic and microbubble-cell parameters. *Sci. Rep.* **2018**, *8*, 3885. [CrossRef] [PubMed]
46. Bonnafous, P.; Vernhes, M.C.; Teissié, J.; Gabriel, B. The generation of reactive-oxygen species associated with long-lasting pulse-induced electropermeabilisation of mammalian cells is based on a non-destructive alteration of the plasma membrane. *Biochim. Biophys. Acta (BBA)-Biomembr.* **1999**, *1461*, 123–134. [CrossRef]
47. Okitsu, K.; Iwasaki, K.; Yobiko, Y.; Bandow, H.; Nishimura, R.; Maeda, Y. Sonochemical degradation of azo dyes in aqueous solution: A new heterogeneous kinetics model taking into account the local concentration of OH radicals and azo dyes. *Ultrason. Sonochem.* **2005**, *12*, 255–262. [CrossRef] [PubMed]
48. Delgado-Povedano, M.M.; De Castro, M.L. A review on enzyme and ultrasound: A controversial but fruitful relationship. *Anal. Chim. Acta* **2015**, *889*, 1–21. [CrossRef] [PubMed]
49. Rojas, M.L.; Hellmeister Trevilin, J.; Augusto, D.; Esteves, P. The ultrasound technology for modifying enzyme activity. *Sci. Agropecu.* **2016**, *7*, 145–150. [CrossRef]
50. Jang, J.H.; Moon, K.D. Inhibition of polyphenol oxidase and peroxidase activities on fresh-cut apple by simultaneous treatment of ultrasound and ascorbic acid. *Food Chem.* **2011**, *124*, 444–449. [CrossRef]

51. Bi, X.; Hemar, Y.; Balaban, M.O.; Liao, X. The effect of ultrasound on particle size, color, viscosity and polyphenol oxidase activity of diluted avocado puree. *Ultrason Sonochem.* **2015**, *27*, 567–575. [CrossRef]
52. Nowacka, M.; Fijalkowska, A.; Dadan, M.; Rybak, K.; Wiktor, A.; Witrowa-Rajchert, D. Effect of ultrasound treatment during osmotic dehydration on bioactive compounds of cranberries. *Ultrasonics* **2018**, *83*, 18–25. [CrossRef] [PubMed]
53. Ścibisz, I.; Mitek, M. Antioxidant activity and phenolics compound capacity in dried highbush blueberries (*Vaccinium corymbosum* L.). *ŻYWNOŚĆ Nauk. Technol. Jakość.* **2006**, *4*, 68–76. (in Polish).
54. Liu, L.; Yang, Y.; Liu, P.; Tan, W. The influence of air content in water on ultrasonic cavitation field. *Ultrason. Sonochem.* **2014**, *21*, 566–671. [CrossRef] [PubMed]
55. Tsai, P.J.; Hsieh, Y.Y.; Huang, T.C. Effect of sugar on anthocyanin degradation and water mobility in a roselle anthocyanin model system using 17O NMR. *J. Agric. Food Chem.* **2004**, *52*, 3097–4009. [CrossRef] [PubMed]
56. Watanabe, Y.; Yoshimoto, K.; Okada, Y.; Nomura, M. Effect of impregnation using sucrose solution on stability of anthocyanin in strawberry jam. *LWT - Food Sci. Technol.* **2011**, *44*, 891–905. [CrossRef]

© 2019 by the authors. Licensee MDPI, Basel, Switzerland. This article is an open access article distributed under the terms and conditions of the Creative Commons Attribution (CC BY) license (http://creativecommons.org/licenses/by/4.0/).

Article

Improved Physicochemical and Structural Properties of Blueberries by High Hydrostatic Pressure Processing

Maria Paciulli [1], Ilce Gabriela Medina Meza [2,*], Massimiliano Rinaldi [1], Tommaso Ganino [1,3], Alessandro Pugliese [1], Margherita Rodolfi [1], Davide Barbanti [1], Michele Morbarigazzi [4] and Emma Chiavaro [1]

1. Department of Food and Drug, University of Parma, Parco Area delle Scienze 27/A, 43124 Parma, Italy
2. Department of Biosystems and Agricultural Engineering, Michigan State University, East Lansing, MI 48824-1323, USA
3. Consiglio Nazionale delle Ricerche, Istituto per la Valorizzazione del Legno e delle Specie Arboree (IVaLSA), 50019 Sesto Fiorentino, Italy
4. HPP Italia, Via E. Carbognani 6, Traversetolo, 43029 PR, Italy
* Correspondence: ilce@msu.edu; Tel.: +1-517-884-1971

Received: 15 June 2019; Accepted: 18 July 2019; Published: 21 July 2019

Abstract: The use of high pressure on fruits and vegetables is today widely studied as an alternative to the traditional thermal preservation techniques, with the aim of better preserving nutritional and organoleptic properties. The use of high hydrostatic pressures (400–600 MPa; 1–5 min; room temperature) was tested on the physicochemical and structural properties of blueberries, in comparison to raw and blanched samples. High hydrostatic pressures led to higher tissue damages than blanching, related to the intensity of the treatment. The cellular damages resulted in leakage of intracellular components, such as bioactive molecules and enzymes. As a consequence, among the high pressure treatments, the resulting antioxidant activity was higher for samples treated for longer times (5 min). Pectinmethyl esterase (PME), deactivated by blanching, but strongly barotolerant, was more active in blueberries treated with the more intense high pressure conditions. Blueberry texture was better retained after high pressure than blanching, probably because of the PME effect. Blueberry color shifted towards purple tones after all of the treatments, which was more affected by blanching. Principal component analysis revealed the mild impact of high pressure treatments on the organoleptic properties of blueberries.

Keywords: high pressure; blanching; fruit; microscopy; pectin methyl esterase; texture; color; antioxidant activity

1. Introduction

Blueberries are nowadays a very popular fruit because of their sensory and health related properties. Blueberries, indeed, in addition to being appreciated for their color, flavor, texture, and juiciness, are also a natural source of bioactive compounds. The so-called polyphenols, including anthocyanins, the molecules responsible for blueberry color, are powerful antioxidants with recognized anti-inflammatory and antihypertensive properties, in addition to being involved in many cell regulation pathways [1].

Blueberries are often consumed fresh, but untreated fruit show short storage life, usually due to both microbial and enzymatic spoilage. The presence of enzymes such as peroxidase (POD), polyphenol oxidase (PPO), lipoxygenase (LOX), and lipase may be responsible for color and flavor changes, while pectin-methylesterase (PME) and polygalacturonase (PG) are involved in texture modifications [2]. The application of preservation processes, in order to guarantee consumer safety and extend shelf life, thus, became necessary. Blueberries are generally sold frozen, dehydrated, or heat treated. Such treatments induce important modifications in the product quality. Traditional thermal preservation methods, such as canning, make them susceptible to significant losses in their natural sensorial (i.e., aroma, flavor) and nutritional (bioactive compounds) properties [3]. Moreover, freezing and drying have a huge effect on fruit structure and texture [4].

Retention of nutritional value and freshness of fruit and vegetables (F&V) is a major challenge for the food industry. Novel processing technologies can address these issues. Among them, high hydrostatic pressure (HHP) is an attractive technology, because of the use of low temperatures during processing (i.e., room temperature), combined with high hydrostatic pressures (100–1000 MPa) and short time (a few seconds or minutes). HHP belongs to the so-called "non-thermal technologies", which are capable of retaining low molecular weight food compounds (i.e., flavoring agents, pigments, and vitamins) by not affecting covalent bonds [5,6]. Additionally, recent attention has been paid to the effects of HHP on the color attributes of F&V, since HHP can increase the intensity of color characteristics because of cell lysis and subsequent leakage of pigments [7]. The effect of high hydrostatic pressure (HHP) on the improvement of the technological functionalities of polymers, such as proteins and polysaccharides, has been recently studied [8,9]. Texture is an important quality attribute in F&V, and indicates freshness from a consumer point-of-view. The structural integrity and texture of F&V is attributed mainly to the primary cell wall, the middle lamella, and the turgor generated within cells by osmosis [10]. The primary cell wall basic structure consists of a cellulose–hemicellulose network [11]. The firmness developed by these two polysaccharides is not affected by processing or storage; however, pectin is affected by both enzymatic and non-enzymatic reactions. Pectin is the main constituent of the middle lamella, and gives firmness and elasticity to tissues [12]. All of these positive results indicate that HHP may be useful for retaining major structural quality characteristics of F&V.

Based on our knowledge, the application of HHP on whole blueberries has mainly been used to investigate microbial inactivation [13] or biomolecule extraction [14]. No studies are available on the effect of high hydrostatic pressure on the physicochemical and structural properties of whole blueberry fruit. The aim of this study was to evaluate the related properties affecting blueberry structure, color quality, and antioxidant capacity after HHP treatments in comparison with conventional thermal processing.

2. Material and Methods

2.1. Sample Preparation

Blueberries (*Vaccinium corymbosum*, cv. Duke) were purchased from a local market (Parma, Italy). After washing and draining, whole samples from the same batch and with similar dimensions (10 ± 0.1 mm) were selected and used for the study.

2.2. Treatments

For each sample, six conditions were analyzed: Raw (R), Blanched (BL), and treated with high hydrostatic pressure under four different conditions, as described later.

2.2.1. Blanching

Blueberries were blanched in a Combi-Steam SL (V-Zug, Zurich, Switzerland) oven, which presented an internal volume of 0.032 m^3, an air speed of 0.5 m s^{-1}, and a steam injection rate of 0.03 kg min^{-1}, at 95 °C for two min in accordance with Sablani et al. [15]. Nine samples were arranged in a circle and one was placed at the center to ensure uniform heating conditions. The treatment was conducted in triplicate.

2.2.2. High Hydrostatic Pressure

The treatments were conducted in a 300 L high pressure plant (Avure Technologies Inc., Kent, WA, USA), at the "HPP Italia" of Traversetolo (Italy). Ten blueberries were vacuum sealed inside flexible (75 mm thickness) plastic pouches (Ultravac Solutions, Kansas City, MO, USA). Cold water (4 °C) was used as pressure medium. HHP treatments were conducted at 400 and 600 MPa, both for 1 and 5 min; the samples were, thus, called 400-1, 400-5, 600-1, and 600-5. These conditions were chosen on the basis of preliminary experiments performed on the same types of samples [16]. Treatments were conducted at room temperature (20 ± 1 °C). Temperature increase due to compression was not higher than 2–3 °C/100 MPa. Three pouches were processed and analyzed for each treatment condition.

2.3. Moisture Content

Moisture content was determined according to the Association of Official Agricultural Chemists (AOAC) method [17] on both raw and treated samples. Almost 5 g of homogenized sample (as triplicate) were dried in a convection oven (ISCO NSV 9035, ISCO, Milan, Italy) at 105 °C for at least 16 h until constant weight was reached. The samples were stored at 4 °C and their moisture content was determined one day after HHP and thermal treatments.

2.4. Histological Analysis

The samples were fixed in formalin, acetic acid, alcohol (FAA) solution (formalin: acetic acid: 60% ethanol solution, 2:1:17 v/v) [18]. After two weeks, they were dehydrated using increasingly concentrated alcoholic solutions. The inclusion was made in a methacrylate resin (Technovit 7100, Heraeus Kulzer and Co., Wehrheim, Germany) and the resulting blocks were sectioned with transversal cuts at 3 µm thickness using a semithin Leitz 1512 microtome (Leitz, Wetzlar, Germany). The sections were stained with Toluidine Blue (TBO) and with a solution containing $FeSO_4$ for the tannin analysis [18]. Four pieces of fruit were sampled for each treatment. Sections were observed with a Leica DM 4000B optical microscope (Leica Imaging Systems Ltd., Wetzlar, Germany) equipped with a Leica DMC 2900 digital camera (Leica Imaging Systems Ltd., Wetzlar, Germany). The tissues were measured using an image analysis system (LAS v4.10.0, Leica Application Suite, Wetzlar, Germany). The microscopic observations were carried out by observing at least ten slides carrying ten sections each, for each specimen. The image analyses were carried out using a manual configuration of the image analysis system.

2.5. DPPH Free Radical Scavenging Activity Test

Antioxidant molecules were extracted from 1 g of ground sample, using 5 mL of a methanol/water (70:30 v/v) solution, kept under motion for 60 min at room temperature, and then paper filtered. The solvent was evaporated, and the extract was then dissolved in 10 mL of methanol, thus centrifuged at 5040× g for 15 min at 4 °C. Analyses were performed in triplicate mixing 100 µL of surnatant and 1 mL of 2,2-Diphenyl-1-picrylhydrazyl (DPPH) methanolic solution (0.2 mm), bringing the mix to a final volume of 2.4 mL with methanol. The absorbance of the solution was recorded at 517 nm by a Perkin Elmer UV-Visible spectrophotometer after an incubation time of 30 min in the dark at room temperature. Blank was prepared and analyzed following the same procedure, using 100 µL of methanol instead of sample. The radical scavenging activity was calculated as follows:

$$I\% = ((Abs0 - Abs1)/Abs0) \times 100, \qquad (1)$$

where Abs0 was the absorbance of the blank and Abs1 was the absorbance of the sample. The Trolox Equivalent Antioxidant Capacity value (TEAC) expressed as µmol Trolox equivalents/gram of dry weight (µmol TE/g of dw) of the samples was calculated from the calibration curve obtained by measuring the absorbance at 517 nm of Trolox methanolic solutions at different concentrations. Three replicates were analyzed for each treatment condition.

2.6. Pectin Methylesterase (PME) Activity

The enzyme extraction and the determination of its activity were conducted according to the method of Vicente et al. [19]. Five grams of sample were ground with 15 mL of 1 M NaCl (1:3, w/v) containing 1% (w/v) of polyvinylpolypyrrolidone (PVPP). The suspension obtained was stirred for 4 h and then centrifuged at 10,000× g for 30 min. The supernatant was collected, adjusted to pH 7.5 with 1 N NaOH, and used for assaying the enzyme activity. The activity was assayed in a mixture containing 600 µL of 0.5% (w/v) pectin, 150 µL of 0.01% bromothymol blue pH 7.5, 100 µL of water pH 7.5, and 150 µL of enzymatic extract. PME activity results in a progressive discoloration of the blue solution. The reduction of the absorbance at 620 nm was measured every 15 s for two minutes. The PME activity was calculated using the slope of a linear segment absorbance-time [20]. Percentage variations were calculated in comparison to the raw sample. Three replicates were analyzed for each treatment condition.

2.7. Texture Analysis

The texture of raw and treated samples was analyzed using a TA.XT2i Texture Analyzer equipped with a 25 kg load cell (Stable Micro Systems, Godalming, UK), a force resolution equal to 0.01 N, and an accuracy value of 0.025%. Puncture test (trigger force 0.05 N) was performed using a 2 mm diameter stainless steel needle probe, driven up in a radial direction to the center of the samples at a speed of 1 mm s^{-1}, following the method proposed by Paciulli et al. [21].

The following parameters were determined from the force vs. time curves: first peak force (F_{P1} given in N), which indicates the resistance opposed by external cell layers to needle penetration [21]; maximum puncture force (F_{max} given in N), which indicates the resistance opposed by the pulp to needle penetration; and Area under the force/time curve (Area given in N*s), which represents the total work carried out by the needle probe to penetrate the sample. The parameters were quantified using the application software provided (Texture Exponent for Windows, version 6.1 10.0). Ten blueberries units were analyzed for each treatment.

2.8. Color

Color determination was carried out using a Minolta Colorimeter (CM2600d, Minolta Co., Osaka, Japan) equipped with a standard illuminant D65, which simulates natural noon daylight in order to mimic the vision of the human eye. The measurement was performed on two opposite points on the blueberries epidermis. The instrument was calibrated before each analysis with white and black standard tiles. L^* (lightness; black = 0, white = 100), a^* (redness > 0, greenness < 0), b^* (yellowness > 0, blue < 0), C (chroma, 0 at the centre of the color sphere), and h° (hue angle, red = 0°, yellow = 90°, green = 180°, blue = 270°) were quantified on each fruit using a 10° position of the standard observer (Commission Internationale de l'eclairage), [22]). The ΔE for all the treated samples in comparison to the raw vegetables was also calculated. 10 determinations were performed for each treatment.

2.9. Statistical Analysis

One-way analysis of variance (ANOVA) among all the different treated samples and two way-ANOVA among the HHP treated samples, using pressure and time as independent variables, were performed using Statistical Package for Social Science (SPSS) software (Version 25.0, SPSS Inc., Chicago, IL, USA). A Least Significant Difference (LSD) post-hoc test at a 95% confidence level ($p \leq 0.05$) was performed to further identify differences among treatments.

Pearson correlation coefficients were calculated among all variables considering 95% and 99% confidence levels ($p < 0.05$ and $p < 0.01$).

Principal component analysis (PCA) was performed using the normalized variables, as reported by Medina Meza et al. [23]. Before running PCA, factor analysis (FA) was applied to exclude the variables that showed low contribution to explain the variance. FA was carried out on 10 independent

3. Results and Discussion

3.1. Histological Analysis

Raw blueberries fruits showed an epidermis composed of a single layer of cells (Figure 1A), with abundant tannin inclusions (Figure 1B). The subdermis, located immediately under the skin, was composed of 2 or 3 layers of cells, and showed multiple solid tannin inclusions (Figure 1B). The epidermis and subdermis exhibited thickened cell walls. These layers contain the pigments. Mesocarp, composed of parenchymatic cells, showed thin layers and large vacuoles (Figure 1A). From the observation of the transversal section stained with Tannin Solution, the presence of tannins was perceived mainly in the epidermis and subdermis (Figure 1B). In these cellular layers, tannins were solid and crystalline, while in the mesocarp parenchymatic tissue, tannins appeared as single inclusions or "crystalline powder". Zifkin and collaborators [24], in a study on blueberries, revealed that flavonols (condensed tannins) are among the major antioxidant compounds in epidermis and subdermis.

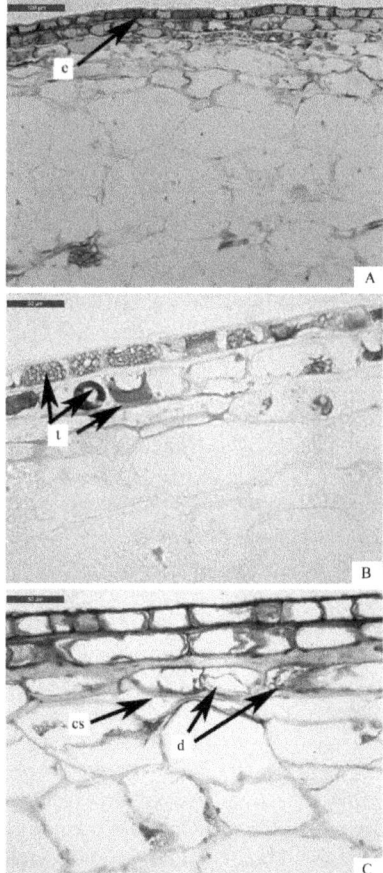

Figure 1. Transverse sections of blueberries samples: (**A**) raw or uncooked stained with Toluidine Blue (TBO); (**B**) raw or uncooked stained with tannin solution; (**C**) blanched stained with Toluidine Blue (TBO). Legend: e = epidermis; t = tannins; cs = cell separation; d = dehydration.

After blanching, blueberry microstructure did not show clear alterations in comparison to raw samples. Subdermis cells showed swollen cell walls due to the absorption of intra or extracellular water; this effect is called gelatinization. The major variations of blueberry tissues after blanching treatment were: cellular dehydration, cell wall gelatinization, and cell separation. Mesocarp cells, after thermal treatment, showed dehydration symptoms, with the detachment of the cellular membrane from the wall. In Figure 1C, it is also possible to observe cell separation with formation of large intracellular spaces between parenchymatic cells. The same observations were previously reported by Fuchigami et al. [25] in a study on carrots subjected to slight cooking, and by Paciulli and collaborators [21] in a study on blanched asparagus. As confirmed by several authors [21,26–28], cell separation is due to the breakage of chemical bonds between the pectic components of the middle lamellae of adjacent cells or to hydrolysis of some other components of the cell wall, such as pectin, hemicellulose, and cellulose. Compared to raw samples, tannins showed shape mild alteration and apparently a higher concentration. These compounds were visible in the epidermis, subdermis, mesocarp, and near the seeds. The blanching treatment seemed to induce leaking of tannins from cellular walls. Zaupa and coworkers [29] observed the same effect in different rice cultivars, where total antioxidant activity of samples increased after thermal treatments.

The high pressure treatment at 400 MPa for 1 min (400-1) induced separation of the external cells layers (epidermis and few layers of mesocarp). Cells seemed to be deformed by the appearance of air bubbles, and as a consequence, elliptical or circular lacunas are shown (Figure 2A). The same effects were previously observed by Prestamo and Arroyo [30] in a study on spinach and broccoli exposed to high pressures. Similarly to our observations, Tangwongchai et al. [31] observed the formation of large cavities in tissues of cherry tomatoes treated by high pressures of between 200 and 400 MPa. These authors hypothesized that during depressurization, the previously compressed air expanded rapidly, aggregating into larger bubbles, which caused the formation of cavities. Moreover, in our experimental conditions, mesocarp cells appeared dehydrated with gelatinized walls; in some cases, cells separation and cell wall breakage are also shown (Figure 2A).

The use of Tannin Solution staining on blueberry sections highlighted the presence of high quantities of tannins mainly in the epidermis and mesocarp (Figure 2B). The literature does not explain the increase of tannins in blueberries after high pressure treatments, but it is clear that the exposition of berries to high pressures increases the extraction of these antioxidant compounds. Serment-Moreno and collaborators [10], in their review, observed that the major role in the variation of polyphenol content after 200–600 MPa treatment is their release after disruption of cellular membranes and possible degradation due to their high susceptibility to oxidation and enzymatic reactions, however, details of the specific reaction taking place still remain unknown.

In Figure 2C the microstructure of blueberries treated at 400 MPa for 5 min (400-5) is shown. The tissue appeared disorganized and composed by shapeless cells that have lost their turgidity and, in comparison to the shorter treatment (400-1), more intercellular spaces were observed. Mesocarp resulted collapsed with formation of circular and elliptic cavities from the subdermis to the inner parenchyma (Figure 2C), probably due to the effect of the air bubbles aggregation during depressurization [31]. Cavities resulted larger if compared to the ones found in 400-1 (Figure 2A). Moreover, epidermis, subdermis and the firsts layer of mesocarp resulted detached; cell walls appeared thickened (gelatinized) all over the microstructure and the entire tissue showed considerable damages. Tannins solution stain highlights the presence of large amount of tannins that cover the entire structure (Figure 2D).

Blueberries treated at 600 MPa for 1 min (600-1) reflected the observations done for 400-1. The degree of parenchyma gelatinization, the dehydration of the external mesocarp tissues, as well as the presence of tannins, resulted comparable for 400-1 and 600-1 (Figure 2E).

After the treatment at 600 MPa for 5 min, blueberry tissues revealed evident damages, especially near the epidermis, where deep cavities, dehydration and gelatinization are present (Figure 2F). This treatment showed effects on the final product similar to those observed for HHP 400-5. Also tannins, as seen for 400-5, resulted distributed in all the structure.

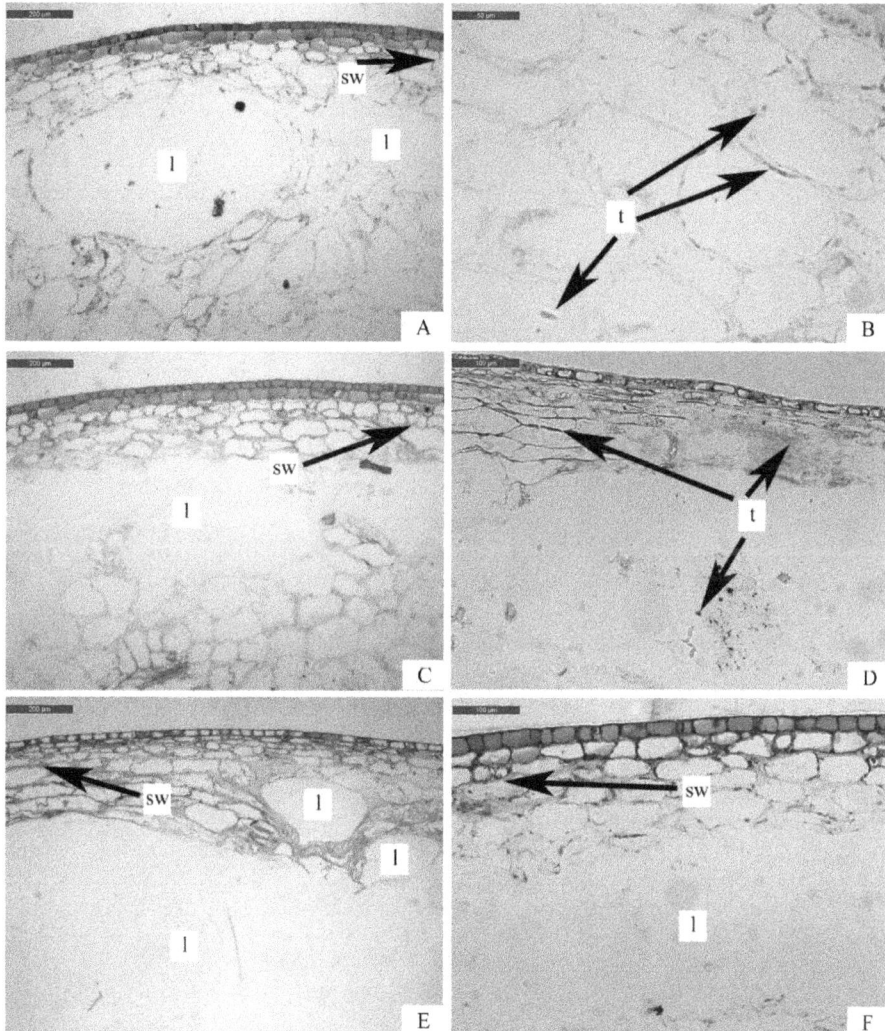

Figure 2. Transverse sections of blueberries samples treated using HHP technology: (**A**) 400-1 stained with Toluidine Blue (TBO); (**B**) 400-1 stained with tannin solution; (**C**) 400-5 stained with Toluidine Blue (TBO); (**D**) 400-5 stained with tannin solution; (**E**) 600-1 stained with Toluidine Blue (TBO); (**F**) 600-5 stained with Toluidine Blue (TBO). Abbreviations: l = lacuna; sw = swelling; t = tannins.

Blanching and high pressures caused changes in the cellular structure of blueberries under the effect of different phenomena: chemical and mechanical, respectively. Indeed, while blanching mainly led to hydrolysis of pectin chemical bonds, causing cell separation, high pressures brought to localized air bubbles explosions, breaking groups of cells. Blanching appeared thus less invasive than high pressure, although it involves widespread changes in the entire structure. On the other hand high pressures treatments, despite their strength, provoked localized damages. The extent of the cellular damages due to high pressures resulted more influenced by treatment time than pressure intensity.

3.2. Antioxidant Activity (DPPH)

In Table 1 are reported the values of blueberries antioxidant activity measured by DPPH assay. According to Zifkin et al. [24], the molecules responsible for antioxidant activity in blueberries are flavonols (condensed tannins), proantocyanidins and anthocyanins in epidermis and subdermis, anthocyanins in mesocarp, flavanols in placenta. These molecules act as radical scavengers [32].

Table 1. Water content, antioxidant activity and residual PME activity of raw and treated blueberries [†].

	R	BL	400-1	400-5	600-1	600-5	P	t	P×t
DPPH (TE µmol/g_{dw})	103.4 ± 18.6 [a]	79.1 ± 5.1 [ab]	67.8 ± 6.5 [b C]	79.0 ± 4.1 [ab AB]	75.5 ± 7.5 [b B]	84.9 ± 2.1 [ab A]	n.s.	*	n.s.
PME (%)	100 [a]	9.1 ± 2.14 [c]	64.8 ± 3.7 [b C]	95.7 ± 5.2 [a A]	83.5 ± 9.0 [ab B]	87.2 ± 14.2 [a AB]	n.s.	*	*
Water content (%)	90.7 ± 1.9 [a]	88.4 ± 0.3 [a]	87.0 ± 0.8 [a C]	88.0 ± 0.7 [a B]	88.2 ± 1.1 [a B]	89.2 ± 0.2 [a A]	*	*	n.s

[†] Data are expressed as means ± standard deviations of 3 samples. Means in row followed by different lowercase letters are significantly different according to the post-hoc analysis after one-way analysis of variance (ANOVA) ($p ≤ 0.05$). Means in row, of high pressure treated samples, followed by different uppercase letters are significantly different according to post-hoc comparisons after two-way ANOVA ($p ≤ 0.05$), performed considering pressure (P) and time (t) as independent variables. The p values were corrected for multiple comparisons use LSD method. Abbreviations: R = raw/untreated; BL = blanched; 400-1 = HHP at 400 MPa for 1 min; 400-5 = HHP at 400 MPa for 5 min; 600-1 = HHP at 600 MPa for 1 min; 600-5 = HHP at 600 MPa for 5 min; TE = Trolox equivalent; PME = pectin methylesterase; n.s. = not significant ($p ≥ 0.05$); * = significant ($p ≤ 0.05$).

Raw samples showed values of about 105 µmolTE/g_{dw}; this value is in line with the results of Lohachoompol et al. [32], with slight differences attributable to the different blueberries varieties.

All the treatments affected blueberries antioxidant activity in comparison to raw samples. Blanching decreased it slightly, despite tannins, resulted better extracted from the cell walls (Par.3.1). Brownmiller et al. [3] registered a decrease of the total antioxidant activity after blueberries blanching, despite they didn't observe any anthocyanins reduction, probably because of other antioxidant molecules losses. Conversely, Rossi et al. [33] observed higher antioxidant activity in blueberries juice after fruits blanching, probably because of the rapid polyphenoloxidase inactivation and/or increase in extraction yield, due to heat induced skin permeability [34].

The high pressure treated samples showed behaviors similar to the blanched ones (Table 1). On the other hand, in comparison to R, 400-1 and 600-1 showed a significantly lower antioxidant activity, with losses of around 34 and 27%, respectively. 600-5 resulted instead the less impacting treatment with around 82% retention of the total activity. Two-way ANOVA, conducted among the HHP treated samples, revealed a time-dependent behavior: the longer the time, the higher the antioxidant activity. The retention of the antioxidant activity with the increase of the pressure holding time may be explained as a better extraction of the bioactive molecules from the broken cells, as suggested by other authors [20] and confirmed by the histological observations (Figure 2). Moreover it's known that the total antioxidant activity in blueberries is due to several classes of molecules, differently distributed among the tissues [24], that can thus be extracted under different conditions. The highest loss of antioxidant activity registered for 400-1 and 600-1 may be also attributable to the poor inactivation of PPO and POD, the main enzymes responsible for phenol decay [35].

3.3. Pectin Methylesterase (PME) Activity

Pectin methyl esterase (PME) is the enzyme responsible for the demethylesterification of plant cell walls pectin. This enzyme, in combination with polygalacturonase, affects texture of fruit and vegetables during postharvest storage [35].

In Table 1, the blueberry PME activity is reported as percentage variation in comparison to the raw sample. Among treatments, blanching, with a residual activity of 10%, was the most effective on PME inactivation, because of the protein thermal denaturation [36]. On the other hand, high pressure treatments had lower effect on PME inactivation. HHP400-1, leading to a residual PME activity of around 65%, resulted the most effective high-pressure treatment. The high PME baro-resistance is reported for many fruit and vegetables [37]. It has been shown that, among many studied products, tomato PME is the most pressure-resistant, even being inactivated at ambient conditions up to 800 MPa [37]. The two-way ANOVA, performed among the four HHP treatments, revealed a time dependence for blueberry PME inactivation by high pressure—the higher the pressure holding time, the lower the inactivation. Similarly, Paciulli et al. [20] reported increased enzymatic activity for longer pressure exposure times on beetroot slices. These authors explained this phenomenon with enzymes leaking from the broken cells, which leading to easier contact with the substrates, results in higher activities. This hypothesis follows the same trend of the antioxidant activity and is supported by the histological observations that show 400-5 and 600-5 as the most damaged tissues (Figure 2C–F). Phenomena of enzyme activation, under the effect of high pressure, due to conformational changes have also been reported [38]. A significant interaction of time and pressure was also revealed by the two-way ANOVA; indeed, 400-5 showed almost no effect on PME. In contrast, in a study on two different varieties of pumpkin, Paciulli et al. [39] reported high-pressure treatment at 400 MPa for 5 min as the most effective against PME, with a residual activity of 20–25%; this demonstrates that the PME barotolerance is specie dependent.

3.4. Texture

Figure 3 reports the texture analysis profiles of blueberries in all the studied conditions, obtained by puncture test. In the same figure, the parameters F_{p1}, F_{max}, and Area are shown, and their values summarized in Table 2. The absence of significant differences in water content (Table 1) between raw and treated samples indicates that the observed texture differences were not related to the cellular turgidity, but they strictly depend on the tissues structure, as modified by the treatments.

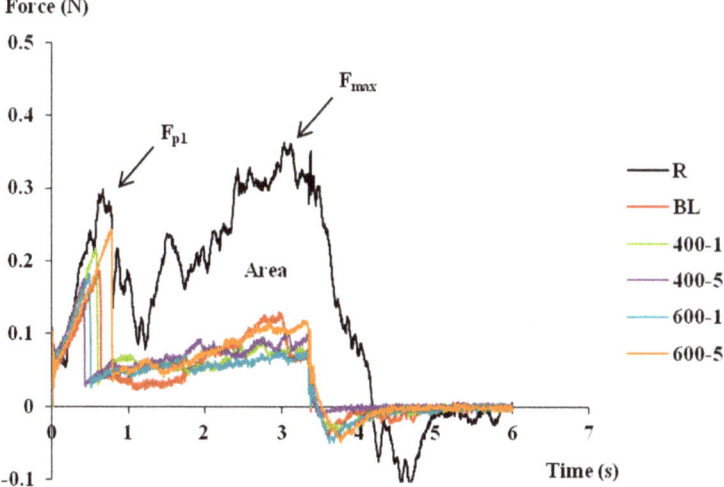

Figure 3. Texture analysis profiles of raw and treated blueberries obtained by puncture test. Abbreviations: R = raw or untreated; BL = blanched; 400-1 = HHP at 400 MPa for 1 min; 400-5 = HHP at 400 MPa for 5 min; 600-1 = HHP at 600 MPa for 1 min; 600-5 = HHP at 600 MPa for 5 min; F_{P1} = maximum force first peak; F_{max} = absolute maximum force; Area = area under the force/time curve.

Table 2. Texture parameters [†] for raw and treated blueberries.

Samples	F_{P_1} (N)	F_{max} (N)	Area (N*s)
R	0.30 ± 0.03 [a]	0.32 ± 0.06 [a]	0.79 ± 0.14 [a]
BL	0.19 ± 0.03 [c]	0.10 ± 0.03 [b]	0.22 ± 0.03 [c]
400-1	0.22 ± 0.04 [bc AB]	0.10 ± 0.03 [b B]	0.26 ± 0.04 [bc B]
400-5	0.20 ± 0.04 [bc B]	0.10 ± 0.03 [b B]	0.28 ± 0.06 [bc B]
600-1	0.20 ± 0.04 [bc B]	0.10 ± 0.03 [b B]	0.27 ± 0.05 [bc B]
600-5	0.25 ± 0.05 [ab A]	0.14 ± 0.04 [b A]	0.36 ± 0.08 [b A]
	P t P×t	P t P×t	P t P×t
	n.s. n.s. *	n.s. n.s. n.s.	n.s. * n.s.

Note: [†] Data are expressed as means ± standard deviations of 10 samples. Means in columns followed by different lowercase letters are significantly different according to the post-hoc analysis after one-way analysis of variance (ANOVA) ($p \leq 0.05$). Means in columns of high-pressure treated samples followed by different uppercase letters are significantly different according to post-hoc comparisons after two-way ANOVA ($p \leq 0.05$), performed considering pressure (P) and time (t) as independent variables. The p values were corrected for multiple comparisons use LSD method. Abbreviations: R = raw/untreated; BL = blanched; 400-1 = HHP at 400 MPa for 1 min; 400-5 = HHP at 400 MPa for 5 min; 600-1 = HHP at 600 MPa for 1 min; 600-5 = HHP at 600 MPa for 5 min; F_{P_1} = maximum force first peak; F_{max} = absolute maximum force; Area = area under the force/time curve; n.s. = not significant ($p \geq 0.05$); * = significant ($p \leq 0.05$).

The first peak force (F_{P_1}) was generated in blueberries by epidermal and subdermal cells, characterized by cutinization and thickened cell walls (Figure 1A, Section 3.1). It is visible how untreated blueberries showed the highest FP_1 (~0.30 N) among all samples. Blanching led to the highest FP_1 drop among all treatments. This phenomenon, already reported for other blanched vegetables [21], may be ascribable to the external cell wall swelling (Figure 1C, Par.3.1) and pectin thermal degradation, with consequent softening. HHP treated samples showed Fp_1 values significantly lower than R, with the exception of 600-5, which showed values of Fp_1 around 0.25 N, slightly higher than BL. Fp_1 reduction under high pressure can be justified by the detachment of the epidermis, subdermis, and first mesocarp layers, as well as cell wall swelling, as observed from the histological analysis (Figure 2). The two-way ANOVA performed among the high-pressure treated samples (Table 1) showed a synergistic effect of pressure and time on Fp_1. Texture recovery during pressure holding time was already reported by other authors [20,40] and associated to the insufficient inactivation of PME, whose reaction product is the low-methoxy pectin, that forms a gel-networkwith divalent ions such as Ca and Mg, contributing to the enhanced hardness value. This hypothesis, more evident at 600 MPa, is supported by the PME activity results (Table 1). At 400 MPa, the effect of the tissue damage prevails.

Comparing F_{max} with Fp_1 (Table 2), it is visible how, after treatments, Fp_1 was almost 50% higher than F_{max}, indicating that the skin had the major effect on blueberries mechanical properties. Moreover, F_{max} of all the treated samples was significantly lower than the raw ones (Table 2); this phenomenon may be justified by an easier penetration of the needle probe across the dehydrated and gelled parenchyma of the blanched (Figure 1C, Section 3.1) and HHP treated blueberries (Figure 2, Par.3.1). Among the HHP treated samples, 600-5 was significantly firmer than the others. Texture recovery after prolonged exposures to high pressure has already been reported [40]. This phenomenon, as discussed for Fp_1, can be justified with the high PME activity.

Area values comprise both the effect of skin and pulp penetration. It has been confirmed that all of the treatments were detrimental for blueberry texture, showing significantly lower Area values in comparison to the untreated samples. Despite BL samples showing the lowest values among all samples (~0.22 N*s), only 600-5 (~0.36 N*s) was significantly higher than BL. The two-way ANOVA confirmed a time dependence of the Area values under the effect of pressure—the higher the exposure time, the higher the Area values. This phenomenon, already reported in previous studies, was justified either with tissue recovery during the holding time [41], or with the still high PME activity [42]. Confirming the last hypothesis, a positive correlation was found between Area values and PME ($R = 0.498$; $p < 0.05$); with increasing PME activity, the firmness of blueberries increased.

3.5. Color

The bluish blueberry skin color is affected by anthocyanin content, as well as by the presence of surface waxes [43]. L*, a*, and b* values, measured on raw samples (Table 3), are in line with data already reported for different varieties of blueberries [43]. After blanching, blueberries were slightly darker than the raw samples, with lower L* values (Table 3); this may be related to melting of blueberries' cutaneous wax [44]. At the same time, a* and b* increase, indicating enhancement of the perceived red and blue colors. The slight increase of the color intensity C may be instead ascribed to alterations of the surface reflecting properties. Similarly, Mazzeo et al. [45] found an enhancement of the asparagus, green bean, and zucchini green color after blanching; they associated it to a change of the color perception due to the air replacement with water and cell juices. The shift of $h°$ from about 270 to 350 degrees confirms the shift to a red-purple color. According to ΔE values, blanching was the more affecting treatment. According to previous studies, thermal treatments are more detrimental than high pressure on the anthocyanin content of fruits, with consequent higher color changes [46]. Focusing on the HHP samples, while a* behaved similar to the blanched samples, L*, b*, and consequently C were influenced by the time/pressure interaction. Among the HHP treatments, HHP600-1 was the most affecting treatment, leading to brighter and more intense blue tones, as also confirmed by the shift of $h°$ to values around 300 degrees. On the other hand, HHP600-5 was the least affecting HHP treatment; this may be due to better anthocyanins extraction [46], associated in our study with the high antioxidant activity of these samples.

Table 3. Color parameters for raw and treated blueberries [†].

	L*	a*	b*	C	h°	ΔE
R	31.50 ± 2.00 [ab]	0.08 ± 0.02 [b]	−1.66 ± 0.53 [ab]	1.66 ± 0.53 [c]	272.72 ± 1.94 [c]	-
BL	28.30 ± 3.96 [b]	2.67 ± 0.98 [a]	−0.41 ± 0.18 [a]	2.72 ± 0.95 [bc]	349.49 ± 7.50 [a]	5.64 ± 1.10 [a]
400-1	30.15 ± 3.46 [ab B]	2.52 ± 0.96 [a A]	−1.74 ± 1.08 [ab A]	3.19 ± 1.08 [ab B]	326.41 ± 18.48 [a A]	4.18 ± 1.66 [abc AB]
400-5	31.05 ± 2.79 [ab B]	2.01 ± 1.07 [a A]	−2.37 ± 1.50 [bc A]	3.42 ± 1.02 [ab B]	312.02 ± 24.83 [ab AB]	3.56 ± 1.23 [bc B]
600-1	34.80 ± 1.69 [a A]	2.30 ± 0.47 [a A]	−3.61 ± 1.04 [c B]	4.43 ± 0.53 [a A]	304.62 ± 14.43 [b B]	4.72 ± 1.01 [ab A]
600-5	30.86 ± 1.44 [ab B]	2.06 ± 0.80 [a A]	−1.95 ± 0.73 [bc A]	2.93 ± 0.73 [b B]	317.06 ± 16.23 [ab AB]	2.60 ± 0.73 [c C]
	P t P×t	P t P×t	P t P×t	P t P×t	P t P×t	P t P×t
	n.s. n.s. *	n.s. n.s. n.s.	n.s. n.s. *	n.s. n.s. *	n.s. n.s. n.s.	n.s * n.s.

Note: [†] Data are expressed as means ± standard deviations of 10 samples. Means in columns followed by different lowercase letters are significantly different according to the post-hoc analysis after one-way analysis of variance (ANOVA) ($p \leq 0.05$). Means in columns of high-pressure treated samples followed by different uppercase letters are significantly different according to post-hoc comparisons after two-way ANOVA ($p \leq 0.05$), and were performed considering pressure (P) and time (t) as independent variables. The p values were corrected for multiple comparisons using LSD method. Abbreviations: R = raw or untreated; BL = blanched; 400-1 = HHP at 400 MPa for 1 min; 400-5 = HHP at 400 MPa for 5 min; 600-1 = HHP at 600 MPa for 1 min; 600-5 = HHP at 600 MPa for 5 min; L = lightness; a* = redness; b* = blueness; C = chroma; $h°$ = hue angle; n.s. = not significant ($p \geq 0.05$); * = significant ($p \leq 0.05$).

3.6. Principal Component Analysis

Based on the obtained results, Principal Component Analysis (PCA) (Figure 4) was performed to enable an overview of the variables that mainly influenced the final quality of blueberries under the effect of the different studied treatments. Starting from 10 variables (DPPH, PME, F_{p1}, F_{max}, Area, L, a*, b*, C, $h°$), factor analysis excluded DPPH because it showed low contribution to the total variance.

Nine variables were selected, generating a score plot in which the first two principal components (PC) explained 86.66% of the total variance. The first and more discriminating component (PC1) was related to texture and color parameters; in particular, F_{p1}, F_{max}, and Area showed positive loadings on PC1, being inversely related to a*. On the second component (PC2), the color parameters L* and b* showed high positive factor loadings. The variables PME and $h°$ resulted in opposite positions on the plan, having respectively positive and negative factor loadings, both on PC1 and PC2. The color parameter C showed positive loadings on PC1 and negative on PC2. Based on this distribution, the samples were grouped into three clusters: Raw, Blanched, and HHP treated samples. Raw samples were directly related with all of the texture parameters, being instead inversely related to the red component a*, indicating a more turgid structure with an opaque red color. Blanched samples were inversely related to PME and with the color parameters L and b*, having instead a positive relation with h°. The wide distance between raw and blanched samples on the factor plan confirmed that the blanched blueberries were the ones that deviated most from raw fruit in terms of organoleptic properties, however being the most effective treatment on PME inactivation. HHP treated samples clustered together in the middle zone between raw and blanched samples, indicating high pressure as a mild treatment if compared to the thermal one. Among the HHP treatments, 600-1 slightly differed from the other samples, mainly in relation to the color parameters L and b*.

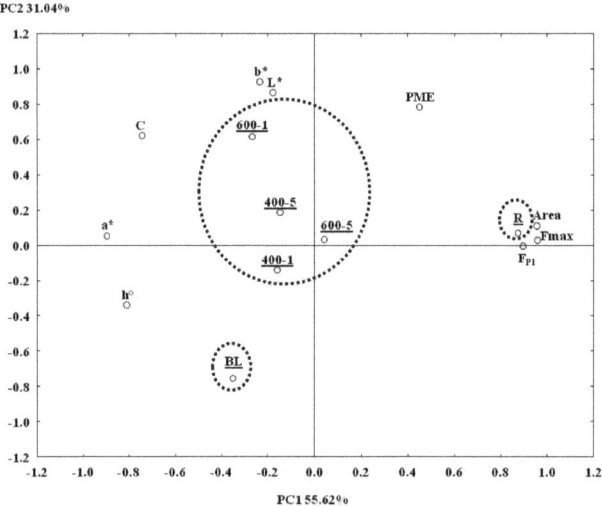

Figure 4. Principal Component Analysis (PCA) results obtained for the two principal components: Projection of the variables and of the cases on the factor plane (1 × 2). Abbreviations: R = raw/untreated; BL = blanched; 400-1 = HHP at 400 MPa for 1 min; 400-5 = HHP at 400 MPa for 5 min; 600-1 = HHP at 600 MPa for 1 min; 600-5 = HHP at 600 MPa for 5 min; F_{P1} = maximum force first peak; Fmax = absolute maximum force; Area = area under the force/time curve; PME = pectin methylesterase; L = lightness; a* = redness; b* = blueness; C = chroma; h° = hue angle; PC = Principal Component.

4. Conclusions

The application of high pressure as a fruit preservation technique has shown mild effects on the organoleptic properties of blueberries if compared to blanching, despite the more severe tissue damage. Focusing on the high-pressure treatments, an interesting effect of the pressure holding time was observed on almost all of the investigated variables. Indeed, less PME deactivation, higher texture and color retention, and better antioxidant activity were observed by increasing the treatment time for both of the tested pressures. These phenomena, probably related to an easier extraction of intracellular molecules from the broken cells, were particularly visible for 600-5, the sample subjected to the most intense high-pressure treatment. Almost no effect of the pressure level was evidenced in this study. The high PME activity, considered a defect of the high pressure treated samples, was, however, related to a better texture recovery during treatment. A shelf life study will be necessary to evaluate the evolution of these parameters over time.

Author Contributions: Conceptualization, P.M., I.G.M.M., R.M., and E.C.; methodology, P.M., R.M., T.G., A.P., M.R., and M.M.; validation, P.M., R.M., T.G., and E.C.; formal Analysis, P.M., A.P., and M.R.; investigation, P.M., A.P., and M.R.; data Curation, P.M. and I.G.M.M.; writing—original draft preparation, P.M., I.G.M.M., and T.G.; writing—review and editing, M.P., I.G.M.M., T.G., and E.C.; visualization, D.B. and E.C.; supervision, D.B., M.M. and E.C.

Funding: This research received no external funding.

Conflicts of Interest: The authors declare no conflict of interest.

References

1. Kalt, W.; Joseph, J.A.; Shukitt-Hale, B. Blueberries and human health: A review of current research. *J. Am. Pomol. Soc.* **2007**, *61*, 151–160.
2. Toivonen, P.M.; Brummell, D.A. Biochemical bases of appearance and texture changes in fresh-cut fruit and vegetables. *Postharvest Boil. Technol.* **2008**, *48*, 1–14. [CrossRef]
3. Brownmiller, C.; Howard, L.R.; Prior, R.L. Processing and storage effects on monomeric anthocyanins, percent polymeric color, and antioxidant capacity of processed blueberry products. *J. Food Sci.* **2008**, *73*, H72–H79. [CrossRef] [PubMed]
4. Zielinska, M.; Sadowski, P.; Błaszczak, W. Freezing/thawing and microwave-assisted drying of blueberries (*Vaccinium corymbosum* L.). *LWT Food Sci. Technol.* **2015**, *62*, 555–563. [CrossRef]
5. Guerrero-Beltrán, J.A.; Barbosa-Cánovas, G.V.; Swanson, B.G. High hydrostatic pressure processing of fruit and vegetable products. *Food Rev. Int.* **2005**, *21*, 411–425. [CrossRef]
6. Medina-Meza, I.G.; Barnaba, C.; Barbosa-Cánovas, G.V. Effects of high pressure processing on lipid oxidation: A review. *Innov. Food Sci. Emerg. Technol.* **2014**, *22*, 1–10. [CrossRef]
7. Medina-Meza, I.G.; Barnaba, C.; Villani, F.; Barbosa-Cánovas, G.V. Effects of thermal and high pressure treatments in color and chemical attributes of an oil-based spinach sauce. *LWT Food Sci. Technol.* **2015**, *60*, 86–94. [CrossRef]
8. Tejada-Ortigoza, V.; García-Amezquita, L.E.; Serna-Saldívar, S.O.; Welti-Chanes, J. The dietary fiber profile of fruit peels and functionality modifications induced by high hydrostatic pressure treatments. *Food Sci. Technol. Int.* **2017**, *23*, 396–402. [CrossRef]
9. Tejada-Ortigoza, V.; Garcia-Amezquita, L.E.; Serna-Saldívar, S.O.; Martín-Belloso, O.; Welti-Chanes, J. High hydrostatic pressure and mild heat treatments for the modification of orange peel dietary fiber: Effects on hygroscopic properties and functionality. *Food Bioprocess Tech.* **2018**, *11*, 110–121. [CrossRef]
10. Serment-Moreno, V.; Jacobo-Velázquez, D.A.; Torres, J.A.; Welti-Chanes, J. Microstructural and physiological changes in plant cell induced by pressure: Their role on the availability and pressure-temperature stability of phytochemicals. *Food Eng. Rev.* **2017**, *9*, 314–334. [CrossRef]
11. Harris, P.J.; Smith, B.G. Plant cell walls and cell-wall polysaccharides: Structures, properties and uses in food products. *Int. J. Food Sci. Technol.* **2006**, *41*, 129–143. [CrossRef]
12. Voragen, A.G.; Coenen, G.J.; Verhoef, R.P.; Schols, H.A. Pectin, a versatile polysaccharide present in plant cell walls. *Struct. Chem.* **2009**, *20*, 263. [CrossRef]
13. Huang, R.; Ye, M.; Li, X.; Ji, L.; Karwe, M.; Chen, H. Evaluation of high hydrostatic pressure inactivation of human norovirus on strawberries, blueberries, raspberries and in their purees. *Int. J. Food Microbiol.* **2016**, *223*, 17–24. [CrossRef] [PubMed]
14. Altuner, E.M.; Tokuşoğlu, Ö. The effect of high hydrostatic pressure processing on the extraction, retention and stability of anthocyanins and flavonols contents of berry fruits and berry juices. *Int. J. Food Sci. Technol.* **2013**, *48*, 1991–1997. [CrossRef]
15. Sablani, S.S.; Andrews, P.K.; Davies, N.M.; Walters, T.; Saez, H.; Syamaladevi, R.M.; Mohekar, P.R. Effect of thermal treatments on phytochemicals in conventionally and organically grown berries. *J. Sci. Food Agric.* **2010**, *90*, 769–778. [CrossRef]
16. Oey, I.; Lille, M.; Van Loey, A.; Hendrickx, M. Effect of high-pressure processing on colour, texture and flavour of fruit-and vegetable-based food products: A review. *Trends Food Sci. Technol.* **2008**, *19*, 320–328. [CrossRef]
17. AOAC. *Official Methods of Analysis*, 16th ed.; Association of Official Analytical Chemists: Arlington, VA, USA, 2002.

18. Ruzin, S. *Plant Microtechnique and Microscopy*; Oxford University Press: Oxford, UK, 1999; Volume 198.
19. Vicente, A.R.; Costa, M.L.; Martínez, G.A.; Chaves, A.R.; Civello, P.M. Effect of heat treatments on cell wall degradation and softening in strawberry fruit. *Postharvest Biol. Technol.* **2005**, *38*, 213–222. [CrossRef]
20. Paciulli, M.; Medina-Meza, I.G.; Chiavaro, E.; Barbosa-Cánovas, G.V. Impact of thermal and high pressure processing on quality parameters of beetroot (*Beta vulgaris* L.). *LWT Food Sci. Tech.* **2016**, *68*, 98–104. [CrossRef]
21. Paciulli, M.; Ganino, T.; Pellegrini, N.; Rinaldi, M.; Zaupa, M.; Fabbri, A.; Chiavaro, E. Impact of the industrial freezing process on selected vegetables—Part I. Structure, texture and antioxidant capacity. *Food Res. Int.* **2015**, *74*, 329–337. [CrossRef]
22. Commission Internationale de l'eclairage (CIE). *Recommendations on Uniform Colourspaces-Colour Equations, Psychometric Colour Terms. Supplement no. 2 to CIE Publ. No. 15 (E-1.3.L) 1971/9TC-1-3*; CIE: Paris, France, 1978.
23. Medina-Meza, I.G.; Aluwi, N.A.; Saunders, S.R.; Ganjyal, G.M. GC–MS profiling of triterpenoid saponins from 28 quinoa varieties (*Chenopodium quinoa* Willd.) grown in Washington State. *J. Agric. Food Chem.* **2016**, *64*, 8583–8591. [CrossRef]
24. Zifkin, M.; Jin, A.; Ozga, J.A.; Zaharia, I.; Schernthaner, J.P.; Gesell, A.; Abrams, S.R.; Kennedy, J.A.; Constabel, C.P. Gene expression and metabolite profiling of developing highbush blueberry (*Vaccinium corymbosum* L.) fruit indicates transcriptional regulation of flavonoid metabolism and activation of abscisic acid metabolism. *Plant Physiol.* **2012**, *158*, 200–224. [CrossRef] [PubMed]
25. Fuchigami, M.; Hyakumoto, N.; Miyazaki, K. Programmed freezing affects texture, pectic composition and electron microscopic structures of carrots. *J. Food Sci.* **1995**, *60*, 137–141. [CrossRef]
26. Sila, D.N.; Smout, C.; Elliot, F.; Loey, A.V.; Hendrickx, M. Non-enzymatic depolymerization of carrot pectin: Toward a better understanding of carrot texture during thermal processing. *J. Food Sci.* **2006**, *71*, E1–E9. [CrossRef]
27. Lecain, S.; Ng, A.; Parker, M.L.; Smith, A.C.; Waldron, K.W. Modification of cell-wall polymers of onion waste—Part I. Effect of pressure-cooking. *Carbohydr. Polym.* **1999**, *38*, 59–67. [CrossRef]
28. Van Marle, J.T.; Stolle-Smits, T.; Donkers, J.; van Dijk, C.; Voragen, A.G.; Recourt, K. Chemical and microscopic characterization of potato (*Solanum tuberosum* L.) cell walls during cooking. *J. Agric. Food Chem.* **1997**, *45*, 50–58. [CrossRef]
29. Zaupa, M.; Ganino, T.; Dramis, L.; Pellegrini, N. Anatomical study of the effect of cooking on differently pigmented rice varieties. *Food Struct.* **2016**, *7*, 6–12. [CrossRef]
30. Prestamo, G.; Arroyo, G. High hydrostatic pressure effects on vegetable structure. *J. Food Sci.* **1998**, *63*, 878–881. [CrossRef]
31. Tangwongchai, R.; Ledward, D.A.; Ames, J.M. Effect of high-pressure treatment on the texture of cherry tomato. *J. Agric. Food Chem.* **2000**, *48*, 1434–1441. [CrossRef]
32. Lohachoompol, V.; Mulholland, M.; Srzednicki, G.; Craske, J. Determination of anthocyanins in various cultivars of highbush and rabbiteye blueberries. *Food Chem.* **2008**, *111*, 249–254. [CrossRef]
33. Rossi, M.; Giussani, E.; Morelli, R.; Scalzo, R.L.; Nani, R.C.; Torreggiani, D. Effect of fruit blanching on phenolics and radical scavenging activity of highbush blueberry juice. *Food Res. Int.* **2003**, *36*, 999–1005. [CrossRef]
34. Kalt, W.; McDonald, J.E.; Donner, H. Anthocyanins, phenolics, and antioxidant capacity of processed lowbush blueberry products. *J. Food Sci.* **2000**, *65*, 390–393. [CrossRef]
35. Cao, S.; Zheng, Y.; Wang, K.; Rui, H.; Tang, S. Effect of methyl jasmonate on cell wall modification of loquat fruit in relation to chilling injury after harvest. *Food Chem.* **2010**, *118*, 641–647. [CrossRef]
36. Duvetter, T.; Sila, D.N.; Van Buggenhout, S.; Jolie, R.; Van Loey, A.; Hendrickx, M. Pectins in processed fruit and vegetables: Part I—Stability and catalytic activity of pectinases. *Compr. Rev. Food Sci. Food Saf.* **2009**, *8*, 75–85. [CrossRef]
37. Terefe, N.S.; Buckow, R.; Versteeg, C. Quality-related enzymes in fruit and vegetable products: Effects of novel food processing technologies, Part 1: High-pressure processing. *Crit. Rev. Food Sci. Nutr.* **2014**, *54*, 24–63. [CrossRef] [PubMed]
38. Hendrickx, M.; Ludikhuyze, L.; Van den Broeck, I.; Weemaes, C. Effects of high pressure on enzymes related to food quality. *Trends Food Sci. Technol.* **1998**, *9*, 197–203. [CrossRef]

39. Paciulli, M.; Rinaldi, M.; Rodolfi, M.; Ganino, T.; Morbarigazzi, M.; Chiavaro, E. Effects of high hydrostatic pressure on physico-chemical and structural properties of two pumpkin species. *Food Chem.* **2019**, *274*, 281–290. [CrossRef] [PubMed]
40. Basak, S.; Ramaswamy, H.S. Effect of high pressure processing on the texture of selected fruits and vegetables. *J. Texture Stud.* **1998**, *29*, 587–601. [CrossRef]
41. Araya, X.I.T.; Hendrickx, M.; Verlinden, B.E.; Van Buggenhout, S.; Smale, N.J.; Stewart, C.; Mawson, A.J. Understanding texture changes of high pressure processed fresh carrots: A microstructural and biochemical approach. *J. Food Eng.* **2007**, *80*, 873–884. [CrossRef]
42. Kaushik, N.; Kaur, B.P.; Rao, P.S. Application of high pressure processing for shelf life extension of litchi fruits (*Litchi chinensis* cv. Bombai) during refrigerated storage. *Food Sci. Technol. Int.* **2014**, *20*, 527–541. [CrossRef]
43. Saftner, R.; Polashock, J.; Ehlenfeldt, M.; Vinyard, B. Instrumental and sensory quality characteristics of blueberry fruit from twelve cultivars. *Postharvest Biol. Technol.* **2008**, *49*, 19–26. [CrossRef]
44. Lacombe, A.; Niemira, B.A.; Gurtler, J.B.; Fan, X.; Sites, J.; Boyd, G.; Chen, H. Atmospheric cold plasma inactivation of aerobic microorganisms on blueberries and effects on quality attributes. *Food Microbiol.* **2015**, *46*, 479–484. [CrossRef] [PubMed]
45. Mazzeo, T.; Paciulli, M.; Chiavaro, E.; Visconti, A.; Fogliano, V.; Ganino, T.; Pellegrini, N. Impact of the industrial freezing process on selected vegetables-Part II. Colour and bioactive compounds. *Food Res. Int.* **2015**, *75*, 89–97. [CrossRef] [PubMed]
46. Patras, A.; Brunton, N.P.; Da Pieve, S.; Butler, F. Impact of high pressure processing on total antioxidant activity, phenolic, ascorbic acid, anthocyanin content and colour of strawberry and blackberry purées. *Innov. Food Sci. Emerg. Technol.* **2009**, *10*, 308–313. [CrossRef]

© 2019 by the authors. Licensee MDPI, Basel, Switzerland. This article is an open access article distributed under the terms and conditions of the Creative Commons Attribution (CC BY) license (http://creativecommons.org/licenses/by/4.0/).

Article

Convective Drying of Fresh and Frozen Raspberries and Change of Their Physical and Nutritive Properties

Zoran Stamenković [1], Ivan Pavkov [1], Milivoj Radojčin [1,*], Aleksandra Tepić Horecki [2], Krstan Kešelj [1], Danijela Bursać Kovačević [3] and Predrag Putnik [3]

1. Faculty of Agriculture, University of Novi Sad, Trg Dositeja Obradovića 8, 21000 Novi Sad, Serbia
2. Faculty of Technology, University of Novi Sad, Bulevarcara Lazara 1, 21000 Novi Sad, Serbia
3. Faculty of Food Technology and Biotechnology, University of Zagreb, Pierottijeva 6, 10000 Zagreb, Croatia
* Correspondence: mradojcin@polj.uns.ac.rs; Tel.: +381-214853431; Fax: +381-21459989

Received: 7 June 2019; Accepted: 4 July 2019; Published: 11 July 2019

Abstract: Raspberries are one of Serbia's best-known and most widely exported fruits. Due to market fluctuation, producers are looking for ways to preserve this fresh product. Drying is a widely accepted method for preserving berries, as is the case with freeze-drying. Hence, the aim was to evaluate convective drying as an alternative to freeze-drying due to better accessibility, simplicity, and cost-effectiveness of Polana raspberries and compare it to a freeze-drying. Three factors were in experimental design: air temperature (60, 70, and 80 °C), air velocity (0,5 and 1,5 m·s^{-1}), and state of a product (fresh and frozen). Success of drying was evaluated with several quality criteria: shrinkage (change of volume), color change, shape, content of L-ascorbic acid, total phenolic content, flavonoid content, anthocyanin content, and antioxidant activity. A considerable influence of convective drying on color changes was not observed, as ΔE was low for all samples. It was obvious that fresh raspberries had less physical changes than frozen ones. On average, convective drying reduced L–ascorbic acid content by 80.00–99.99%, but less than 60% for other biologically active compounds as compared to fresh raspberries. Convective dried Polana raspberry may be considered as a viable replacement for freeze-dried raspberries.

Keywords: raspberry; convective drying; freeze-drying; bioactive compounds; shrinkage; color change

---

1. Introduction

Raspberries (Rubus idaeus) are one of the most important fruits in Serbian agriculture. It has been recently reported that raspberry production in 2017 was 109,742 t, which positions Serbia as one from the three leading countries in raspberry world production [1]. Over 90% of produced Serbian raspberries are commonly frozen, while only 10% is immediately used for processing or sold on open markets [2]. At their full maturity, raspberries have high moisture content (84% w.b.), L-ascorbic acid content, and potassium, as well as proteins, fibers, and minerals [3–7]; moreover, they contain various biologically active compounds (BACs) with high antioxidant activity. BACs found in raspberries include anthocyanins (cyanidin-3-sambubioside, cyanidin-3-glucoside, cyanidin-3-xylosylrutinoside, and cyanidin-3-rutinoside), ellagic acid, hydrolysable tannins (derivatives of gallic and ellagic acid), proanthocyanidins, vecetin, quercetin and rutin [8–10], carotenoids (lutein, zeaxanthin, alpha carotene, and beta carotene), chlorophyll derivatives, and tocopherols [11], with antioxidative and anti-inflammatory potential [12,13].

In Serbia, raspberries are often processed to jelly, diary, or confectionery products able to be stored for a longer time [6,14–17], however it is difficult to preserve this berries with high level of natural properties. In industry, this problem is commonly tackled with some form of drying, where one of the most popular procedures is convective drying that is done at high temperatures and for a long processing time. This often leads to chemical and biochemical changes and loss of native quality of this fruits with changes of color, taste, aroma, and nutritive value [18]. Previous studies showed that convective drying of red raspberries changed their volume, color, shape, and the content of BACs, this is especially true for contents of L–ascorbic acid, flavonoids, anthocyanins, and others [6,19–21]. Therefore, freeze-drying has been established as a standard process in the industry, to preserve nutritive value and extend the shelf life. A downside, however, is that freeze-drying is an expensive technology, not only in terms of the initial investment, but also during processing, even though previous reports showed that freeze-dried fruits, in comparison to convective dried ones, have better preserved their physiochemical, nutritive, and sensory properties [22,23]. The choice of the drying method for a particular food product is a crucial step as the drying procedure and operating conditions have impact on the quality of the dried product and its cost [24]. Despite its simplicity and low investment cost, convective drying is the most common dehydration technique in the food industry with a focus on minimizing economic and environmental impacts. Energy consumption and energy saving potential of convective drying may be tempered by combining new technologies with traditional drying procedures. Bórquez et al. [6] evaluated quality changes during osmotic dehydration of raspberries in sucrose solution with vacuum pretreatment, and followed by microwave-vacuum drying. Obtained results showed acceptable results regarding the preservation of color, taste and structure of these berries. Unfortunately, L-ascorbic acid content decreased 5-fold (220 to 41 mg/100 g), therefore reutilization of sucrose solution with higher initial concentration of L-ascorbic acid is recommended to reduce the losses to 60%.

Kowalski et al. [20] tested different drying techniques of raspberries, i.e., hot air drying vs. combined hybrid drying consisting of simultaneous hot air, microwave, and ultrasound drying, on the kinetics, energy consumption, and product quality. Results revealed that combined hybrid drying significantly improved the drying kinetics and the energy utilization. However, the total CIELab color difference (ΔE) ranged from 12 to 15, which implies significant degradation of color in dried raspberries. In addition, when final product was compared to fresh raspberry, considerable volumetric shrinkage was recorded with slight changes in the shape, but only after convective drying.

Bustos et al. [25] studied the impact of convective drying at various temperatures on phenolic characterization in both, raspberries (*Rubus idaeus* var. Autumn Bliss) and boysenberries (*Rubus ursinus* × Rubus idaeus var. Black Satin). Berries were applied to different convective drying conditions: 50 °C for 48 h, 65 °C for 20 h or 130 °C for 2 h until a moisture content was below 15%. Obtained results indicated that drying regime at 65 °C during 20 h was optimum for the best preservation of color, polyphenol content and antioxidant activity in dried berries. Moreover, authors argued that conventional drying is more economical than freeze-drying, and with considerable increase of total polyphenolic content due to depolymerization of native polyphenols which lead to improved antioxidant activity. Different raspberry cultivars, have different sensitivity to convective drying mostly evident by changes in their physical (e.g., reduced rehydration), mechanical (e.g., initial shape), nutritive (e.g., loss of nutrients), and sensory properties (e.g., formation of unpleasant aroma). Furthermore, Pavkov et al. [17] found that physiochemical properties and rehydration capacity of Polana and Polka varieties could be partially preserved after convective drying.

Therefore, the aim of this research was to investigate the influence of convective drying on physiochemical properties and quality parameters of red Polana raspberry for fresh and frozen samples. Evaluated quality parameters included changes in volume and shape, color, L-ascorbic acid, total phenolic content, flavonoid content, anthocyanin content, and antioxidant activity. All results were controlled against freeze-dried raspberries.

2. Materials and Methods

2.1. Plant Material

Polana variety samples of red raspberries at full maturity were taken from the local farmers during August and October, 2017, at area of Novi Sad, Republic of Serbia. The raspberry fruits were harvested few hours prior to each experiment. The selected samples were approximately equal in size, volume, color, mass, and humidity. Average values of these properties were obtained from raspberry fruits subsample ($n = 500$) that were as following; (i) moisture content $X_{d.b} = 5.45 \pm 0.45$ kg$_w$/kg$_{d.b.}$; (ii) mass m = 3.30 ± 0.24 g; (iii) length a = 21.26 ± 0.74 mm, width b = 20.08 ± 0.56 mm, and thickness c = 18.63 ± 0.62 mm; (iv) volume V = 3.17 ± 0.22 cm^3; and (iv) water activity $a_w = 0.979 \pm 0.001$, at the temperature of 20 °C, total soluble solids of 6.6% and pH = 3.26 ± 0.02.

2.2. Convective Drying and Freeze-Drying of Raspberries

Raspberry fruits were dried as fresh or as frozen (with commercial freezer at −20 °C) with lab-scale convective dryer, designed and constructed to control for air flow (through the layer of processing material), air drying temperature, and continuous monitoring of the sample mass in processing [26]. The dryer chamber door was made up of glass, hence light could have some minimal impact on the quality drying products. In order to compare all results with controls, part of fresh fruits was freeze-dried with Martin Christ, Alpha 2-4 LDplus, without heating of the plates. Drying conditions were −83.8 °C on an ice condenser, with vacuum pressure of 0.0088 mbar in a drying chamber. Samples were dried for 48 h until average humidity at the end of the drying process of $X_{d.b} = 0.07$ kg$_w$/kg$_{d.b.}$.

2.3. Experimental Design and Statistical Analysis

Convective raspberry drying was three-factor experiment with one qualitative and two quantitative factors. The qualitative factor represented the initial state of the raspberry fruit before drying with two levels (fresh and frozen raspberries). Quantitative factors of the experiment were drying temperatures (60, 70, to 80 °C) and air velocity (0.5 to 1.5 m·s^{-1}). The absolute air humidity was approximately constant in the experiments with average value 0.011 ± 0.002 kg$_w$/kg$_{d.air}$. For each experimental run, the initial mass of the raspberries was approximately 500 g. Raspberry samples were set on perforated sieve in a thin stagnant layer and placed in a drying chamber. Air flew along or across the surface of the material in the dryer, and the samples were dried until the same value of moisture content of approximately $X_{d.b} = 0.152$ kg$_w$/kg$_{d.b.}$. The experiment was conducted with three repetitions for each experimental run.

2.4. Measuring of Volume and Shrinkage Determination

When drying some biomaterial, volume shrinkage (V_{sh}) is one of the most common physical and quality indicators of the final product. Shrinkage is expressed by the ratio between the change of volume after drying and volume of sample before drying. The samples volume ($n = 45$) were measured by immersing the raspberries into 96% concentration of ethanol [27] according to

$$V_0 = (m_0 - m_l)/\rho_t \qquad (1)$$

where m_0 is the mass of liquid and the immersed sample (kg), m_l is the mass of liquid, and ρ_t is the liquid density (kg/m^3). The volumetric shrinkage of raspberries (V_{sh}) was based on the following equation [28].

$$V_{sh} = ((V_0 - V_i)/V_0) \times 100 \qquad (2)$$

where V_0 is initial average volume and V_i is volume of each raspberry after drying.

2.5. Determination of Heywood Shape Factor

If observed independently, the volumetric shrinkage is not a sufficient indicator of the changes in dried material. For this reason, an additional indicator was used to monitor changes of shape, i.e., Heywood shape factor (k), able to detect the changes after drying [29–31]. This factor k = 0.523 and was calculated from the relation

$$k = V_p/d_a^3 \quad (3)$$

where V_p is the particle volume with equivalent diameter of the projected area of the particles. This was obtained by assigning the area of an equivalent circle with the same greater diameter as that of the fruit [32].

2.6. Color Measurement and Total Color Difference

Before and after each drying regime, CIELab color parameters were assessed for raspberry samples (n = 45) by colorimeter Konica Minolta CR400(C-light source and the observer angle of 2°). Where L* was whiteness/brightness, a* was redness/greenness, and b* represented yellowness/blueness. The total color difference (ΔE), hue angle (h_o) and chromaticity (C*) were calculated by Equations (4)–(6) [33,34]:

$$\Delta E = ((L^* - L_0)^2 + (a^* - a_0)^2 + (b^* - b_0)^2)^{1/2} \quad (4)$$

$$h_o = \arctan(b^*/a^*) \quad (5)$$

$$C^* = ((a^*)^2 - (b^*)^2)^{1/2} \quad (6)$$

where, L_0, a_0, and b_0 are the color values before drying, while L*, a*, and b* are the color values after drying.

2.7. Analysis of Nutritiveproperties

2.7.1. Extraction Procedure

Methanol extract was prepared from the fresh/dried raspberry samples to determine the contents of total phenols, flavonoids, and radical scavenging capacity. Briefly, 50 mL of an extraction solvent (methanol, 99.8%) (Fisher Scientific, UK) was poured over the raspberry sample into an Erlenmeyer flask. The flasks were covered and placed on a laboratory stirrer for 24 h (in a dark place). After the extraction, the samples were transferred to volumetric flasks of determined volume, filtered, and stored in a dark and cool place until the analysis was carried out.

2.7.2. Determination of Total Phenolic Content (TPC)

The TPC in methanol extracts of fresh and dried raspberries was determined by Folin–Ciocalteu spectrophotometric method [35]. In a 50 mL volumetric flask with V = 0.5 mL of extract, 0.25 mL of Folin–Ciocalteu was mixed and 0.75 mL of 20% Na_2CO_3 (m/v). After 3 min of stirring, distilled water was added and made up to volume of 50 mL. Reaction mixture was left to stand at room temperature for 2h and absorbencies were measured at 765 nm by UV–Vis spectrophotometer. Based on the measured absorbance, the concentration (mg/mL) of TPC was calculated from the calibration curve of the standard solution of gallic acid. The results are expressed in g of gallic acid equivalents (GAE) per 100 g of fruit dried basis (gGAE/100$g_{d.b}$).

2.7.3. Determination of Total Flavonoids Content (TFL)

The TFL was determined by previously described colorimetric method [36]. In short, the reaction mixture was prepared by mixing 1 mL of an extract with 4 mL of distilled water and 0.3 mL of a 5% $NaNO_2$ solution (m/v). Then mixture was incubated at room temperature for five minutes, and then 0.3 mL of 10% $AlCl_3$ (m/v) was added. After six minutes, when the solution became very yellow, 2 mL of NaOH was added. Distilled water was added to the reaction mixture and made up the volume to 10 mL in a volumetric flask. The absorbance was measured at 510 nm. The TF were calculated according to the catechin standard calibration curve and expressed in mg of catechin equivalents (CAE) per 100 g of fruit dried basis (mgCAE/100$g_{d.b}$).

2.7.4. Determination of Radical Scavenging Capacity

The free radical scavenging capacity (RSC) of raspberry extracts was determined using a simple and fast spectrophotometric method described by Espin et al. [37]. Briefly, the prepared extracts were mixed with methanol (95%) and 90 µM 2,2-diphenyl-1-picryl-hydrazyl (DPPH) to give different final concentrations of extract. After 60 min at room temperature, the absorbance was measured at 517 nm. RSC was calculated according to Equation (7) and expressed as IC50 value, which represents the concentration of extract solution required for obtaining 50% of RSC.

$$RSC\ (\%) = 100 - (A_{sample} \times 100)/A_{blank} \qquad (7)$$

where A_{blank} is the absorbance of the blank and A_{sample} is the absorbance of the sample. The obtained results were presented as a mass of dry sample material that is necessary for inhibition of 50% of DPPH (IC50 (mg$_{d.b}$/mL)).

2.7.5. Determination of Monomeric Anthocyanin Content (AC)

The sample preparation for the content of total AC was conducted by previously described method [38]. Here an extraction solvent (ethanolic acid solution) [39] was poured over the samples (fresh or dried raspberries) and the mixture was thoroughly homogenized in a glass beaker. Afterwards, the beaker was covered with paraffin film and left to sit at 4 °C. After 24 h, the extraction mixture was kept at room temperature, filtered, and transferred to volumetric flask and made up to the volume of 100 mL with an extraction solvent. An aliquot of an extract was transferred into two volumetric flasks with added buffers at pH = 1.0 and pH = 4.5. After 15 min, the absorbencies were measured at 510 and 700 nm against distilled water as a blank. The content of AC was recalculated to cyanidin-3-glucoside by

$$AC = (A \times M_w \times D_f \times V_m \times 1000)/(\epsilon \times m) \qquad (8)$$

$$A = (A_{510} - A_{700})pH_{1.0} - (A_{510} - A_{700})pH_{4.5} \qquad (9)$$

where AC = anthocyanin content (mg/100g); A_{510} = sample absorbance at λ = 510nm; A_{700} = sample absorbance at λ = 700nm; M_w = molecular weight of cyanidin-3-glucoside (449.2), D_f = dilution factor = original solution volume; ϵ = molar extraction coefficient of cyanidin-3-glucoside (26900); and m = sample weight (g). The content of AC was expressed in mg per 100 g of fruit dried basis (mg/100$g_{d.b}$).

2.7.6. Determination of Vitamin C Content

Separations and quantifications of vitamin C were performed by HPLC equipment (Thermo Scientific™ UltiMate 3000) on Nucleosil 100-5C_{18}, 5 µm (250 × 4.6 mm I.D.) column (Phenomenex, Los Angeles, CA). Separation was performed with standard method BS EN 14130:2003 (Foodstuffs. Determination of vitamin C by HPLC). The content of AC was expressed in mg ascorbic acid (the sum of ascorbic acid and its oxidative form of dehydroascorbic acid) per 100 g of fruit dried basis (mg/100$g_{d.b}$).

2.8. Statistical Analysis

For the purposes of statistical tests, analysis of variance was performed (ANOVA) with Statistica13 (Stat Soft, Inc., Oklahoma, United States). In order to define homogenous groups of samples an additional Duncan test was performed with statistical significance at $p < 0.05$.

3. Results and Discussion

3.1. Volume Shrinkage

Comparison of convective and freeze-drying technique for fresh vs. frozen samples revealed that the least changes in volume had freeze-drying (V_{sh} = 16.49 ± 2.75%). For convective drying, the least changes in volume had fresh raspberry samples dried at T = 60 °C and air velocity of 1.5 m·s^{-1} (V_{sh} = 35.74 ± 6.78%). Pavkov et al. 2017 [17] reported results of air drying red Polana raspberry, dried at air temperature of 50, 60, 70, and 80 °C and constant air velocity of 1 m·s^{-1}. Judging by the volume shrinkage during convective drying, air temperature of 50 °C will lead to a total collapse and loss of the product shape. Interestingly, the least volume shrinkage (23.17%) was achieved with air temperature at 70 °C. Drying with air temperature at 60 °C provoked shrinkage of 28.74%, what is still lower than it was obtained in the current study. Samples dried with air temperature of 80 °C reached volume shrinkage of 43.13%. Results obtained from these two experiments revealed that higher air temperatures do not necessarily lead to higher changes in volume shrinkage. This may be explained by the fact that higher air temperature have tendency to lean towards mechanical stabilization of the raspberry surface, thus limiting the degree of shrinkage.

Air temperature and initial state of the raspberry, prior to convective drying significantly changed the volume of dried raspberry. On the other hand, the air velocity did not have any impact on the change in raspberry volume (Table 1). The most considerable changes in volume occurred when drying frozen raspberries at T = 80 °C and air velocity of 0.5 m·s^{-1} (V_{sh} = 79.07 ± 4.07%). As previously reported, drying temperature of 80 °C had similar trend on volume shrinkage [17].

Table 1. Statistical analysis of convective drying factor effect on all determined quality indicators.

Depend. Value	Statistical Indicators	Intercept	AT	AV	RS	AT*AV	AT*RS	AV*RS	AT*AV*RS	Error
Volume shrinkage	SS	1687039	96255	11	3834	2874	1686	2743	103	28630
	MS	1687039	48127	11	3834	1437	843	2743	51	59
	F	28814.82	822.02	0.20	65.49	24.54	14.40	46.85	0.88	
	p	0.0000	0.0000	0.6587	0.0000	0.0000	0.0000	0.0000	0.4166	
Heywood shape factor	SS	0.208363	0.0044	0.0001	0.0015	0.0000	0.0000	0.0001	0.0001	0.0094
	MS	0.208363	0.0022	0.0001	0.0015	0.0000	0.0000	0.0001	0.0000	0.0000
	F	10773.36	115.30	10.08	79.20	1.59	1.16	7.79	4.55	
	p	0.0000	0.0000	0.0015	0.0000	0.2057	0.3129	0.0054	0.0110	
Color change	SS	17759.28	576.41	16.78	68.15	31.92	56.31	0.57	1.77	654.85
	MS	17759.28	288.20	16.78	68.15	15.96	28.15	0.57	0.89	2.74
	F	6481.604	105.186	6.126	24.871	5.825	10.275	0.207	0.323	
	p	0.000000	0.000000	0.014017	0.000000	0.00330	0.000052	0.649312	0.724239	
Ascorbic acid	SS	1689.955	1357.504	640.217	6.840	430.577	91.156	0.008	7.249	8.575
	MS	1689.955	678.752	640.217	6.840	215.289	45.578	0.008	3.625	0.357
	F	4729.853	1899.694	1791.841	19.144	602.551	127.564	0.024	10.144	
	p	0.000000	0.000000	0.000000	0.000203	0.000000	0.000000	0.879336	0.000641	
Total phenolic content	SS	40829272	96596	58924	52597	8089	23366	65802	46258	10129
	MS	40829272	48298	58924	52597	4044	11683	65802	23129	422
	F	96744.51	114.44	139.62	124.63	9.58	27.68	155.92	54.80	
	p	0.0000	0.0000	0.0000	0.0000	0.0008	0.0000	0.0000	0.0000	
Total flavonoid content	SS	3799457	6538	2505	1149	8042	1435	6	4362	5100
	MS	3799457	3269	2505	1149	4021	718	6	2181	213
	F	17879.73	15.38	11.79	5.40	18.92	3.38	0.03	10.26	
	p	0.000000	0.000050	0.002170	0.028864	0.000012	0.051007	0.863581	0.000602	
Anthocyanin content	SS	2163673	22925	715	2014	711	2315	782	9184	18331
	MS	2163673	11463	715	2014	355	1157	782	4592	764
	F	2832.821	15.008	0.936	2.637	0.465	1.515	1.024	6.012	
	p	0.0000	0.0000	0.3428	0.117456	0.6334	0.2400	0.3215	0.0076	
Radical scavenging	SS	0.696911	0.003357	0.000115	0.011585	0.027292	0.008212	0.009029	0.015719	0.00044
	MS	0.696911	0.001678	0.000115	0.011585	0.013646	0.004106	0.009029	0.007860	0.00001
	F	37390.89	90.05	6.18	621.55	732.14	220.29	484.40	421.69	
	p	0.000000	0.000000	0.020259	0.000000	0.000000	0.000000	0.000000	0.000000	

AT—Air Temperature; AV—Air Velocity; RS—Raspberry state before drying; SS—Sum of Squares; MS—Mean square; F—Fisher's F-ratio; p— p-value.

Additional for convective drying, some reports indicated variations in volume shrinkages with regards to raspberry varieties. Sette et al. [29] dehydrated with convective drying previously frozen Autumn Bliss raspberry, at T = 60 °C and air velocity of 1–1.5 m·s^{-1}. Here they found higher volume shrinkage (V_{sh} = 81 ± 3%) than what was reported in the current study.

Initial raspberry state effected the shrinkage, which was not surprising as creation of the ice crystals tends to destabilize cellular structures and this is particularly emphasized with drying air velocity of 0.5 m·s^{-1}. On the contrary, Duncan's test revealed that there are no significant differences in the volume shrinkage of the samples which were dried at the same temperature and with velocity of 1.5 m·s^{-1}. The reason for this may be the faster drying rate in the first drying period, which can lead to a faster mechanical stabilization of the surface, hence the preservation of the volume. As expected, initial raspberry state had an effect on volume shrinkage, and the results after drying are presented in Figure 1. Figure 2 shows the shrinkage and the changes in fruit size after convective and freeze-drying of fresh raspberry at T = 60 °C and air velocity of 1.5 m·s^{-1}.

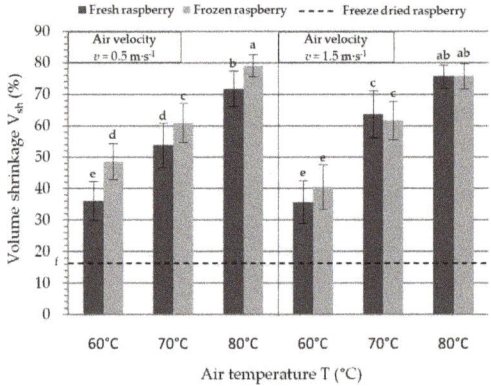

Figure 1. Volume shrinkage after convective drying of raspberry variety Polana. Different lowercase letters indicate significant differences ($p < 0.05$).

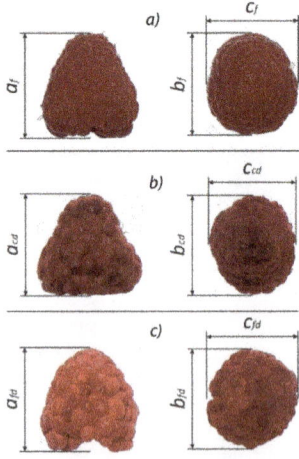

Figure 2. Red raspberry, variety Polana: (**a**) fresh (a_f = 20.74±2.45mm; b_f = 20.04 ± 1.96 mm; c_f = 18.88 ± 1.72 mm), (**b**) after convective drying of fresh raspberry at T = 60 °C and air velocity of 1.5 m·s^{-1} (a_{cd} = 16.97 ± 1.86 mm; b_{cd} = 15.37 ± 1.65 mm; c_{cd} = 15.34 ± 1.43 mm), and (**c**) after freeze-drying (a_{fd} = 20.62 ± 1.98 mm; b_{fd} = 20.62 ± 2.66 mm; c_{fd} = 18.58 ± 1.78 mm).

3.2. Heywood Shape Factor Results

The referent Heywood shape factor before drying of fresh raspberry was k = 0.3323 (Figure 3), and all three experimental factors were significant for the changes in Heywood shape factor. As compared to k of a fresh raspberry, the factor after freeze-drying equaled to k = 0.27. In case of convective drying, the lowest deviation from k occurred for fresh raspberries at T = 60 °C and air velocity of 1.5 m·s^{-1} (k = 0.2694 ± 0.003). Results showed that under the same experimental conditions, the convective dried frozen raspberries had greater deviation of size as compared to the fresh raspberries. Hence, Heywood shape factor corresponded with the results for volume shrinkage.

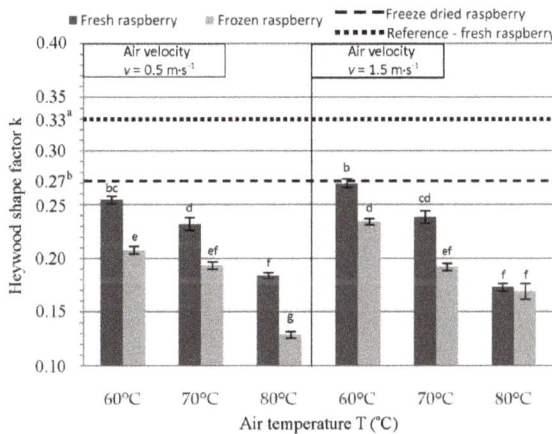

Figure 3. Of Heywood shape factor after convective drying of Polana raspberry. Different lowercase letters indicate significant differences ($p < 0.05$).

3.3. Color Change

CIE Lab color parameters L*, a*, b*, C*, h*, and ΔE* measured on fresh and dried raspberries at different drying conditions are shown in Table 2. Any considerable influences on color caused by the drying of raspberry was not detected, as total color change was roughly 10, except for freeze dried samples, and essentially lightness (L) remained similar to those of a fresh samples. Hence, changes in color were driven by the parameters a* and b*. Generally, convective dried raspberry samples slightly shifted towards maroon color, which can originate from decomposition of carotenoid pigments. Moreover, high temperature induces nonenzymatic Maillard browning with formation of brownish pigmentations [40]. Alternatively, this may be the consequences of high concentrations of preserved anthocyanins in dried samples [3]. A slight increase of a* and b* will have positive repercussions, as it will lean towards more saturated color of products, which corresponds well with increased chroma values.

Air temperature, air velocity, and initial state of raspberry before drying had statistically significant effect on color change ($p < 0.05$) (Table 1). The least color change (ΔE = 5.18) was observed with convective drying at T = 60 °C and air velocity of 1.5 m·s^{-1}. This temperature remained optimal choice regarding ΔE, as it was not modified by different air velocities and initial state of material (fresh and frozen). The largest color change for this drying type was at T = 80 °C and air velocity of 0.5 m·s^{-1} for both frozen and fresh raspberry when ΔE was 11.17 and 10.12, respectively. Unexpectedly, freeze-dried raspberries had the largest color changes (ΔE = 19.60) that were caused by increase in all of the three-color parameters, and especially for a* (Δa = 18.04). Bustos et al. [25] reported similar findings for freeze-dried berries with higher values for redness (Δa = 25.62) as compared to convective samples. Also, study by Sette et al. [3] reported an increase of a* and emphasized that besides pigmentation, differences of internal structures should be considered among convective and freeze dried raspberries.

For instance, after freeze-drying, free water from raspberry is replaced by air, so shifts in red color and lightness can be a consequence of different diffusion of light that passes throughout a material. This effect is likely more pronounced for fruits with defined and vibrant hues, as for the raspberries [3,41]. During the conventional air-drying, increasing drying temperatures reduce the drying time, whereas shorter drying times may result in reduced risks of food quality deterioration [23]. Increasing hot air temperature for convective drying of *Cassia alata* from 40 °C to 60 °C reduced drying time from 180 min to 120 min [42]. Consequently, from data obtained, it can be assumed that as the temperature and the drying time increase, the color change of dried raspberries will increased too.

Table 2. Results of CIE Lab parameters before and after all drying treatments: L* (whiteness/brightness), a* (redness/greenness), b* (yellowness/blueness), and ΔE (color change).

Experiment Factors			Measured Values									ΔE
			Before Drying (Fresh Samples)					After Drying (Dried Samples)				
			L_0	a_0	b_0	$h_0°$	C_0*	L*	a*	b*	h°	C*	
Fresh dried	1.5 m·s⁻¹	Drying air temperature [°C]	80	24.829	22.172	11.091	26.575	24.791	23.848	27.570	11.924	23.388	30.038	5.549 ± 1.67 [h]
			70	24.768	20.932	10.134	25.833	23.256	25.799	29.650	13.448	24.397	32.557	9.383 ± 1.66 [def]
			60	26.225	22.251	9.5562	23.242	24.216	22.608	30.589	14.021	24.625	33.649	10.12 ± 1.10 [bc]
	0.5 m·s⁻¹		80	24.407	22.009	10.553	25.617	24.408	23.058	26.982	11.101	22.363	29.176	5.181 ± 0.95 [h]
			70	24.372	22.513	10.886	25.805	25.006	24.194	30.224	13.100	23.433	32.940	8.024 ± 1.16 [efg]
			60	25.212	22.125	10.084	24.502	24.314	22.029	30.390	14.193	25.033	33.540	9.763 ± 1.21 [cde]
Frozen dried	1.5 m·s⁻¹		80	24.491	21.032	10.397	26.305	23.461	24.264	28.325	12.834	24.375	31.096	7.692 ± 1.23 [g]
			70	24.078	20.509	9.414	24.655	22.566	24.094	29.314	12.897	23.747	32.025	9.468 ± 1.62 [def]
			60	26.368	22.380	10.172	24.442	24.583	26.329	32.674	14.529	23.973	35.758	11.17 ± 1.38 [b]
	0.5 m·s⁻¹		80	25.147	21.106	9.865	25.051	23.297	24.514	28.904	12.892	24.038	31.648	8.38 ± 1.43 [fg]
			70	25.122	21.477	10.321	25.667	23.828	23.603	29.665	13.733	24.841	32.689	8.99 ± 0.80 [def]
			60	25.785	22.719	10.593	24.997	25.067	21.991	30.368	13.977	24.714	33.430	9.18 ± 1.27 [cd]
Freeze dried				24.557	22.367	9.779	23.615	24.411	28.144	40.412	16.562	22.285	43.674	19.608 ± 1.63 [a]

Different lowercase letters indicate significant differences ($p < 0.05$).

3.4. Ascorbic Acid Reduction

The average amount of L-ascorbic acid in fresh samples before drying was 118.27 mg/100g$_{d.b.}$ (18.92 mg/100g$_{w.b.}$) (Table 4), which was similar to quantities reported by Bobinaite et al. [12]. This content of L-ascorbic acid was significantly reduced during convective drying under all experimental conditions. This is expected as prolonged exposure to heightened temperatures and oxygen has tendency to reduce the content of this acidin fruits [21,43–47]. Figure 4 shows temperature kinetics of raspberry samples during convective drying from a fresh state. Type K thermocouple probes were used to monitor and control product temperature during the process, by placing probes inside the drupelet. For all experiments, the temperature at the beginning of the process is approximately 35 °C, but after 10 minutes of the drying, the raspberry temperature can reach 50 °C. Due to the reduced moisture content, during the last quarter of drying all samples have the same temperature as drying air. This means that drying time at T = 80 °C is 6–8 h, and depending of the air velocity can last almost three times longer at T = 60 °C. However, the highest content of L-ascorbic acid was after the shortest convective drying with air velocity of 1.5 m·s⁻¹. This was regardless of the fact that the raspberry temperature reached T = 80 °C, and equaled to 27.46 ± 1.12 mg/100g$_{d.b.}$ and 22.54 ± 1.28 mg/100$_{d.b.}$, after drying of frozen and fresh raspberry, respectively (Table 4). Conversely, the degradation of 99% L-ascorbic acid was detected for longest drying with T = 60 °C and air velocity of 0.5 m·s⁻¹. Accordingly, this might mean that L-ascorbic acid degradation is more induced by longer exposure to higher oxygen levels during the convective drying than to the drying temperature itself.

This reasoning is in accordance with Verbeyst et al. [43] research with thermal and high-pressure effects on vitamin C degradation in strawberries and raspberries. Here it was shown that ascorbic acid degradation from strawberry and raspberry is slightly temperature dependent for temperature range of 80 to 90 °C, and that oxygen presence plays the key role. As expected, the highest levels of

L-ascorbic acid preservation was achieved by freeze-drying (115.48 ± 2.29 mg/100g$_{d.b.}$), since there was neither thermal nor oxygen degradation involved.

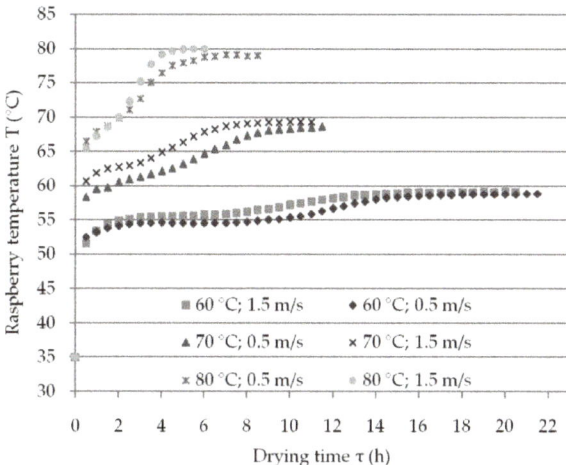

Figure 4. Raspberry temperature kinetics during convective drying from a previously fresh state.

3.5. Total Phenols Reduction

The average values for relevant nutritive profile of fresh raspberries are presented in Table 3. Average total amount of polyphenols in fresh raspberry was 1.63 g GAE/100g$_{d.b.}$. All three individual experimental factors had influence on the content of total phenols (Table 1).

When these samples were dried convectively the best preserved polyphenolic content was at T = 70 °C and air velocity of 1.5 m·s^{-1} (1.28 gGAE/100g$_{d.b.}$). On the contrary, they were least preserved at 60 °C and air velocity of 0.5 m·s^{-1} (0.92 gGAE/100g$_{d.b.}$). Freeze-drying preserved 1.10 g GAE/100g$_{d.b.}$ of total phenols, and, as expected, convective drying reduced polyphenolic content in the samples. The exceptions were the samples freshly dried at air velocity of 1.5 m·s^{-1} and drying temperatures of 70 °C; and 80 °C in which higher total phenolic content was observed in comparison to freeze-dried samples. Similar results were recently reported where higher phenolic content was found in convectively hot air-dried *Cassia alata* in comparison to freeze-dried samples [42]. Hossain et al. has suggested that freeze-drying may not have completely deactivated degradative enzymes due to the low-temperature process. Therefore, reactivation of this degradative enzymes could be further occurred in freeze-dried samples thus result in lower phenolic content [48].

Vasco et al. [49] made classification of 17 fruit types from Ecuador based on their content of total phenols and according to this classification there are three main groups: one with low levels of total phenols (<0.1 gGAE/100g$_{w.b.}$), one with medium level (0.2–0.5 gGAE/100g$_{w.b.}$), and the third with high levels (>1.0 gGAE/g100g$_{w.b.}$). This classification was accepted by others [50,51], and states that fresh Polana raspberry belongs to a high content group, as do freeze-dried and convectively dried samples from this study (under all experimental conditions).

Table 3. The content of BACs and radical scavenging capacity in fresh raspberry sample.

Compound	mg/100g$_{d.b.}$
L-ascorbic acid	118.27
Total phenolic content	1,635.60
Flavonoid content	386.19
Anthocyanin content	513.55
Antioxidative activity IC50 [mg$_{d.b.}$/mL]	0.0534

3.6. Total Anthocyanin Reduction

Temperature had significant influence on the content of anthocyanin (Table 4), however air velocity had no effect on this group of compounds. Amount of anthocyanin in fresh raspberry was 511.7 mg/100g$_{d.b.}$. After convective drying, anthocyanin content was preserved from 40–56%. Anthocyanin content (287.0 mg/100$_{d.b.}$) was best preserved with drying of fresh raspberries at T = 70 °C and air velocity of 0.5 m·s^{-1}. Their least retention occurred after convective drying of frozen samples at T = 70 °C and air velocity of 0.5 m·s^{-1} (205.3 mg/100$_{d.b.}$). Thermal degradation of anthocyanins and complementary oxidization is the origin of the maroon color that was detected with the CIELab analysis. After freeze-drying, the content of anthocyanin in raspberry was 410.4 mg/100g$_{d.b.}$ which is expected due to minimized considerable influence of temperature and oxygen.

Table 4. Results of biologically active compounds in fresh raspberry and dried raspberry after all drying treatments.

Experimental Factors			Measured Values				
			Ascorbic Acid (mg/100g$_{d.b.}$)	Total Phenolic Content (gGAE/100g$_{d.b.}$)	Total Flavonoid Content (mgCAE/100g$_{d.b.}$)	Anthocyanin Content (mg/100g$_{d.b.}$)	Radical Scavenging IC50 (mg$_{d.b.}$/Ml)
Fresh dried	0.5 m·s^{-1}	80	<0.25 [d]	0.921 ± 0.010 [k]	315.1 ± 8.9 [cde]	215.3 ± 8.6 [hi]	0.097 ± 0.0011 [g]
		70	2.45 ± 0.22 [cd]	0.985 ± 0.018 [ij]	298.8 ± 5.4 [de]	287.0 ± 9.4 [c]	0.218 ± 0.0015 [a]
		60	4.04 ± 0.34 [cd]	1.152 ± 0.020 [d]	317.4 ± 6.5 [cde]	256.1 ± 14.6 [ef]	0.198 ± 0.0011 [b]
	1.5 m·s^{-1}	80	2.53 ± 0.33 [cd]	1.075 ± 0.003 [ef]	299.3 ± 16.0 [de]	235.5 ± 6.4 [fg]	0.210 ± 0.0065 [b]
		70	7.65 ± 0.27 [cd]	1.281 ± 0.028 [b]	352.7 ± 14.7 [ab]	242.8 ± 12.5 [fg]	0.101 ± 0.0025 [g]
		60	22.54 ± 1.28 [b]	1.202 ± 0.003 [c]	331.8 ± 15.6 [bc]	248.5 ± 5.7 [ef]	0.124 ± 0.0010 [de]
Frozen dried	0.5 m·s^{-1}	80	<0.25 [d]	0.994 ± 0.001 [ij]	290.0 ± 15.0 [e]	206.2 ± 16.7 [i]	0.135 ± 0.0055 [d]
		70	<0.25 [d]	1.080 ± 0.002 [ef]	302.1 ± 20.1 [cde]	205.3 ± 5.9 [i]	0.095 ± 0.0010 [g]
		60	8.87 ± 0.90 [c]	1.011 ± 0.006 [hi]	375.5 ± 3.2 [a]	276.0 ± 11.7 [cd]	0.084 ± 0.0012 [h]
	1.5 m·s^{-1}	80	<0.25 [d]	0.976 ± 0.019 [j]	322.4 ± 7.0 [bcd]	227.1 ± 12.2 [gh]	0.182 ± 0.0075 [c]
		70	6.58 ± 0.46 [cd]	1.029± 0.015 [h]	361.7 ± 3.5 [a]	238.3 ± 16.6 [fg]	0.111 ± 0.0043 [f]
		60	27.46 ± 1.12 [b]	1.067 ± 0.019 [f]	331.1 ± 9.3 [bc]	263.9 ± 3.4 [de]	0.124 ± 0.0077 [ef]
Freeze dried			115.48 ± 2.29 [a]	1.103 ± 0.019 [e]	327.8 ± 1.24 [cde]	410.4 ± 9.4 [b]	0.064 ± 0.0001 [i]
Fresh raspberry			118.27 ± 2.88 [a]	1.635 ± 0.025 [a]	386.1 ± 21.1 [a]	511.7 ± 5.0 [a]	0.053 ± 0.0005 [j]

* Different lowercase letters indicate significant differences ($p < 0.05$).

3.7. Radical Scavenging Capacity

As expected, all experimental factors influenced the radical scavenging capacity (Table 4). As a smaller IC50 means higher radical scavenging capacity, the majority of convectively dried raspberries exhibited lower radical scavenging capacity in comparison to fresh or freeze-dried samples. The IC50 value of fresh raspberry was IC50 = 0.0534 mg$_{d.b.}$/mL. Freeze-dried samples had highly preserved radical scavenging capacity that was equal to 0.0641 mg$_{d.b.}$/mL, likely due to high preservation of all bioactive compounds. The lowest IC50 value (e.g. the highest radical scavenging capacity) had convective drying for frozen samples, of IC50 = 0.0845 mg$_{d.b.}$/mL which was obtained at T = 80 °C and air velocity of 0.5 m·s^{-1}. The main reason for this may be the high preservation of total flavonoid content (0.97%) in dried samples with same convective drying regime. Raspberry belongs to a group of biomaterial with high radical scavenging capacity [12,50]. It is also believed that almost 20% of its total radical scavenging capacity is secured by the content of L-ascorbic acid [52]. As previously reported, heat and oxygen have influence on almost all bioactive compounds with some form of degradation, so it is not surprising that to find the loss of radical scavenging capacity due to convective drying.

4. Conclusions

Using physical properties, contents of various biologically active compounds and radical scavenging capacity proved to be useful in selecting alternatives for preservation of raspberries as in the case of convective and freeze-drying. For Polana variety, the most desirable results against freeze-drying as standard in terms of color, volume shrinkage, and Heywood shape factor change was achieved with convective drying of fresh raspberry at T = 60 °C with air velocity of 1.5 m·s^{-1}. Convective drying of raspberry had influenced all measured biologically active compounds. In comparison to fresh samples, in convectively dried raspberries 60–78% of total phenols was preserved as well as 75–97% of flavonoids and 40–56% of anthocyanins. Consequently, lower radical scavenging capacity was found in convectively dries samples as compared to fresh or freeze-dried. The largest shortcoming for convective drying was observed in difference between freeze-dried for preservation levels of L-ascorbic acid. Freeze-drying preserved more than 97% of L–ascorbic acid, while convective drying samples had degradation of over 80% of this compound. This might not be as relevant where L–ascorbic acid is added in processing of raspberries (e.g., confectionery products, biscuits, cookies, dairy product etc.). In conclusion, Polana raspberry dried convectively with air temperature of 60 °C and air velocity of 1.5 m·s^{-1}, may be considered as sufficient alternative to freeze-drying.

Author Contributions: Conceptualization, I.P., M.R., and Z.S.; Methodology, I.P., M.R., A.T.H., and Z.S.; Formal analysis, Z.S. and K.K.; Contributions to sample and analysis experiments, Z.S. and K.K.; writing—original draft preparation, Z.S.; writing—review and editing, I.P., M.R., A.T.H., P.P., and D.B.K.; supervision, P.P. and D.B.K.

Funding: This manuscript is a result of the research within the national project TR31058, 2011-2020, supported by the Ministry of Education, Science and Technology, Republic of Serbia.

Conflicts of Interest: The authors declare no conflicts of interest.

References

1. Statistical Office of the Republic of Serbia. *Statistical Yearbook*; Statistical Office of the Republic of Serbia: Belgrade, Serbia, 2018; 51, p. 214.
2. Paraušić, V.; Simeunović, I. Market analysis of Serbia's raspberry sector and cluster development initiatives. *Econ. Agricult.* **2016**, *63*, 1417–1431.
3. Sette, P.; Franceschinis, L.; Schebor, C.; Salvatori, D. Fruit snacks from raspberries: Influence of drying parameters on colour degradation and bioactive potential. *Int. J. Food Sci. Technol.* **2017**, *52*, 313–328. [CrossRef]
4. Szadzińska, J.; Łechtańska, J.; Pashminehazar, R.; Kharaghani, A.; Tsotsas, E. Microwave- and ultrasound-assisted convective drying of raspberries: Drying kinetics and microstructural changes. *Dry Technol.* **2018**, *37*, 1–12. [CrossRef]
5. Rodriguez, A.; Bruno, E.; Paola, C.; Campañone, L.; Mascheroni, R.H. Experimental study of dehydration processes of raspberries (*Rubus Idaeus*) with microwave and solar drying. *Food Sci. Technol.* **2018**, *39*. [CrossRef]
6. Bórquez, R.M.; Canales, E.R.; Redon, J.P. Osmotic dehydration of raspberries with vacuum pretreatment followed by microwave-vacuum drying. *J. Food Eng.* **2010**, *99*, 121–127. [CrossRef]
7. Wang, S.Y.; Lin, H.-S. Antioxidant activity in fruits and leaves of blackberry, raspberry, and strawberry varies with cultivar and developmental stage. *J. Agr. Food Chem.* **2000**, *48*, 140–146. [CrossRef]
8. Verbeyst, L.; Crombruggen, K.V.; Van der Plancken, I.; Hendrickx, M.; Van Loey, A. Anthocyanin degradation kinetics during thermal and high pressure treatments of raspberries. *J. Food Eng.* **2011**, *105*, 513–521. [CrossRef]
9. Summen, M.A.; Erge, H.S. Thermal degradation kinetics of bioactive compounds and visual color in raspberry pulp. *J. Food Process Preserv.* **2014**, *38*, 551–557. [CrossRef]
10. Nile, S.H.; Park, S.W. Edible berries: Bioactive components and their effect on human health. *Nutrition* **2014**, *30*, 134–144. [CrossRef]
11. Carvalho, E.; Fraser, P.D.; Martens, S. Carotenoids and tocopherols in yellow and red raspberries. *Food Chem.* **2013**, *139*, 744–752. [CrossRef]
12. Bobinaitė, R.; Viškelis, P.; Venskutonis, P.R. Variation of total phenolics, anthocyanins, ellagic acid and radical scavenging capacity in various raspberry (*Rubus* spp.) cultivars. *Food Chem.* **2012**, *132*, 1495–1501. [CrossRef] [PubMed]

13. Szymanowska, U.; Baraniak, B.; Bogucka-Kocka, A. Antioxidant, anti-inflammatory, and postulated cytotoxic activity of phenolic and anthocyanin-rich fractions from polana raspberry (*Rubus idaeus* L.) fruit and juice—In vitro study. *Molecules* **2018**, *23*, 1812. [CrossRef] [PubMed]
14. Giuffrè, A.M.; Louadj, L.; Rizzo, P.; Poiana, M.; Sicari, V. Packaging and storage condition affect the physicochemical properties of red raspberries (*Rubus idaeus* L., cv. Erika). *Food Control* **2019**, *97*, 105–113. [CrossRef]
15. Tamer, C.E. A research on raspberry and blackberry marmalades produced from different cultivars. *J. Food Process Preserv.* **2012**, *36*, 74–80. [CrossRef]
16. Rajkovic, A.; Smigic, N.; Djekic, I.; Popovic, D.; Tomic, N.; Krupezevic, N.; Uyttendaele, M.; Jacxsens, L. The performance of food safety management systems in the raspberries chain. *Food Control* **2017**, *80*, 151–161. [CrossRef]
17. Pavkov, I.; Stamenković, Z.; Radojčin, M.; Babić, M.; Bikić, S.; Mitrevski, V.; Lutovska, M. Convective and freeze drying of raspberry: Effect of experimental parameters on drying kinetics, physical properties and rehydration capacity. In Proceedings of the INOPTEP 5th International Conference Sustainable Postharvest and Food Technologies, Vršac, Serbia, 23–28 April 2017; pp. 261–266.
18. Mierzwa, D.; Szadzińska, J.; Pawłowski, A.; Pashminehazar, R.; Kharaghani, A. Nonstationary convective drying of raspberries, assisted by microwaves and ultrasound. *Dry Technol.* **2019**, *37*, 988–1001. [CrossRef]
19. Ratti, C. Hot air and freeze-drying of high-value foods: A review. *J. Food Eng.* **2001**, *49*, 311–319. [CrossRef]
20. Kowalski, S.J.; Pawłowski, A.; Szadzińska, J.; Łechtańska, J.; Stasiak, M. High power airborne ultrasound assist in combined drying of raspberries. *Innov. Food Sci. Emerg.* **2016**, *34*, 225–233. [CrossRef]
21. López, J.; Uribe, E.; Vega-Gálvez, A.; Miranda, M.; Vergara, J.; Gonzalez, E.; Di Scala, K. Effect of air temperature on drying kinetics, vitamin c, antioxidant activity, total phenolic content, non-enzymatic browning and firmness of blueberries variety O´Neil. *Food Bioprocess Technol.* **2010**, *3*, 772–777. [CrossRef]
22. Zorić, Z.; Pedisić, S.; Kovačević, D.B.; Ježek, D.; Dragović-Uzelac, V. Impact of packaging material and storage conditions on polyphenol stability, colour and sensory characteristics of freeze-dried sour cherry (*Prunus cerasus* var. Marasca). *J. Food Sci. Technol.* **2015**, *53*, 1247–1258. [CrossRef]
23. Sun, Y.; Zhang, M.; Mujumdar, A. Berry drying: Mechanism, pretreatment, drying technology, nutrient preservation, and mathematical models. *Food Eng. Rev.* **2019**, *11*, 61–77. [CrossRef]
24. Sabarez, H.T. Airborne ultrasound for convective drying intensification. In *Innovative Food Processing Technologies —Extraction, Separation, Component Modification and Process Intensification*; Knoerzer, K., Juliano, P., Smithers, G., Eds.; CSIRO Food and Nutrition: Werribee, Australia, 2016; pp. 361–386.
25. Bustos, M.C.; Rocha-Parra, D.; Sampedro, I.; de Pascual-Teresa, S.; León, A.E. The influence of different air-drying conditions on bioactive compounds and antioxidant activity of berries. *J. Agric. Food Chem.* **2018**, *66*, 2714–2723. [CrossRef] [PubMed]
26. Pavkov, I. Combined Technology of Fruit Tissue Drying. Ph.D. Thesis, Faculty of Agriculture, University of Novi Sad, Serbia, Novi Sad, 2012.
27. Mohsenin, N.N. *Physical Properties of Plant and Animal Materials*; Gordon and Breach Sci. Publ.: New York, NY, USA, 1986.
28. Radojčin, M.; Babić, M.; Pavkov, I.; Stamenković, Z. Osmotic drying effects on the mass transfer and shrinkage of quince tissue. *PTEP J. Process Energy Agric.* **2015**, *19*, 113–119.
29. Sette, P.; Salvatori, D.; Schebor, C. Physical and mechanical properties of raspberries subjected to osmotic dehydration and further dehydration by air- and freeze-drying. *Food Bioprod. Process.* **2016**, *100*, 156–171. [CrossRef]
30. Panyawong, S.; Devahastin, S. Determination of deformation of a food product undergoing different drying methods and conditions via evolution of a shape factor. *J. Food Eng.* **2007**, *78*, 151–161. [CrossRef]
31. Michelis, A.D.; Pirone, B.N.; Vullioud, M.B.; Ochoa, M.R.; Kesseler, A.G.; Márquez, C.A. Cambios de volumen, área superficial y factor de forma de Heywood durante la deshidratación de cerezas (*Prunus avium*). *Ciência Tecnol. Aliment.* **2008**, *28*, 317–321. [CrossRef]
32. De Michelis, A. Effect of structural modifications on the drying kinetics of foods: Changes in volume, surface area and product shape. *Int. J. Food Stud.* **2013**, *2*, 188–211. [CrossRef]
33. Radojčin, M.; Babić, M.; Babić, L.; Pavkov, I.; Stojanović, Č. Color parameters change of quince during combined drying. *PTEP J. Process Energy Agric.* **2010**, *14*, 81–84.

34. Maskan, M. Kinetics of colour change of kiwifruits during hot air and microwave drying. *J. Food Eng.* **2001**, *48*, 169–175. [CrossRef]
35. Singleton, V.L.; Rossi, J.A.J. Colorimetry of total phenolics with phosphomolybdic-phosphotungstic acid reagents. *Am. J. Enol. Viticult.* **1965**, *16*, 144–158.
36. Harborne, J.B.; Williams, C.A. Advances in flavonoid research since 1992. *Phytochemistry* **2000**, *55*, 481–504. [CrossRef]
37. Espín, J.C.; Soler-Rivas, C.; Wichers, H.J. Characterization of the total free radical scavenger capacity of vegetable oils and oil fractions using 2,2-diphenyl-1-picrylhydrazyl radical. *J. Agric. Food Chem.* **2000**, *48*, 648–656. [CrossRef] [PubMed]
38. Giusti, M.M.; Wrolstad, R.E. Characterization and measurement of anthocyanins by UV-visible spectroscopy. *Curr. Protoc. Food Anal. Chem.* **2001**, *00*, F1.2.1–F1.2.13. [CrossRef]
39. Fuleki, T.; Francis, F.J. Quantitative methods for anthocyanins. 1. Extraction and determination of total anthocyanin in cranberries. *J. Food Sci.* **1968**, *33*, 72–77. [CrossRef]
40. Si, X.; Chen, Q.; Bi, J.; Yi, J.; Zhou, L.; Wu, X. Infrared radiation and microwave vacuum combined drying kinetics and quality of raspberry. *J. Food Process Eng.* **2016**, *39*, 377–390. [CrossRef]
41. Saarela, J.M.S.; Heikkinen, S.M.; Fabritius, T.E.J.; Haapala, A.T.; Myllylä, R.A. Refractive index matching improves optical object detection in paper. *Meas. Sci. Technol.* **2008**, *19*, 055710. [CrossRef]
42. Chua, L.Y.; Chua, B.L.; Figiel, A.; Chong, C.H.; Wojdyło, A.; Szumny, A.; Lech, K. Characterisation of the convective hot-air drying and vacuum microwave drying of *Cassia alata*: Antioxidant activity, essential oil volatile composition and quality studies. *Molecules* **2019**, *24*, 1625. [CrossRef]
43. Verbeyst, L.; Bogaerts, R.; Van der Plancken, I.; Hendrickx, M.; Van Loey, A. Modelling of vitamin C degradation during thermal and high-pressure treatments of red fruit. *Food Bioprocess. Technol.* **2012**, *6*, 1015–1023. [CrossRef]
44. Rodríguez, Ó.; Gomes, W.; Rodrigues, S.; Fernandes, F.A.N. Effect of acoustically assisted treatments on vitamins, antioxidant activity, organic acids and drying kinetics of pineapple. *Ultrason Sonochem* **2017**, *35*, 92–102. [CrossRef]
45. Herbig, A.-L.; Renard, C.M.G.C. Factors that impact the stability of vitamin C at intermediate temperatures in a food matrix. *Food Chem.* **2017**, *220*, 444–451. [CrossRef]
46. Santos, P.H.S.; Silva, M.A. Retention of vitamin C in drying processes of fruits and vegetables—A Review. *Dry Technol.* **2008**, *26*, 1421–1437. [CrossRef]
47. Arancibia-Avila, P.; Namiesnik, J.; Toledo, F.; Werner, E.; Martinez-Ayala, A.L.; Rocha-Guzmán, N.E.; Gallegos-Infante, J.A.; Gorinstein, S. The influence of different time durations of thermal processing on berries quality. *Food Control* **2012**, *26*, 587–593. [CrossRef]
48. Hossain, M.B.; Barry-Ryan, C.; Martin-Diana, A.B.; Brunton, N.P. Effect of drying method on the antioxidant capacity of six *Lamiaceae* herbs. *Food Chem.* **2010**, *123*, 85–91. [CrossRef]
49. Vasco, C.; Ruales, J.; Kamal-Eldin, A. Total phenolic compounds and antioxidant capacities of major fruits from Ecuador. *Food Chem.* **2008**, *111*, 816–823. [CrossRef]
50. De Souza, V.R.; Pereira, P.A.P.; da Silva, T.L.T.; de Oliveira Lima, L.C.; Pio, R.; Queiroz, F. Determination of the bioactive compounds, antioxidant activity and chemical composition of Brazilian blackberry, red raspberry, strawberry, blueberry and sweet cherry fruits. *Food Chem.* **2014**, *156*, 362–368. [CrossRef] [PubMed]
51. Alibabić, V.; Skender, A.; Bajramović, M.; Šertović, E.; Bajrić, E. Evaluation of morphological, chemical, and sensory characteristicsof raspberry cultivars grown in Bosnia and Herzegovina. *Turk. J. Agric. For.* **2018**, *42*, 67–74. [CrossRef]
52. Beekwilder, J.; Hall, D.R.; Ric de Vos, C.H. Identification and dietary relevance of antioxidants from raspberry. *Biofactors* **2005**, *23*, 197–205. [CrossRef]

© 2019 by the authors. Licensee MDPI, Basel, Switzerland. This article is an open access article distributed under the terms and conditions of the Creative Commons Attribution (CC BY) license (http://creativecommons.org/licenses/by/4.0/).

Article

Phenolic and Antioxidant Analysis of Olive Leaves Extracts (*Olea europaea* L.) Obtained by High Voltage Electrical Discharges (HVED)

Irena Žuntar [1], Predrag Putnik [2], Danijela Bursać Kovačević [2], Marinela Nutrizio [2], Filip Šupljika [2], Andreja Poljanec [2], Igor Dubrović [3], Francisco J. Barba [4] and Anet Režek Jambrak [2,*]

1. Faculty of Pharmacy and Biochemistry, University of Zagreb, 10000 Zagreb, Croatia
2. Faculty of Food Technology and Biotechnology, University of Zagreb, 10000 Zagreb, Croatia
3. Teaching Institute for Public health of Primorje-Gorski Kotar County, 51000 Rijeka, Croatia
4. Nutrition and Food Science Area, Preventive Medicine and Public Health, Food Sciences, Toxicology and Forensic Medicine Department, Faculty of Pharmacy, Universitat de València, Avda. Vicent Andrés Estellés, s/n, Burjassot, 46100 València, Spain
* Correspondence: anet.rezek.jambrak@pbf.hr; Tel.: +385-1460-5287

Received: 6 June 2019; Accepted: 5 July 2019; Published: 8 July 2019

Abstract: Background: The aim of this study was to evaluate high voltage electrical discharges (HVED) as a green technology, in order to establish the effectiveness of phenolic extraction from olive leaves against conventional extraction (CE). HVED parameters included different green solvents (water, ethanol), treatment times (3 and 9 min), gases (nitrogen, argon), and voltages (15, 20, 25 kV). Methods: Phenolic compounds were characterized by ultra-performance liquid chromatography-tandem mass spectrometer (UPLC-MS/MS), while antioxidant potency (total phenolic content and antioxidant capacity) were monitored spectrophotometrically. Data for Near infrared spectroscopy (NIR) spectroscopy, colorimetry, zeta potential, particle size, and conductivity were also reported. Results: The highest yield of phenolic compounds was obtained for the sample treated with argon/9 min/20 kV/50% (3.2 times higher as compared to CE). Obtained results suggested the usage of HVED technology in simultaneous extraction and nanoformulation, and production of stable emulsion systems. Antioxidant capacity (AOC) of obtained extracts showed no significant difference upon the HVED treatment. Conclusions: Ethanol with HVED destroys the linkage between phenolic compounds and components of the plant material to which they are bound. All extracts were compliant with legal requirements regarding content of contaminants, pesticide residues and toxic metals. In conclusion, HVED presents an excellent potential for phenolic compounds extraction for further use in functional food manufacturing.

Keywords: high voltage electrical discharge; olive leaves extracts; green solvents; eco-extraction; sustainability

---

1. Introduction

Plant extracts obtained from fruits, leaves, flowers, woods, roots, resins or seeds of known medicinal properties are responsible for many health benefits, like reduction of hypertension, prevention of cardiovascular disease, suppression of different types of cancer and viral disease, etc. Classical/conventional extraction (CE) methods are not only time and energy demanding, but are also prone to usage of solvents, resulting with overall hazardous/toxic effects on the environment. Therefore, there is a growing interest to utilize plant extracts with "green chemistry" in science-based production of functional foods, food supplements, cosmetics, perfumes, nutraceuticals, and pharmaceuticals in a sustainable ways [1–7].

Olive tree (*Olea europaea* L.) is cultivated from antiquity in many parts of the world, but Mediterranean region is the main agricultural area for its production. Mediterranean region is also known for its diet with olives containing nutritional, health and antioxidant benefits, while being rich in with phenols naturally originated from olive plant [8–10]. Olive leaves (OLs) which may be considered as both, waste from olive oil production and medicinal and aromatic herbs, contain a large variety of phenols including oleuropeosides (oleuropein and verbascoside); flavonols (rutin); flavones (luteolin-7-glucoside, apigenin-7-glucoside, diosmetin-7-glucoside, luteolin and diosmetin); flavan-3-ols (catechin), substituted phenols (tyrosol, hydroxytyrosol, vanillin, vanilic acid and caffeic acid), oleoside and secoiridoid glycoside (oleuricine A and oleuricine B) [1,2,9–13].

It is well known that polyphenols as antioxidants may prevent or minimize oxidative damage/stress at cellular levels, resulted from generated concentration imbalance between reactive oxygen (ROS) and nitrogen (RNS) species and cell antioxidants. Numerous phenolics in OLs have strong radical scavenging activity, showing that olive phenolics exhibit more beneficial effects in form of a mixture (e.g., olive leaves extract; OLE) than isolated as a single phenolic compound. Olive leaves extract have a synergistic capacity in the elimination of free radicals, that is more superior to the antioxidant capacity of the vitamin C and E [3,7–12]. Consequently, OLE as an antioxidant may reduce the risks for harmful health effects [3,10,11]. Also, recent studies on OLE confirmed potential coadjuvant use for cervical cancer treatments [12], inhibitory effects on the obesity [13], suppression of inflammatory cytokine production and the activation of NLRP3 inflammasomes in human placenta [14], potential as preventive therapy for neurodegenerative diseases [15], protective quality of testis and sperms [16], protection against intoxication of liver by carbon tetrachloride, kidney by cadmium, and brain by lead poisoning [17–19].

Furthermore, OLE polyphenols are useful as natural food additives that have antimicrobial and antioxidant properties. For instance, OLE are good for fermentation and oxidative processes of table olives and for improving their nutritional properties [20], i.e., as replacement of synthetic additives and the shelf life extension of fish patties [21] and salmon burgers [22]. OLs as by-products/waste generated during olive oil production contains even higher amounts of polyphenols than appreciated olive oil. Therefore, there is a great interest for their extraction from OLs and especially with application of eco-friendly and sustainable processes [2,23–25].

However, green chemistry only applies in a part of the life cycle of a single natural product [4] which also includes food safety aspects. Therefore, OLEs must be safe with respect to contaminants, pesticide residues and toxic metals (Pb, Hg, Cd, etc) [26,27]. Nonetheless, it is important to note that OLs are a rich source of various minerals (Ca, K, Na, P, S, Cl, Mg, Al, Si, K, Mn, Fe) and trace elements (Ni, Cu, Sr, Ba, Cr, Zn, Rb, Th, Co, Ng, Cs, La, Ce, V, Nd) [28].

In a context of the Chemat's principles of green chemistry, an extraction is defined as:

"Green extraction is based on the discovery and design of extraction processes which will reduce energy consumption, allows use of alternative solvents and renewable natural products, and ensure a safe and high-quality extract/product" [4].

Current regulations including REACH (Registration, Evaluation, Authorization and Restriction of Chemicals, CLP (Classification, Labelling and Packaging of substances and mixtures), IPPC (Integrated Pollution Prevention and Control) [29], and notation of BAT (Best Available Technology) have direct progressive impact in diminishing the consumption of organic solvents and Volatile Organic Compounds (VOC) [4]. Hence, they aim to save human health and environment, both in research and development (R&D), as well as in production. Among green solvents, the agro- or bio-solvents play an important role for the replacement of organic solvents. Ethanol is a well-known, common bio-solvent, classified as environmentally preferable. It is obtained by the fermentation of sugar-rich materials such as sugar beet and cereals. Although it is flammable and potentially explosive, ethanol is used on a large scale due to synthesis in high purity, low price, and complete biodegradability [4].

In order to improve CE, modern non-conventional extractions were developed to reduce the mass transfer limitations and to increase yields in shorter time with minimal consumption of solvents. These advanced techniques include ultrasound-assisted extraction (UAE), microwave-assisted extraction (MAE), sub- and super-critical fluid extraction (SFE), pressurized liquid extraction (PLE), pulsed electric fields (PEF), and high voltage electrical discharges (HVED) [1,30,31]. Although there are limitations of innovative technologies for industrial implementations (e.g., high costs, control of process variables, lack of regulatory approval, and consumer acceptance) [32], HVED (cold plasma) is promising green technique for the extraction of biologically active compounds (BACs) with green solvents. Thus, it has the potential for development of green chemical engineering and sustainable production [25]. PLE, in example, would yield the greatest possible amounts of flavonoids (25.66 mg/g) in one extraction cycle, that lasts for 15 min at 100 °C. If extraction, for PLE, would be-taken to obtain the most of flavonols (9.22 mg/g), that can be achieved with one extraction cycle, that lasts for 5 min at 87 °C [23].

Developments of non-thermal technologies, such PEF and HVED, have been advancements for industry and academia in meeting the challenges of producing safe and high-quality food [1,2]. HVED is an emerging technology and broad designation for pulsed mode plasma systems, but it is mainly known as "corona discharge" due to its discharge reminiscent of crown that surrounds the cathode wire with pulsed DC power supply [33].

It is well known that HVED extraction is one of the applications of the liquid phase discharge technology where a phenomenon of electrohydraulic discharge might occur in the water by the effect of high voltage pulsed electric field which is accompanied by several secondary phenomena. Additionally, UV light can lead cell inactivation by damaging to the DNA, shock waves and strong liquid turbulence can cause products fragmentation as well as mechanical destruction of cell tissues, and the high density of radicals can damage to the cell by cell oxidation. These are reasons why HVED may destruct cellular structure and enhance mass transfer from the cell to the solution, thus greatly improving the yields of BACs [34]. Pioneers of HVED extraction summarized many aspects of HVED extraction, but recent review puts emphasis on introducing different extraction devices, including batch, continuous and circulating extraction systems, generalizing the critical processes factors and recent applications, discussing the advantages and disadvantages of HVED extraction as well as the future trends of HVED assisted extraction technique. Although HVED is very potent extraction method with so many advantages, the authors also discussed some disadvantages and problems needed to be solved for its future development that will benefit various fields such as food and medical industries [34].

Therefore, the aim of this study is to evaluate HVED in the extraction efficiency of phenolic components from autochthonous Mediterranean OLs by varying: (i) concentrations of ethanol; (ii) voltages applied; and (iii) treatment times. Phenolic content in OLE was characterized by ultra-performance liquid chromatography-tandem mass spectrometer (UPLC-MS/MS), while antioxidant tests were conducted to analyze the antioxidant potency. Near infrared spectroscopy (NIR) was used for the prediction of an extract quality and analyzed by principal component analysis (PCA) and all results were controlled against CE as standard.

2. Materials and Methods

2.1. Plant Material

Dried olive leaves (*Olea europaea* L.) were provided locally from Mediterranean area (Zadar county, Croatia). Herbs were stored in polyethylene bags in a dark and dry place until extractions. Herbs were milled using knife mill (Grindomix GM 300 – RETSCH; Retsch GmbH, Haan, Germany) before HVED treatment. Plant particle size distribution was: $d(0.9) \leq 330.563$ µm; $d(0.5) \leq 118.540$ µm; $d(0.1) \leq 23.105$ µm measured by the laser particle size analyzer (Malvern, Mastersizer 2000, Germany). Prior to HVED extraction, leaves were weighted (1 g) and mixed with 50 mL of extracting solvent, as distilled water, 25% and 50% aqueous ethanol (v/v) at room temperature (22 °C).

2.2. Conventional and HVED Assisted Extraction

The CE was used as a control procedure to compare the efficiency of HVED under the same conditions with respect to solvent type and treatment time. Therefore, the CE was conducted by varying: (i) solvent type (0, 25 and 50% of ethanol (v/v)); (ii) stirring time (3 and 9 min) at room temperature (22 °C). Magnetic stirring was used to provide effective stirring of herb and extraction solvent during CE. HVED assisted extraction (plasma) generator "IMP-SSPG-1200" (Impel group, Zagreb, Croatia) was used for rectangular pulses from direct current (DC) as high voltage (HV) generator. Maximum adjustable current was 30 mA up to voltage of 25 kV. Frequency was 100 Hz; pulse width was 400 µs; voltage was 15 kV and 20 kV for argon gas; and 20 kV and 25 kV for nitrogen.

Parameters were chosen based on conducted preliminary experiments with different HVED parameters (frequency, voltage, pulse length, distance between electrodes), as well as mass to solvent ratio. From more than 100 experiments, mentioned parameters were chosen. Regarding gases: it is important to have significant voltage to obtain discharge, if too low, there will be no discharge. For that reason, 15 kV and 20 kV were chosen for argon, and 20 kV and 25 kV for nitrogen. For ethanol concentration, pharmacopeia regulations were followed. The mixture of leaves and solvent was transferred to beaker shaped reactor V = 100 mL. This reactor was opened on both sides and fitted with silicone tops (diameter 1 cm). They were used for easier mounting of the electrode from the top and needle form the bottom. Both, argon or nitrogen were introduced through the needle with a flow of 5 L min^{-1}. The gap between electrodes was 15 mm, while the set-up of generator and reactor is shown in Figure 1. The high voltage probe (Tektronix P6015A) connected to the oscilloscope (Hantek DS05202BM) was used to measure the output voltage (data not shown). Physical properties as pH, conductivity (µs/cm), temperature and power consumption of the instrument were monitored before and after HVED treatment and modified CE.

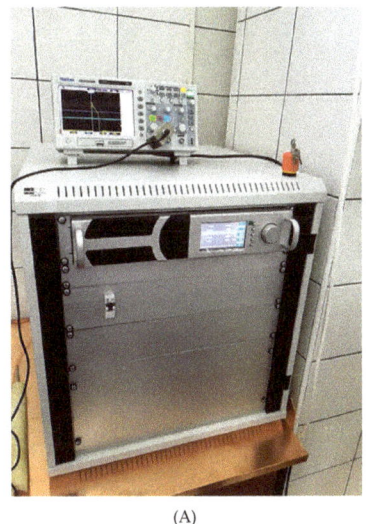
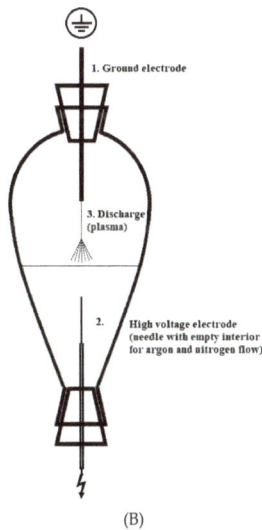

Figure 1. Set-up of generator and reactor for high voltage electrical discharges (HVED) treatments. (**A**) HVED and plasma generator "IMP-SSPG-1200" (Impel group, Zagreb, Croatia); (**B**) Beaker shaped reactor.

Experimental Design and Statistical Analysis

The experiment was designed in STATGRAPHICS Centurion (StatPoint Technologies, Inc, Warrenton, VA, USA) software. Multifactorial design consisting of 12 experimental trials using per each gas (argon and nitrogen). The three chosen independent variables were: (i) concentration of ethanol (0, 25 or 50%), voltage applied (15 kV or 20 kV for argon, and 20 kV or 25 kV for nitrogen) and treatment time (3 and 9 min) (Table 1). Both extractions, HVED and CE were performed in duplicates. The model was fitted by multiple linear regressions (MLR). Calculations were done at 95% of confidence level.

Table 1. Denotation of sample for high voltage electrical discharges (HVED) treatments.

ID	Sample	Treatment Time (min)	Voltage (kV)	Ethanol Content (%)	Stirring (min)	Extraction Type
1	3 OL0	0	0	0	3	
2	9 OL0	0	0	0	9	
3	3 OL25	0	0	25	3	CE
4	9 OL25	0	0	25	9	
5	3 OL50	0	0	50	3	
6	9 OL50	0	0	50	9	
7	OLN1	3	20	50	/	
8	OLN2	9	20	0	/	
9	OLN3	3	20	0	/	
10	OLN4	3	25	0	/	
11	OLN5	9	25	25	/	
12	OLN6	9	20	25	/	
13	OLN7	9	20	50	/	
14	OLN8	9	25	50	/	
15	OLN9	3	25	25	/	
16	OLN10	9	25	0	/	
17	OLN11	3	25	50	/	
18	OLN12	3	20	25	/	HVED
19	OLA1	3	15	50	/	
20	OLA2	9	15	0	/	
21	OLA3	3	15	0	/	
22	OLA4	3	20	0	/	
23	OLA5	9	20	25	/	
24	OLA6	9	15	25	/	
25	OLA7	9	15	50	/	
26	OLA8	9	20	50	/	
27	OLA9	3	20	25	/	
28	OLA10	9	20	0	/	
29	OLA11	3	20	50	/	
30	OLA12	3	15	25	/	

OL = olive leaf, N = nitrogen, A = argon. For HVED, numbers 1–12 are the order of conducted treatment. For CE treatments, 3 and 9 are referred to treatment time (min) while 0, 25, and 50 stands for concentration of an ethanol solvent (%).

Statistical analysis was done in STATGRAPHICS Centurion software (StatPoint Technologies, Inc, Warrenton, VA, USA).

2.3. Determination of Total Phenolic Content (TPC)

Total Phenolic Content (TPC) of OLE was determined using Folin-Ciocalteu method (FC) as previously described [35] with slight modification. Briefly, a volume of 0.1 mL of extract (appropriately diluted) was mixed with 0.2 mL of FC reagent. After 3 minutes 1 mL of 20% Na_2CO_3 (m/v) was added. After thorough mixing by vortex, the reaction mixtures were incubated at 50 °C for 25 min, followed by absorbance reading at 765 nm against blank. The blank contained 0.1 mL of extraction solvent instead of an extract. The calibration curve was prepared by gallic acid standard solutions (50–500 mg/L) solubilized in ethanol. The absorbance was measured for each standard solution following the same procedure as for extracts. The concentration of TPC was expressed in mg of gallic acid equivalents per g of sample (mg GAE/g of sample).

2.4. Determination of Antioxidant Capacity (AOC)

2.4.1. DPPH (2,2-Diphenyl-2-Picrylhydrazyl) Free Radical Assay

DPPH assay of OLEs was determined according to the previously reported procedure [36]. An aliquot (0.75 mL) of OLEs (appropriately diluted) was mixed with 1.5 mL of 0.5 mM DPPH methanolic solution. After mixing, the solutions were stored in the dark for 20 min at room temperature and then, the absorbance was measured at 517 nm against 100% methanol as a blank. The methanol solution of Trolox (25–200 µM) was used for the calibration curve. The absorbance values for the extracts were subtracted from the control sample (0.75 mL of 100% methanol and 1.5 mL of 0.5 mM DPPH). The results were calculated using the calibration curve for Trolox and expressed as µmol of Trolox equivalents per gram of samples (µmol TE/g of sample).

2.4.2. Ferric Reducing Antioxidant Power (FRAP) Assay

The FRAP assay was conducted according to the literature [37]. Prior to analysis, the FRAP reagent was prepared by mixing 0.3 M acetate buffer (pH = 3.6), 10 mM 2,4,6-tri(2-pyridyl)-1,3,5-triazine (TPTZ) solution in 40 mM hydrochloric acid and 20 mM $FeCl_3$ in ratio 10:1:1. Furthermore, the FRAP reagent (2080 µL) was mixed with 240 µL of distilled water and 80 µL of appropriately diluted OLE. The mixture was vortexed and allowed to stand for 5 min at 37 °C prior to absorbance measurement at 595 nm. The amount of extract was substituted by the same amount of extraction solvent in blank. The calibration curve was made by preparing a standard aqueous solution of $FeSO_4 \cdot 7H_2O$ (25–750 µM) where absorbance was measured following the same procedure as described for extracts. FRAP values were calculated according to the calibration curve for $FeSO_4 \cdot 7H_2O$ and expressed as µmol of Fe^{2+} equivalents (FE) per g of sample (µmol FE/g of sample).

2.5. NIR

NIR spectroscopy were conducted using the Control Development Inc. (South Bend, IN, USA), NIR-128-1.7-USB/6.25/50 µm to record extract spectra using the SPEC 32 Control Development software. NIR spectra was recorded in the wavelength range of 904 nm to 1699 nm. Each sample was recorded in triplicate and afterwards was calculated the average spectrum which was used for further processing. Since NIR spectra provide a large amount of data of one sample, PCA analysis was used to recognize and extract the most important information from the measurements, thus reducing the amount of data [38]. The PCA analysis was carried out on a part of spectrum that shows the difference between the samples (absorbance from 1350 to 1699 nm). This method was implemented in XLStat (MS Excel 2010, Microsoft, Redmond, WA, USA; XLStat by Addinsoft, Paris, France).

2.6. UPLC-MS/MS

Determination of phenolic compounds was carried out on UPLC-MS/MS system Eskigent Expert Ultra LC 110 and SCIEX 4500 QTRAP. Separation of the phenolic fraction of olive leaves were performed by a Luna Omega 3µm Polar C18 100Å, 100 × 4.6 mm (column), thermostat column temperature 40 °C, automatic sampling temperature 4 °C, and injection volume 10 µL. Mobile phases consisted of: A 100% H2O with 0.1% HCOOH (v/v) and B 100% ACN with 0.1% HCOOH (v/v) with mobile phase flow 0.40 mL/min. Gradient was set as follows: 1 min 10% B, 2 min 10% B, 15 min 90% B, 25 min 90% B, 27 min 10% B, 30 min 10% B. Determination conditions for MS/MS detector were: negative atmospheric pressure ionization mode (API); ionization temperature: 500 °C, Ion Spray voltage: −4500V, drying gas temperature 190 °C and drying gas flow 9.0 L/min. Determination of phenolic compounds content in extracts was carried out once. Quantitation of phenolic compounds was processed using Multiquant 3.6 (SCIEX, Darmstadt, Germany) software. Partial validation of method was carried out by measuring 6 replicas of added (spiked) standard (10 ng/mL) of oleuropein and oleanolic acid in two different extracts with no trace of these compounds. Average recovery was 92.3% for oleuropein, 91.1% for oleanolic acid, standard deviation 1.108 ng/mL for oleuropein, standard deviation 0.956 ng/mL for oleanolic acid, RSD 1.20% for oleuropein and RSD 1.05% for oleanolic acid.

2.7. Colorimetric Evaluation of OLEs

Color parameters for all trials (pure extract) was measured by Konica Minolta colorimeter (Model CM 3500d, Konica Minolta, Tokyo, Japan) at CIE Standard Illuminant D65 by 8 mm thick plate. All measurements were conducted in the Specular Component Included (SCI) mode as previously reported [39]. Colorimetric values (L*, a*, b*) were measured and color change after HVED treatment (ΔC, ΔH and ΔE) against untreated extracts (CE) was calculated. ΔL, Δa and Δb presents the difference between colorimetric value of HVED extracts and untreated (CE) extracts. Hue (H) was calculated from:

$$H = \arctan\left(\frac{b^*}{a^*}\right) \qquad (1)$$

Chroma (C) and the differences in tone color (ΔC) are calculated based on the following formula:

$$C = \sqrt{a^2 + b^2} \qquad (2)$$

$$\Delta C = C_{\text{HVED extract}} - C_{\text{untreated extract}} \qquad (3)$$

The total color difference (ΔE) was calculated on the basis of the measured parameters:

$$\Delta E = \sqrt{\Delta L^2 + \Delta a^2 + \Delta b^2} \qquad (4)$$

Saturation (ΔH) was determined based on the formula below:

$$\Delta H = \sqrt{\Delta E^2 - \Delta L^2 - \Delta C^2} \qquad (5)$$

2.8. Electrophoretic Light Scattering (ELS)

ELS measures the electrophoretic mobility of particles in dispersion or molecules in solution in the way that combines light scattering with electrophoresis. The OLE is introduced into the cell containing two electrodes. When an electric field is applied across the electrodes every charged particle or molecule will migrate towards its oppositely charged electrode with velocity which is dependent upon its charge. The measured electrophoretic mobility is converted to zeta potential using established theories. Zeta-potential measurements of all extracts were made on a Malvern Zetasizer Ultra (Malvern Panalytical, Malvern, UK) in a disposable folded capillary cells, thermostatted to 25 °C, at the forward angle (13°).

2.9. Dynamic Light Scattering (DLS)

DLS is a non-invasive analytical technique for measuring the size of particles and molecules in suspension which undergo Brownian motion (random movement of particles). It requires accurate and stable temperature because of sample viscosity. The velocity of the Brownian motion is defined through a property known as the translational diffusion coefficient (D). The size of a particle (hydrodynamic diameter) is calculated from the translational diffusion coefficient by using the Stokes-Einstein equation. Measurements of all extracts were made on a Malvern Zetasizer Ultra from Malvern Panalytical, UK in a disposable folded capillary cells, thermostatted to 25 °C, at the non-invasive back scatter (NIBS, 173°).

2.10. Determination of Pesticides and Metals in OLs

The content of the pesticides was measured by modified procedures with following national regulations HRN EN ISO 12393-1,12393-2 and 12393-3: 2013, i.e., extraction with petroleum ether/dichloromethane and determination using the GC-ECD Varian CP-3800 instrument. Metal trace content was determined according to the HRN EN ISO 14084: 2005 procedure, or by wet sample digestion by HNO_3 (microwave digestion) with Microwave reaction system Anton Paar, Multiwave 3000. Determination of metals were conducted on the Perkin Elmer AAS Analyst 800 and ICP-MS Perkin Elmer NexION 300X (PerkinElmer, Inc. 940, Waltham, MA, USA), while Hg traces were determined by the Leco AMA 254 Hg analyzer (LECO, St. Joseph, MI, USA).

3. Results and Discussion

3.1. Influence of HVED Treatment on Physical Parameters of OLEs

HVED was applied as an emerging non-thermal technology, with the aim to reduce extraction time and enhance extraction efficiency at lower temperatures than it is usual for CE. Data for pH, electrical conductivity, temperature, and power are given in Table 2.

Table 2. Average values of pH, conductivity (µS/cm) for untreated and HVED treated samples, starting temperature (°C), final temperature (°C) and power (kW) after HVED treatments.

ID	Sample	pH	Conductivity (µS/cm)	Starting Temperature (°C)	Final Temperature (°C)	Power (kW)	Extraction Type
1	3 OL0	5.54 ± 0.21	516.0 ± 4	20.8 ± 0.9	20.8 ± 0.5	/	
2	9 OL0	5.45 ± 0.11	349.0 ± 3	19.7 ± 0.4	19.7 ± 0.4	/	
3	3 OL25	6.07 ± 0.09	139.2 ± 2.3	20.1 ± 0.2	20.1 ± 0.6	/	CE
4	9 OL25	6.21 ± 0.15	141.6 ± 3.2	20.0 ± 0.5	20.0 ± 0.6	/	
5	3 OL50	6.88 ± 0.12	36.6 ± 1.1	21.0 ± 0.4	21.0 ± 0.4	/	
6	9 OL50	6.51 ± 0.11	41.3 ± 1.2	21.0 ± 0.7	21.0 ± 0.2	/	
7	OLN1	6.46 ± 0.22	51.4 ± 1.2	26.4 ± 0.8	26.7 ± 0.8	12 ± 1	
8	OLN2	5.80 ± 0.16	242.4 ± 3.3	24.2 ± 0.6	24.9 ± 0.3	12 ± 1	
9	OLN3	5.87 ± 0.14	272.1 ± 2.3	25.2 ± 0.5	25.6 ± 0.6	12 ± 1	
10	OLN4	5.65 ± 0.21	224.3 ± 3.2	24.8 ± 0.6	26.2 ± 0.6	19 ± 1	
11	OLN5	5.93 ± 0.19	124.7 ± 1.2	23.4 ± 0.3	23.6 ± 0.5	17 ± 0	
12	OLN6	5.85 ± 0.18	126.5 ± 2.5	24.3 ± 0.4	23.8 ± 0.4	12 ± 1	
13	OLN7	6.11 ± 0.17	63.4 ± 1.3	23.5 ± 0.5	23.3 ± 0.6	10 ± 1	
14	OLN8	6.09 ± 0.21	57.9 ± 1.1	22.7 ± 0.7	23.3 ± 0.3	20 ± 1	
15	OLN9	6.00 ± 0.15	93.4 ± 1.2	25.5 ± 0.3	25.7 ± 0.4	19 ± 1	
16	OLN10	5.78 ± 0.11	275.6 ± 3.1	23.3 ± 0.2	28.4 ± 0.6	22 ± 1	
17	OLN11	6.10 ± 0.12	51.9 ± 1.1	28.2 ± 0.5	26.7 ± 0.7	19 ± 0	
18	OLN12	5.99 ± 0.17	104.6 ± 1.2	25.9 ± 0.4	26.1 ± 0.6	14 ± 1	HVED
19	OLA1	6.45 ± 0.16	53.2 ± 1.1	23.7 ± 0.6	23.8 ± 0.6	7 ± 1	
20	OLA2	5.30 ± 0.15	268.3 ± 3.4	23.6 ± 0.7	25.6 ± 0.6	9 ± 1	
21	OLA3	5.21 ± 0.12	250.6 ± 4.5	23.6 ± 0.7	24.7 ± 0.5	10 ± 1	
22	OLA4	5.28 ± 0.19	233.3 ± 3.2	23.7 ± 0.5	25.1 ± 0.4	14 ± 1	
23	OLA5	5.77 ± 0.12	126.4 ± 2.3	24.3 ± 0.3	26.1 ± 0.3	16 ± 0	
24	OLA6	5.88 ± 0.11	127.5 ± 2.3	24.0 ± 0.4	24.2 ± 0.6	7 ± 0	
25	OLA7	6.32 ± 0.12	61.9 ± 1.6	23.9 ± 0.6	23.1 ± 0.6	9 ± 1	
26	OLA8	6.25 ± 0.13	59.0 ± 1	23.2 ± 0.4	24.4 ± 0.5	13 ± 1	
27	OLA9	6.00 ± 0.15	94.6 ± 2.0	23.8 ± 0.6	24.5 ± 0.6	14 ± 1	
28	OLA10	5.18 ± 0.16	285.7 ± 3.8	24.5 ± 0.7	29.9 ± 0.4	16 ± 1	
29	OLA11	6.36 ± 0.17	49.6 ± 1.3	24.5 ± 0.6	24.6 ± 0.7	12 ± 0	
30	OLA12	5.88 ± 0.19	105.9 ± 2.8	24.2 ± 0.5	24.0 ± 0.6	7 ± 1	

OL = olive leaf, N = nitrogen, A = argon. For HVED, numbers 1–12 are the order of conducted treatment. For CE treatments, 3 and 9 are referred to treatment time (min) while 0, 25, and 50 stands for concentration of an ethanol solvent (%).

3.2. Influence of HVED Treatment on TPC and Antioxidant Activity of OLEs

The influence of individual process parameters (treatment time, applied voltage, ethanol content) and their interaction were evaluated for significance ($p \leq 0.05$). It is important to emphasize that during 3 and 9 min of treatment time; temperature did not exceed 30 °C. In Table 3, results for extraction yield, TPC and antioxidant activity (DPPH and FRAP) for CE and HVED extracts were given. It can be seen that by using 50% ethanol (v/v) and HVED with argon for 9 min, high extraction yields were achieved as compared to CE-extracts. Also, 20 kV treatment for 3 min, with nitrogen gas gave 2,5-fold higher yield in comparison to CE extracts. A recent study confirmed HVED as an extraction technique with significant increase in yield due to disruptions of sample cells under electrical breakdown and enhanced mass transfer [40].

All HVED treated samples resulted with higher TPC values than CE samples which implies that HVED could be considered as an effective extraction tool at room temperature. The phenomena behind HVED is electroporation, where strength of electric field is directly proportional to the poration of plant cell membrane [1]. Moreover, it was found that higher ethanol/water ratio promote better polyphenolic extraction, thus gave extracts with higher TPC content. Similar results were observed by Garcia-Castello et al., who found that higher ethanol concentration favored flavonoids' extraction from grapefruit (*Citrus paradisi* L.) solid wastes [41]. Authors explained that reduced dielectric constant of the solvent caused by ethanol could led to enhanced solubility and diffusion of polyphenols from plant matrix. With respect to pure water as an extraction solvent, it is clear that 2-6 times higher TPC values were determined for HVED extraction systems (OLN2, 3, 4 and 10; OLA2, 3, 4 and 10) in comparison to CE samples.

Table 3. Concentrations of total phenolic content (TPC) values, antioxidant capacity (AOC) (2,2-Diphenyl-2-Picrylhydrazyl - DPPH) and ferric reducing antioxidant power (FRAP)) and extraction yield in extracts produced by conventional extraction (CE) and HVED.

ID	Sample	TPC (mg GAE/g)	DPPH (μmol TAE/g)	FRAP (μmol FE/g)	Yield (%)	Extraction Type
1	3 OL0	5.32 ± 0.02	27.06 ± 0.01	63.36 ± 2.24	0.53 ± 0.00	CE
2	9 OL0	15.85 ± 0.11	31.10 ± 0.08	146.93 ± 3.54	1.58 ± 0.01	
3	3 OL25	13.35 ± 0.08	24.78 ± 0.05	150.50 ± 4.57	1.33 ± 0.01	
4	9 OL25	16.04 ± 0.10	30.46 ± 0.07	179.07 ± 6.09	1.60 ± 0.01	
5	3 OL50	14.06 ± 0.09	32.71 ± 0.06	159.79 ± 5.42	1.41 ± 0.01	
6	9 OL50	20.61 ± 0.14	33.31 ± 0.08	221.93 ± 6.99	2.06 ± 0.01	
7	OLN1	49.21 ± 0.10	31.49 ± 0.07	266.21 ± 7.53	4.92 ± 0.01	HVED
8	OLN2	29.82 ± 0.07	28.60 ± 0.05	208.36 ± 4.28	2.98 ± 0.01	
9	OLN3	18.08 ± 0.08	25.96 ± 0.06	197.64 ± 2.00	1.81 ± 0.01	
10	OLN4	33.12 ± 0.12	27.46 ± 0.08	199.79 ± 5.22	3.31 ± 0.01	
11	OLN5	24.17 ± 0.05	31.31 ± 0.04	374.79 ± 6.74	2.42 ± 0.01	
12	OLN6	26.95 ± 0.08	30.14 ± 0.09	308.36 ± 4.45	2.69 ± 0.01	
13	OLN7	45.47 ± 0.06	30.49 ± 0.10	209.79 ± 2.01	4.55 ± 0.01	
14	OLN8	47.21 ± 0.09	31.81 ± 0.08	229.79 ± 5.82	4.72 ± 0.01	
15	OLN9	29.99 ± 0.07	28.35 ± 0.01	301.21 ± 4.68	3.00 ± 0.01	
16	OLN10	28.95 ± 0.11	29.99 ± 0.08	256.21 ± 5.74	2.89 ± 0.01	
17	OLN11	45.82 ± 0.16	27.96 ± 0.07	561.93 ± 9.11	4.58 ± 0.02	
18	OLN12	35.03 ± 0.10	31.53 ± 0.02	284.07 ± 3.21	3.50 ± 0.01	
19	OLA1	39.56 ± 0.08	30.78 ± 0.15	234.07 ± 2.66	3.96 ± 0.01	
20	OLA2	32.69 ± 0.05	26.53 ± 0.08	145.50 ± 2.87	3.27 ± 0.00	
21	OLA3	9.82 ± 0.04	29.71 ± 0.11	97.64 ± 1.15	0.98 ± 0.00	
22	OLA4	26.69 ± 0.02	25.53 ± 0.08	169.07 ± 5.92	2.67 ± 0.00	
23	OLA5	36.17 ± 0.04	29.60 ± 0.17	315.50 ± 4.00	3.62 ± 0.00	
24	OLA6	31.21 ± 0.09	30.64 ± 0.02	321.21 ± 6.38	3.12 ± 0.01	
25	OLA7	53.64 ± 0.14	29.85 ± 0.12	443.36 ± 5.21	5.36 ± 0.01	
26	OLA8	65.99 ± 0.06	31.53 ± 0.01	237.64 ± 3.73	6.60 ± 0.01	
27	OLA9	30.77 ± 0.05	26.81 ± 0.08	326.21 ± 2.16	3.08 ± 0.01	
28	OLA10	21.21 ± 0.04	29.67 ± 0.05	196.21 ± 2.97	2.12 ± 0.00	
29	OLA11	42.60 ± 0.11	32.53 ± 0.10	343.36 ± 3.01	4.26 ± 0.01	
30	OLA12	26.43 ± 0.01	30.03 ± 0.08	354.07 ± 5.36	2.64 ± 0.00	

OL = olive leaf, N = nitrogen, A = argon. For HVED, numbers 1–12 are the order of conducted treatment. For CE treatments, 3 and 9 are referred to treatment time while 0, 25, and 50 stands for concentration of an ethanol solvent (%). Percentage of yield was calculated as (g GAE/g of sample) × 100.

3.3. Influence of HVED Treatment on Soluble Phenolic Compounds in OLEs Analyzed by UPLC-MS/MS

A UPLC-MS/MS analysis of individual phenolic compounds has shown that main constituents of phenolics in OLEs were as follows: apigenin, diosmetin, hydroxytyrosol, luteolin, oleanolic acid, oleuropein, and quercentin (Table 4). It is obvious that there is an increase in content of phenolics with increase in ethanol concentration, likely due to increased solubility of phenolic compounds [42]. Oleuropein is the main constituent of OLE. Similar trend is shown for HVED treatment, with 50% ethanol (v/v), where higher concentrations of hydroxytyrosol and oleuropein were observed Hydroxytyrosol, which is a precursor of oleuropein, is scavenger of superoxide anions, and inhibitor of neutrophils and hypochlorous acid-derived radicals [29]. It is also important to emphasize that the highest values of phenolics in OLEs for HVED treated samples, were shown for OLA8 (argon gas, 9 min treatment, 20 kV, 50% ethanol) sample and OLN1 (nitrogen gas, 3 min, 20 kV, 50% ethanol). HVED extraction at the highest voltage (25 kV) and 9 min extraction time, resulted in degradation of oleanolic acid in OLE. This HVED technology with pulsed rapid discharge voltages (from 20 to 80 kV/cm electric field intensity) is based on the phenomenon of electrical breakdown in liquids which induces physical and chemical processes that affect both the cell walls and the membranes, while freeing intracellular components [43]. Moreover, HVED generates hot and localized plasma during photonic dissociation of water, with emission of the UV light and -OH radicals. At the same time, HVED will create shockwaves and pyrolysis caused by electrohydraulic cavitation [44]. HVED or other electrically assisted extractions are less thermally destructive than CE, and they are useful for extraction of specific thermolabile BACs. With increased effectiveness, such extracts are obtained at lower temperatures in a shorter period [45]. Another study found HVED as and efficient pre-treatment technique that intensifies pectin recovery from sugar beet pulp without modification of pectin structure and chemical composition [46]. For extraction of proteins and polyphenols from olive pit, HVED was found to be faster as compared to UAE or PEF [47]. This treatment retained more proteins and polyphenols in processing, thus demonstrating a promising technique for production of essential oils from oilseeds and herbs [48]. With generating electric fields and electrical discharges of up to 25 kV, there is additional formation of free radical species (ROS and RNS). This can be negative and destabilizing effect for sensitive bioactive compounds, especially in a long non-controlled treatment, where they can deteriorate [30]. This can be avoided by optimizing HVED treatment and shorter treatment time, as well to use gases in HVED treatments to reduce oxidation of phenolics and another compound of antioxidant potential.

UPLC-MS/MS analysis chromatograms of untreated and HVED treated OLs samples are shown in Figure 2a,b, Figure 3a,b, Figure 4a,b).

Figures 2a, 3a and 4a represent total ion chromatogram (TIC) of certain extract and in these chromatograms all MRM (multiple reaction monitoring) transitions of the method are included. On the x axis is intensity and on the y axis is time of elution or retention time (RT) of certain analyte.

Figures 2b, 3b and 4b represent extracted ion chromatogram (XIC) of certain extract and in these chromatograms only certain MRM-s transitions of the compounds are included. On the x axis is intensity and on the y axis is time of elution or retention time (RT) of certain analyte. In Figure 2b (Apigenin, Hydroxytyrosol, Oleanolic acid and Oleuropein), Figure 3b (Apigenin, Hydroxytyrosol, Oleanolic acid, Oleuropein and Luteolin) and Figure 4b Apigenin, Hydroxytyrosol, Diosmetin, Oleuropein and Luteolin).

Table 4. Concentration of individual phenolic compounds in CE and HVED olive leaves extracts (OLEs) (ng/mL).

ID	Sample	Apigenin	Diosmetin	Hydroxytyrosol	Luteolin	Oleanolic Acid	Oleuropein	Quercentin	Extraction Type
1	3 OL0	156	124	9.17	2356.6	4.45	50.4	N/A	CE
2	9 OL0	172	121	11.6	71.5	2.12	56.2	0.054	
3	3 OL25	125	112	494	550	0.66	3959	0.140	
4	9 OL25	144	132	125	580	0.79	825	0.124	
5	3 OL50	31.9	14.5	4120	154	3918	23485	247	
6	9 OL50	193	154	13.0	80.8	8.37	108	3.01	
7	OLN1	58.6	47.0	3016	353	3386	17646	9.07	HVED
8	OLN2	157	118	86.6	138	8.84	99.3	N/A	
9	OLN3	109	101	97.5	96.3	3.65	76.1	0.048	
10	OLN4	127	98.1	96.5	122	N/A	74.2	0.042	
11	OLN5	116	119	80.4	361	1.83	163	0.109	
12	OLN6	134	127	67.5	424	2.92	165	0.115	
13	OLN7	38.9	30.0	3090	255	2803	17311	7.17	
14	OLN8	44.3	33.0	3294	265	2705	18413	9.0	
15	OLN9	149	159	108	375	8.57	168	0.179	
16	OLN10	111	94.2	192	252	3.76	178	0.103	
17	OLN11	106	106	97.5	403	2.00	424	0.121	
18	OLN12	56.8	49.2	3426	386	2851	17388	8.54	
19	OLA1	53.6	59.9	2964	357	2004	16103	5.27	
20	OLA2	198	76.6	10.1	29.8	5.74	57.4	0.043	
21	OLA3	124	130	112	215	2.41	97.9	0.032	
22	OLA4	113	100	64.0	113	N/A	52.5	0.021	
23	OLA5	110	123	128	487	N/A	247	0.146	
24	OLA6	115	122	145	457	N/A	315	0.218	
25	OLA7	60.1	60.6	3660	431	1844	17127	6.43	
26	OLA8	48.2	41.5	3682	334	2043	17895	15.6	
27	OLA9	111	116	149	480	6.22	465	0.189	
28	OLA10	94.2	85.3	139	166	N/A	102	0.061	
29	OLA11	51.4	51.0	2967	372	2193	16695	5.79	
30	OLA12	142	141	77.2	459	7.51	166	0.191	

OL = olive leaf, N = nitrogen, A = argon. For HVED, numbers 1–12 are the order of conducted treatment. For CE treatments, 3 and 9 are referred to treatment time (min) while 0, 25, and 50 stands for concentration of an ethanol solvent (%).

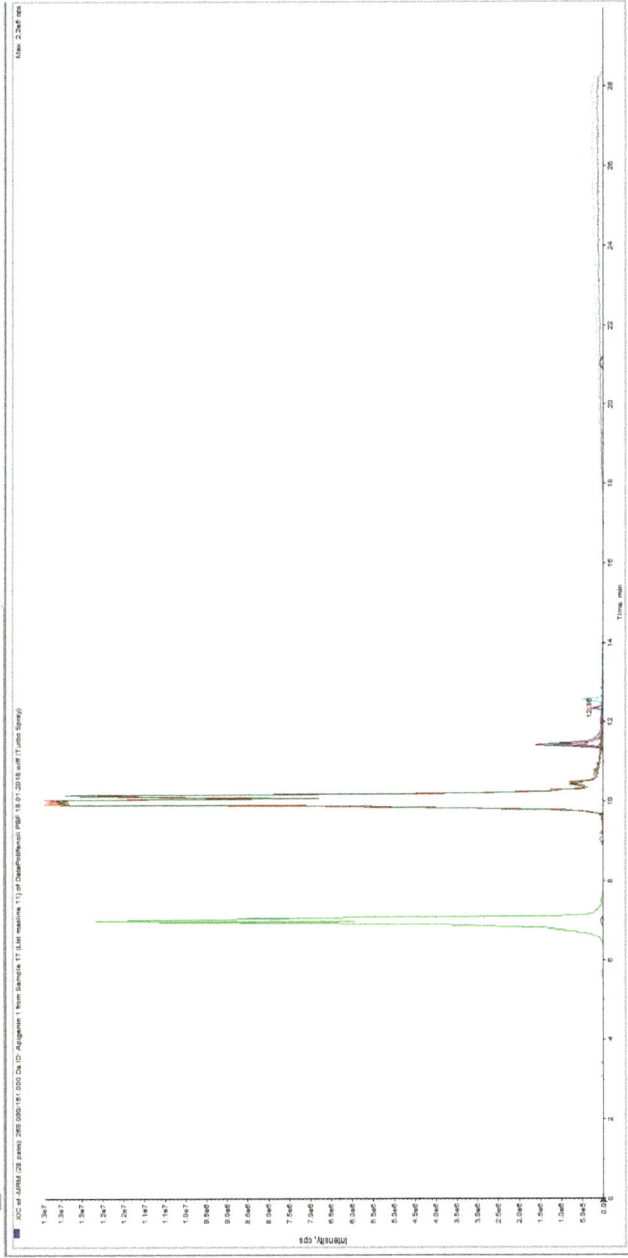

Figure 2. *Cont.*

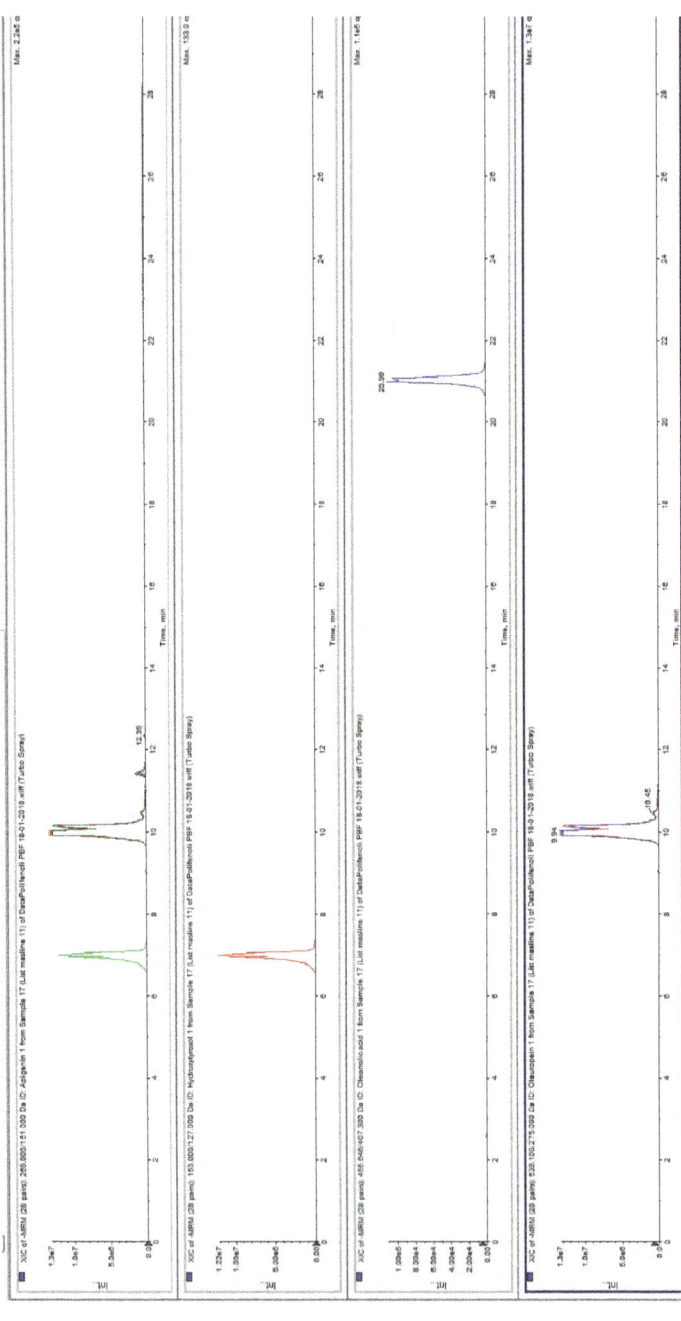

Figure 2. (a) UPLC-MS/MS analysis chromatogram of untreated OLE (3 OL50)—total ion current (TIC). (b) UPLC-MS/MS analysis chromatogram of untreated OLE (3 OL50).

Figure 3. *Cont.*

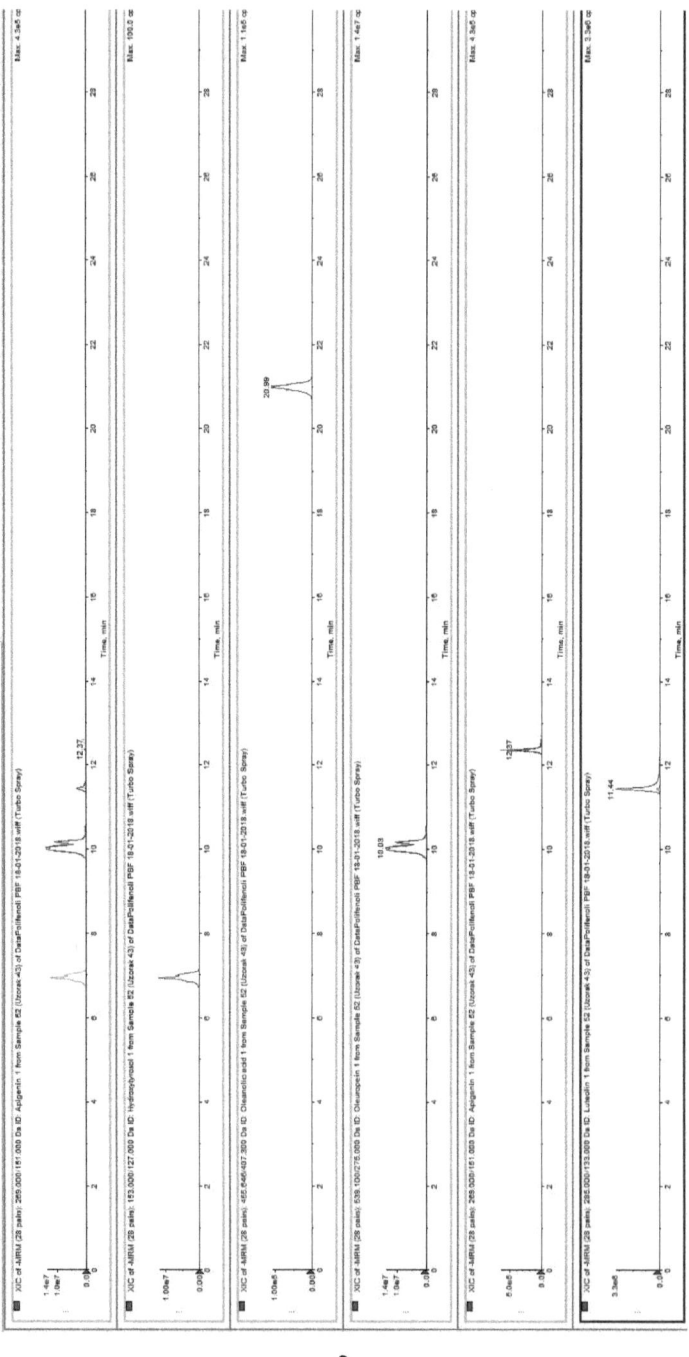

Figure 3. (a) UPLC-MS/MS analysis chromatogram of HVED treated (with nitrogen) OLE (OLN1)—total ion current (TIC). (b) UPLC-MS/MS analysis chromatogram of HVED treated (with nitrogen) OLE (OLN1).

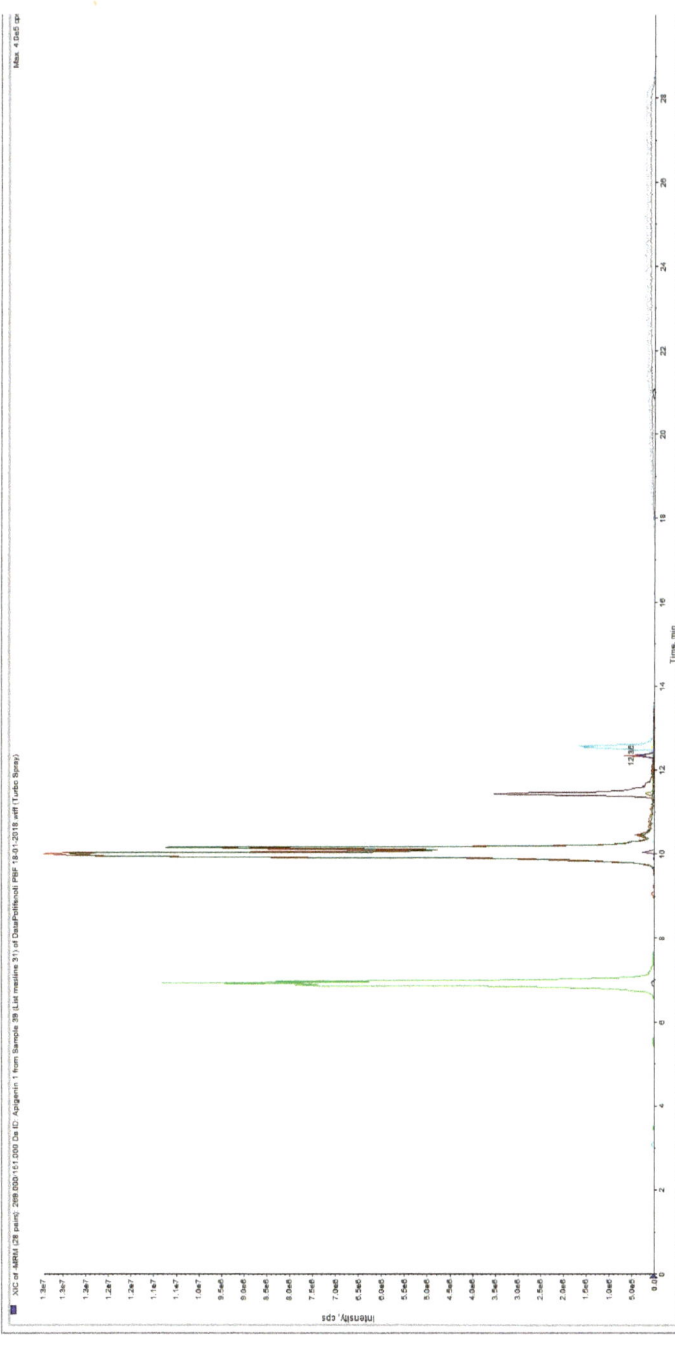

Figure 4. *Cont.*

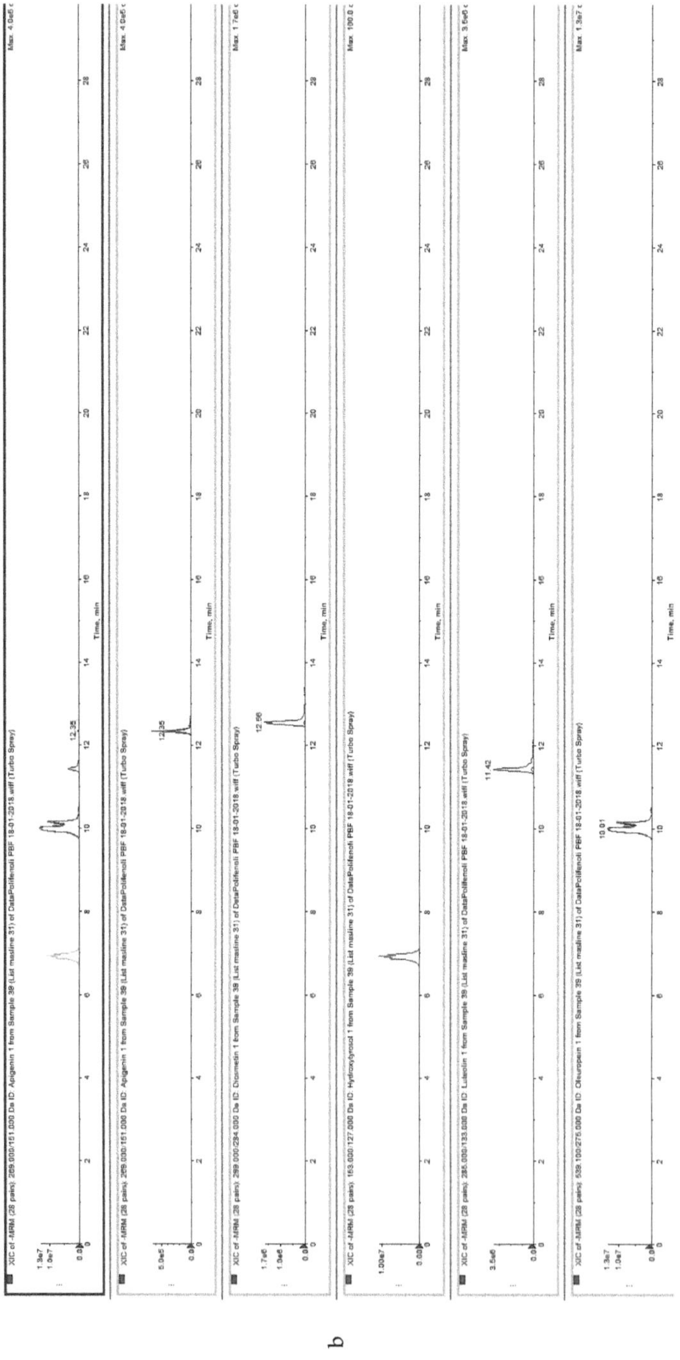

Figure 4. (a) UPLC-MS/MS analysis chromatogram of HVED treated (with argon) OLE (OLA1)—total ion current (TIC). (b) UPLC-MS/MS analysis chromatogram of HVED treated (with argon) OLE (OLA1).

3.4. Influence of HVED Treatment on Color Parameters of OLEs

OLEs color parameters did not change significantly after the HVED treatments (Table 5). HVED treated OLEs had higher values of colorimetric variables a* (more red), b* (more yellowish), C (more tone color) and H (more saturated) than CE extracts, while average L* parameter was lower (darker samples) for HVED OLEs. Total difference in tone color (ΔC), saturation (ΔH) and total color (ΔE) after HVED treatment was lower when using nitrogen gas comparing to argon. The highest difference in tone color (ΔC) and total color difference (ΔE) was observed for HVED samples with 50% ethanol (v/v) (OLN1, OLA1) and 9 min treatment (OLN6 and OLA5). The highest change in saturation (ΔH) was noted for HVED treated samples for 9 min (OLN7 and OLA5). Due to the electroporation effect of plant cells, there is a possibility of an extraction of different molecules depending on the diameter of the pore on electroporated plant membrane [29].

Table 5. CIELab color parameters of CE and HVED treated OLEs.

ID	Sample	L*	a*	b*	C	H	ΔC	ΔH	ΔE	Extraction Type
1	3 OL0	81.86	4.96	42.70	42.99	1.46	/	/	/	CE
2	9 OL0	82.06	6.59	46.76	47.22	1.43	/	/	/	
3	3 OL25	89.56	0.06	30.53	30.53	1.57	/	/	/	
4	9 OL25	91.02	−0.37	31.26	31.26	−1.56	/	/	/	
5	3 OL50	94.12	−3.82	27.89	28.15	−1.43	/	/	/	
6	9 OL50	91.03	−3.82	38.68	38.87	−1.47	/	/	/	
7	OLN1	80.34	−2.69	45.25	45.33	−1.51	17.18	2.74	22.19	HVED
8	OLN2	79.75	7.78	44.90	45.57	1.40	−1.65	1.46	3.20	
9	OLN3	81.28	6.06	42.60	43.03	1.43	0.04	1.10	1.25	
10	OLN4	80.38	4.01	40.60	40.80	1.47	−2.19	0.72	2.74	
11	OLN5	82.35	1.17	29.06	29.08	1.53	−2.18	1.57	9.08	
12	OLN6	81.25	−2.59	45.68	45.75	−1.51	14.49	1.69	17.56	
13	OLN7	84.44	2.87	36.30	36.41	1.49	−2.45	6.66	9.69	
14	OLN8	86.03	−3.34	42.78	42.91	−1.49	4.04	0.84	6.48	
15	OLN9	83.26	0.59	26.29	26.30	1.55	−4.23	0.58	7.61	
16	OLN10	80.90	6.73	47.19	47.67	1.43	0.45	0.08	1.25	
17	OLN11	88.01	−3.70	40.73	40.90	−1.48	12.75	1.54	14.22	
18	OLN12	85.42	0.55	27.78	27.79	1.55	−2.74	0.52	4.99	
19	OLA1	84.82	−2.01	41.69	41.74	−1.52	13.59	3.01	16.74	
20	OLA2	78.28	10.44	52.28	53.31	1.37	6.09	2.86	7.72	
21	OLA3	86.90	3.20	41.42	41.54	1.49	−1.44	1.63	5.49	
22	OLA4	80.40	8.10	48.30	48.97	1.40	5.99	2.32	6.58	
23	OLA5	80.45	4.78	47.39	47.63	1.47	16.37	4.33	19.96	
24	OLA6	83.55	2.67	41.24	41.33	1.51	10.06	2.75	12.83	
25	OLA7	83.12	−1.37	43.09	43.11	−1.54	4.24	2.73	9.38	
26	OLA8	85.95	−2.71	38.36	38.46	−1.50	−0.41	1.08	5.21	
27	OLA9	82.62	2.21	37.57	37.63	1.51	7.10	1.92	10.12	
28	OLA10	78.98	9.06	50.87	51.67	1.39	4.45	1.79	5.70	
29	OLA11	84.57	−2.25	36.37	36.44	−1.51	8.29	2.38	12.87	
30	OLA12	82.27	2.50	36.80	36.88	1.50	6.35	2.21	9.92	

OL = olive leaf, N = nitrogen, A = argon. For HVED, numbers 1–12 are the order of conducted treatment. For CE treatments, 3 and 9 are referred to as treatment time while 0, 25, and 50 stands for concentration of an ethanol solvent (%). L*—lightness from black to white; a* from green to red, and b* from blue to yellow; C—chroma/tone color; H—hue angle; ΔC—total tone color difference; ΔH—total saturation difference; ΔE—total color difference.

3.5. Influence of HVED Treatment and Electroporation Phenomena on Particle Size and Zeta Potential of OLEs

Immediately after electroporation, plant cells swell in four stages with leakage of ions to the outer cellular area. That is the period for the extraction of different molecules (including phenolics) that depends on the pore size. HVED treatment inflicts damage to the membranes by generating shock waves that are characteristic for the discharge. Hence, HVED has an important contribution to the mechanical damages, disintegration of cellular walls, homogenization and aggregation of cells. The results from electroporation on OLs are shown in Table 6.

Table 6. Results of particle size and zeta potential analysis for CE and HVED treated samples.

ID	Sample	Z Average Mean (d·nm)	PI	Zeta potential (mV)	Extraction Type
1	3 OL0	541.55 ± 6.38	0.62 ± 0.01	−9.31 ± 0.47	CE
2	9 OL0	370.60 ± 4.57	0.69 ± 0.06	−7.96 ± 0.25	
3	3 OL25	270.45 ± 3.65	0.63 ± 0.11	−11.30 ± 2.44	
4	9 OL25	303.25 ± 4.56	0.68 ± 0.08	−8.98 ± 1.12	
5	3 OL50	333.10 ± 4.58	0.54 ± 0.00	−10.43 ± 0.81	
6	9 OL50	166.50 ± 2.92	0.56 ± 0.06	−18.07 ± 0.12	
7	OLN1	394.60 ± 3.81	0.41 ± 0.03	−27.13 ± 1.46	HVED
8	OLN2	672.00 ± 7.83	0.48 ± 0.06	−15.77 ± 0.30	
9	OLN3	751.35 ± 9.64	0.73 ± 0.04	−17.79 ± 0.66	
10	OLN4	1545.50 ± 15.01	0.65 ± 0.04	−18.87 ± 1.45	
11	OLN5	139.30 ± 2.45	0.32 ± 0.04	−53.77 ± 0.27	
12	OLN6	203.70 ± 3.65	0.53 ± 0.05	−45.99 ± 5.52	
13	OLN7	653.15 ± 6.83	0.95 ± 0.14	−22.21 ± 0.37	
14	OLN8	378.15 ± 3.63	0.59 ± 0.07	−33.25 ± 1.46	
15	OLN9	279.15 ± 3.72	0.28 ± 0.01	−28.09 ± 0.88	
16	OLN10	952.40 ± 9.43	0.71 ± 0.10	−16.43 ± 0.10	
17	OLN11	193.50 ± 1.23	0.36 ± 0.08	−31.94 ± 2.48	
18	OLN12	304.05 ± 4.72	0.32 ± 0.04	−36.44 ± 2.44	
19	OLA1	452.90 ± 6.74	0.60 ± 0.06	−23.81 ± 0.75	
20	OLA2	1551.50 ± 16.63	0.85 ± 0.13	−19.76 ± 0.37	
21	OLA3	318.20 ± 3.74	0.42 ± 0.05	−16.96 ± 0.40	
22	OLA4	846.95 ± 8.72	0.71 ± 0.15	−17.53 ± 0.23	
23	OLA5	177.25 ± 1.53	0.43 ± 0.04	−54.14 ± 2.72	
24	OLA6	183.25 ± 1.74	0.26 ± 0.00	−36.95 ± 0.26	
25	OLA7	447.40 ± 2.82	0.50 ± 0.11	−29.91 ± 0.03	
26	OLA8	369.40 ± 2.76	0.63 ± 0.06	−35.07 ± 0.73	
27	OLA9	198.45 ± 1.34	0.31 ± 0.03	−41.08 ± 4.01	
28	OLA10	506.95 ± 5.45	0.69 ± 0.06	−19.41 ± 0.10	
29	OLA11	522.55 ± 6.28	0.48 ± 0.10	−23.29 ± 1.53	
30	OLA12	164.70 ± 1.28	0.46 ± 0.00	−35.08 ± 5.12	

OL = olive leaf, N = nitrogen, A = argon. For HVED, numbers 1–12 are the order of conducted treatment. For CE treatments, 3 and 9 are referred to treatment time (min) while 0, 25, and 50 stands for concentration of an ethanol solvent (%).

Physical properties of obtained extracts were tested for polydispersity index (PI) and zeta potential (mV). Data indicate possible usage of HVED technology in industry for simultaneous extraction and nanoformulation, i.e., for production of stable emulsion systems. From values of zeta potential, it is shown that the initial average diameter of particles in CE samples with ethanol are >300 nm. For HVED samples with argon or nitrogen and 25% ethanol content (v/v) (OLN 5, 6, 9, 12 and OLA5, 6, 9, 12) droplet diameter decreases, which is directly linked with negative zeta potential of −30 mV, indicating good stability against coalescence [49]. Zeta potential is often used as an indicator for droplet stability, where positive values above +30 mV and below negative −30 mV indicate good stability against coalescence are electrochemical balance. Therefore, by using OLs, in systems with 25% ethanol content (v/v) it is possible to produce stable emulsion systems and nanoparticles.

The PI, which is a measure of homogeneity, was above 0.3. Ideally, PI should be <0.3, while values higher than 0.3 implicate broad distribution size and suggested particles are not monodispersed. Values for statistical analysis (Tables 7 and 8) are indicating there was a statistically significant effect of ethanol content on ELS for argon samples, and for DLS for nitrogen. Also, it should be mentioned that

HVED samples with water (OLA2, 3, 4, 10 and OLN2, 4, 10) had increased zeta potential and average diameter of particles, where surface charge of particles indicates formation of unstable particle systems.

3.6. Statistical Analysis of HVED Treatment Parameters and Ethanol Influence in Extraction of Bioactive Compound Form OLEs

Table 7. Statistical significance for pH, conductivity, temperature difference, power, dynamic light scattering (DLS), polydispersity index (PI), electrophoretic light scattering (ELS), L*, a*, b*, total phenolic content (TPC), ferric reducing antioxidant power (FRAP), DPPH and yield of extraction. The MANOVA table decomposes the variability of each parameter into contributions due to various factors for OLE treated with argon.

	Source	Main Effects			Interactions		
		A: Treatment Time	B: Voltage	C: Ethanol Content	AB	AC	BC
p-value	pH	0.1751	0.4805	0.0037	0.2276	0.5162	0.7081
	Conductivity	0.0431	0.5992	0.0017	0.2722	0.3039	0.8873
	Temperature difference	0.0867	0.0588	0.0564	0.1082	0.1968	0.6541
	Power	0.2380	0.0099	0.2097	0.3828	0.8125	0.2500
	DLS	0.6704	0.7696	0.3177	0.3569	0.6945	0.8880
	PI	0.4634	0.6328	0.2114	0.9126	0.7919	0.9259
	ELS	0.0703	0.1162	0.0163	0.2681	0.4337	0.1984
	L*	0.3500	0.5867	0.3631	0.5392	0.4922	0.6159
	a*	0.2793	0.6684	0.0474	0.5779	0.5319	0.7388
	b*	0.1225	0.8636	0.1252	0.8647	0.5679	0.3522
	TPC	0.1972	0.4714	0.1067	0.6437	0.6604	0.9387
	FRAP	0.6781	0.9773	0.1791	0.4250	0.8373	0.7394
	DPPH	0.7467	0.8024	0.2450	0.2872	0.6684	0.4903
	Yield of extraction	0.1982	0.4710	0.1072	0.6450	0.6633	0.9392

Where A determines treatment time, B stands for voltage and C stands for ethanol content. The p-values present the statistical significance of each of the factors.

Table 8. Statistical significance for pH, conductivity, temperature difference, power, DLS, PI, ELS, L*, a*, b*, TPC, FRAP, DPPH and yield of extraction. The MANOVA table decomposes the variability of each parameter into contributions due to various factors for OLs treated with nitrogen.

	Source	Main Effects			Interactions		
		A: Treatment Time	B: Voltage	C: Ethanol Content	AB	AC	BC
p value	pH	0.1610	0.1516	0.0255	0.1209	0.2870	0.2384
	Conductivity	0.3678	0.7244	0.0128	0.4040	0.8518	0.9879
	Temperature Difference	0.1782	0.1677	0.0829	0.1081	0.2583	0.1489
	Power	0.5736	0.0046	0.4091	0.1835	0.1957	0.1552
	DLS	0.4132	0.3820	0.0277	0.3044	0.1810	0.0969
	PI	0.2785	0.4608	0.2353	0.7913	0.3141	0.5596
	ELS	0.2004	0.3594	0.0295	0.2676	0.1189	0.4675
	L*	0.6960	0.4382	0.2718	0.9382	0.6644	0.4497
	a*	0.4285	0.4968	0.0706	0.9572	0.5046	0.4027
	b*	0.4367	0.5694	0.2359	0.9952	0.4904	0.5976
	TPC	0.7034	0.8342	0.0652	0.7084	0.5300	0.5030
	FRAP	0.5573	0.2974	0.3888	0.6255	0.3050	0.6206
	DPPH	0.1798	0.8050	0.1974	0.1946	0.6913	0.4720
	Extraction Yield	0.7026	0.8341	0.0650	0.7089	0.5312	0.5062

Where A determines treatment time, B stands for voltage and C stands for ethanol content. The p-values present the statistical significance of each of the factors.

3.7. NIR and PCA Analysis of HVED Treated OLEs

NIR spectra were recorded in the wavelength range of 904–1699 nm for OLEs treated with argon and nitrogen for 3 and 9 min (Figures 5 and 6). Based on their absorbance, they show no significant differences. However, in the ranges 904–928 and 1350–1699 nm, wavelength shifts are visible, indicating changes in the third and second overtone of the C–H and O–H relations. These relations are also associated with the hydroxyl group (–OH) bound directly to aromatic hydrocarbon group. The PCA was used for identifying patterns and highlighting similarities and differences in the data for an individual set of experiments. The goal of PCA was to extract the important information from the data table and express it as a set of new orthogonal variables, called principal components or factors. Figures 7 and 8 show spatial projections defined by the first two main components labeled with F1 and F2 for HVED treated and untreated samples. Although several input variables described the variability of the whole system, often a large part of that variability is described by a small number of variables that are the main components. If this is met, the main components contain the same amount of information as input variables. In this case, the main components F1 and F2 comprise 93.36% variance of the original data for all the samples. Analysis of major components reveals connectivity among variables and allows interpretation that is otherwise difficult. The PCA analysis implies finding the values of the covariance matrix samples. The input data for the analysis of the main components are p variables and n observations (individual) and have a matrix shape $p \times n$. The analysis begins with p variables data for n measurements (observations) using k main components ($k < p$) without losing system information. Main components represent the direction of maximum variability and provide a simpler description of the structure data set. The first principal component is the linear combination of the Y variables that accounts for the greatest possible variance, i.e., F1 (61.31%), while the other contains 32.05% variance.

Figure 7 shows the distribution of samples with respect to the extraction parameters for argon treated and untreated samples, while Figure 8 shows the distribution for nitrogen treated and untreated samples. With higher proportion of ethanol in the solvent and with the longer treatment time, larger influence on the chemical compounds and origin of changes can be detected by NIR analysis (3th and 4th quadrant). For both PCA analysis, in the 1st and 2nd quadrant, there are samples extracted with pure water (0% ethanol) and 25% ethanol (v/v) or extracted for a shorter treatment time with or without HVED treatment. It can be seen the trend of motion patterns by the "strength" of the treatment in clockwise direction. Thus, in Figure 7, in 1st quadrant, samples were subdivided into 0 or 25% ethanol (v/v), treated for 3 or 9 minutes and a voltage of 15 or 20 kV. In the 4th quadrant there are samples extracted with 25 or 50% ethanol solution, treated for 3 or 9 minutes, 15 or 20 kV, while in 3rd quadrant there are samples that had powered treatment, meaning the extraction with 50% ethanol, untreated or HVED treated for 3 minutes at 20 kV. In Figure 8, in the 1st quadrant, samples were subdivided into 0 or 25% ethanol, treated for 3 minutes and with a voltage of 20 or 25 kV. In the 4th quadrant there are samples extracted with 25 or 50% ethanol, treated for 3 or 9 minutes, and 20 or 25 kV. In the 3rd quadrant is placed powered treated samples, meaning extraction with 25% or 50% ethanol, CE or HVED treatment for 3 or 9 minutes and voltages of 20 kV or 25 kV. The conclusion is that ethanol and applied HVED voltage have strong association with the extraction and corresponding changes in plant material. When ethanol is combined with HVED this destroys the linkage between phenolic compounds and components of the plant material to which they are bound.

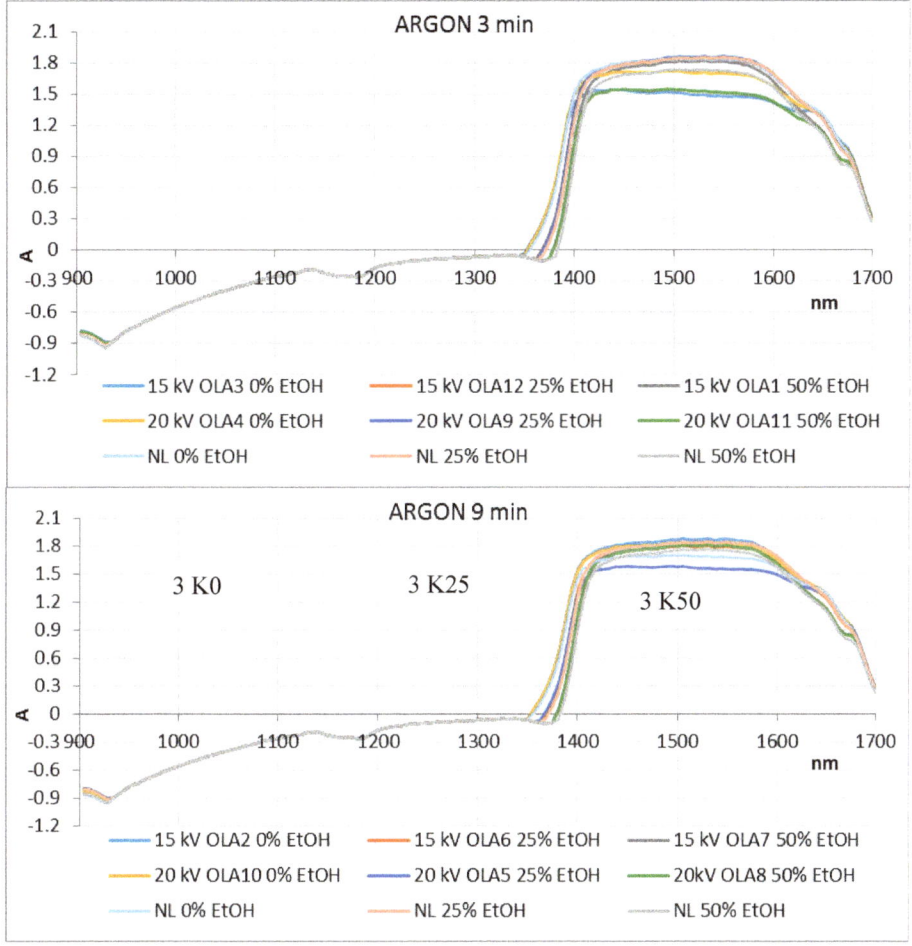

Figure 5. NIR spectra recorded in the wavelength range of 904 nm to 1699 nm for OLEs treated with argon for 3 and 9 min (NL: untreated olive leaves).

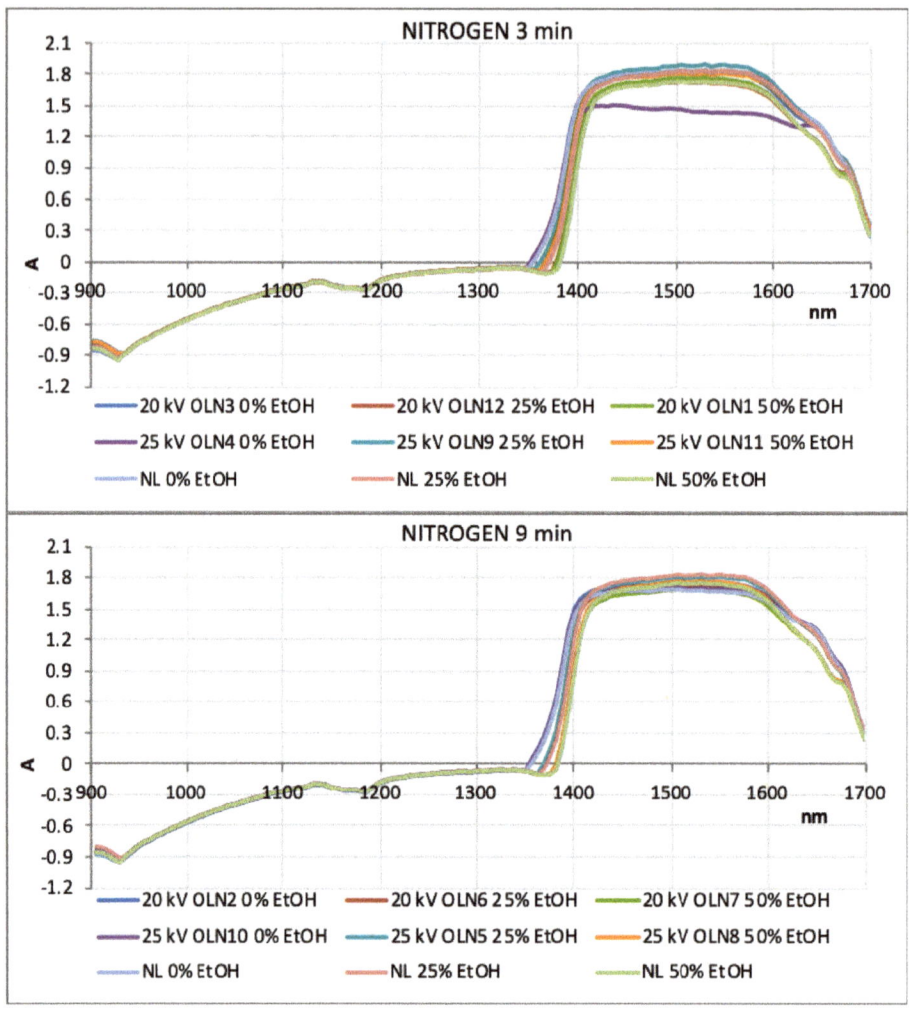

Figure 6. NIR spectra recorded in the wavelength range of 904 nm to 1699 nm for OLEs treated with nitrogen for 3 min and 9 min (NL: untreated olive leaves).

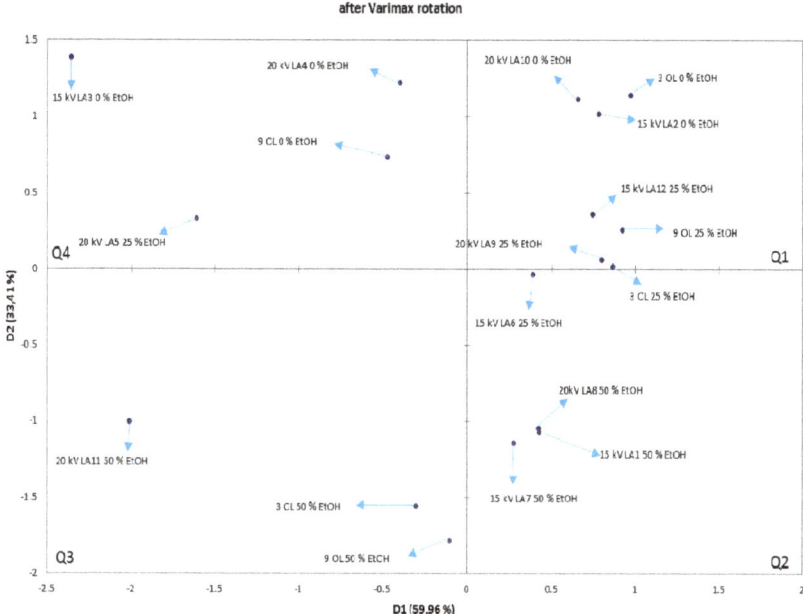

Figure 7. The PCA of untreated and HVED treated samples. The data is denoting untreated OLEs (CE), and OLEs treated with HVED using argon. Q stands for quadrant.

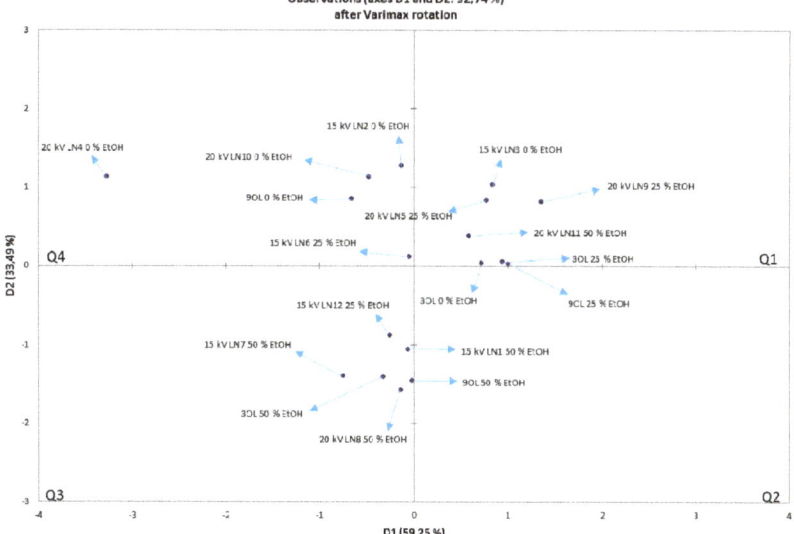

Figure 8. The PCA of untreated and HVED treated samples. The data is denoting untreated OLEs (CE), and OLEs treated with HVED using nitrogen. Q stands for quadrant.

3.8. Analysis of Pesticides and Metals in OL Samples

Tables 9 and 10 show the content of pesticides and metals in initial samples taken for extraction. It is evident that samples have lower levels than set as maximum levels in REGULATION (EC)

No 396/2005 on maximum residue levels (MRLs) of pesticides in or on food and feed of plant and animal origin and amending Council Directive 91/414/EEC and for metals according to COMMISSION REGULATION (EC) No 1881/2006. All extracts were compliant regarding content of contaminants, pesticide residues and toxic metals.

Table 9. Residue levels and maximum residue levels (MRLs) of pesticides (mg/kg) in olive leaves (OLs) samples.

Pesticides	MRL mg/kg	Content mg/kg
Alachlor	0.02	<0.005
Aldrin and Dieldrin (Aldrin and dieldrin combined expressed as dieldrin)	0.01	<0.002
Captan (Sum of captan and THPI, expressed as captan)	0.06	<0.020
DDT (sum of p,p´-DDT, o,p´-DDT, p-p´-DDE and p,p´-TDE (DDD) expressed as DDT)	0.05	<0.004
Endosulphan (sum of alpha- and beta-isomers and endosulphan-sulphate expresses as endosulphan)	0.05	<0.002
Endrin	0.01	<0.004
Heptachlor (sum of heptachlor and heptachlor epoxide expressed as heptachlor)	0.01	<0.002
Hexachlorobenzene	0.01	<0.002
Hexachlorocyclohexane (HCH), alpha-isomer	0.01	<0.002
Hexachlorocyclohexane (HCH), beta-isomer	0.01	<0.002
Iprodione	20	<0.010
Lindane (Gamma-isomer of hexachlorocyclohexane (HCH))	0.01	<0.002
Methoxychlor	0.01	<0.010
Tolylphluanid (Sum of tolylphluanid and dimethylaminosulphotoluidide expresses as tolylphluanid)	0.05	<0.020
Vinclozolin	0.02	<0.002

Table 10. Residue levels and MRLs of metals.

Metals	MRL mg/kg	Content mg/kg
Lead (Pb)	3.00	<0.050
Cadmium (Cd)	1.00	0.011
Mercury (Hg)	0.50	0.023
Arsenic (As)	-	<0.005
Chromium (Cr)	-	0.240
Nickel (Ni)	-	1.82
Manganese (Mn)	-	52.0
Iron (Fe)	-	101
Copper (Cu)	-	9.70
Zinc (Zn)	-	15.0

3.9. Critical Consideration on Costs of Equipment and Application of HVED for Extraction

The nonthermal processing market, was valued at USD 760.7 Million in 2016. It is projected to reach USD 1224.2 million by 2022, at a CAGR of 8.4% from 2017 reported by Markets and Markets™.

This aim of this paper and extensive analysis was to propose the possible application of HVED for the extraction of bioactive compounds from OLs. The aim was also, to emphasize use of water and ethanol as green solvents for efficient extraction of phenolic compounds. However, there should be critical discussion of the obtained results regarding costs of equipment and analysis, as well as short- and long-term improvements and possibilities of use of nonthermal technology like HVED. Regarding, nonthermal processing techniques like ultrasound, high pressure processing, pulsed electric fields, moderate electric fields, supercritical CO_2 and subcritical water extraction, as extraction techniques, the high installation cost is a major restraining factor for this market. Inability to shift from conventional thermal technologies by large players is a major challenge for the nonthermal processing market.

Therefore, there should be extensive research studies of environmental, economic and energy efficient nonthermal processing.

The most used extraction techniques are the conventional techniques, like Soxhlet, heat reflux, boiling and distillation. These techniques are mostly based on the use of mild/high temperatures (50–90 °C) that can cause thermal degradation. The extraction efficiency is dependent of the correct solvent choice, and the agitation intensity in order to increase the solubility of materials and the mass transfer rate. CE is being reflected on long extraction times, high costs, usage of organic solvents and low extraction efficiency consequent low extraction yields. Also, there is potential for overheating of the matrix (herbs), high energy consumption and general cost.

The principal advantages of HVED against thermal conventional extraction procedures are related with the increased mass transfer, improved extraction yield, decreased processing time, decreased intensity of the conventional extraction parameters (i.e., extraction temperature, selection of solvents and solvent concentration), reduction of heat-sensitive compounds degradation (flavors etc.), facilitation of purified extract and reduction of energy costs and environmental impact.

For the highest energy inputs, in this study, there is a significant increase in conductivity, antioxidant activity and total phenolic compounds when compared with the control (a mixture of solvent in the same conditions as treated samples).

Therefore, a short-term overview is that HVED is an effective, but expensive technology, regarding the cost of equipment, high input cost due to the input feed gas (Ar, N_2, He), and chemical residues and toxicological effects have not been studied yet. Possible application to scale-up of HVED to industry requires high capital cost and high energy consumption. Also, there are unresearched risks of corrosion rates of electrodes (the low-cost stainless steel has the highest corrosion rates).

However, long term potential is that there is low operational cost about electricity consumption (i.e., 90 kW/h × 0.05$/kW/h = 4.5 $/h). Advanced application of nonthermal techniques, like HVED, is that they can operate at room temperature ensuring that compound denaturation is avoided or at least decreased. HVED technology involves the application of short duration pulses (from several nanoseconds to several milliseconds) of high electric field strengths (20–80 kV cm^{-1}) and relatively low energy (1–10 kJ/kg). HVED equipment comprises a high-voltage pulse generator, a treatment chamber with a suitable fluid handling system, system of gas delivery and a monitoring and controlling system. The herb (sample) is exposed to the electric field pulses in a static or continuous chamber with at least two electrodes, one on high voltage and the other at ground potential. The sample is submitted to a force per unit charge (the electric field) that is responsible for the cell disintegration (electroporation phenomena).

Therefore, there is long term sustainable application of electrotechnologies like HVED in pharma, food, cosmetics industry etc.

4. Conclusions

The purpose of this extensive study was to evaluate and standardize advanced food technology, as HVED cold plasma-based tech, while accounting for green chemistry and modern sustainability and using autochthonous Mediterranean OLs (*Olea europaea* L.), so their extracts can be further available for industrial production. Such green extracts are abundant in antioxidants and phenolic components and are beneficial for human health and industrial production (food additives).

HVED extraction resulted in effective extraction of phenolic components from leaves when comparing to CE. Highest values of phenolics in OLEs, were shown for HVED treatment with argon gas, 9 min treatment, 20 kV and 50% ethanol; and HVED treatment with nitrogen gas, 3 min, 20 kV and 50% ethanol. HVED extraction at the highest voltage (25 kV) and 9 min extraction time, resulted in the degradation of oleanolic acid in OLEs. This study confirmed that HVED is an extraction technique with significant yield due to disruptions of sample cells under electrical breakdown and enhanced mass transfer. For total phenolic content, all HVED samples had higher values (mg GAE/g) than untreated samples at the same temperature. The conclusion is that ethanol and applied HVED voltage, have strong linkage with changes caused by the treatment. Ethanol in combination with HVED, destroys the bond between phenolic compounds and plant material to which they are bound. From values of zeta potential, it is shown that the initial average diameter of particles in untreated samples with ethanol are around or below 300 nm. Hence, data indicate possible usage of HVED technology for simultaneous extraction and production of nanoemulsions.

In conclusion, green OLEs obtained by green solvents (water/ethanol) and green technology with energy saving (HVED), matched the principles of green engineering and modern sustainability while having high nutritive value, e.g., they were rich in antioxidants and phenolic components [2,4,50].

Author Contributions: I.Ž.—researched prior studies, prepared graphic, drafted and revised the manuscript, P.P.—researched prior studies, revised the manuscript, D.B.K.—researched prior studies, revised the manuscript, M.N.—drafted and revised the manuscript, carried out the experimental work, F.Š.—carried out the experimental work, A.P.—carried out the experimental work, I.D.—carried out the experimental work, F.J.B.—researched prior studies, revised the manuscript, A.R.J.—designed the study, researched prior studies, prepared graphic, drafted and revised the manuscript.

Funding: This work was supported by grant from the Croatian Science Foundation: "High voltage discharges for green solvent extraction of bioactive compounds from Mediterranean herbs (IP-2016-06-1913)". The work of doctoral student Marinela Nutrizio has been fully supported by the "Young researchers' career development project – training of doctoral students" of the Croatian Science Foundation funded by the European Union from the European Social Fund.

Acknowledgments: Authors would like to thank Mara Banović for herb donation and professional support. Authors would like to thank Croatian Science Foundation for funding the project "High voltage discharges for green solvent extraction of bioactive compounds from Mediterranean herbs (IP-2016-06-1913)". Marinela Nutrizio and Anet Režek Jambrak would like to thank to "Young researchers' career development project—training of doctoral students" of the Croatian Science Foundation funded by the European Union from the European Social Fund.

Conflicts of Interest: The authors declare no conflict of interest.

References

1. Putnik, P.; Lorenzo, J.; Barba, F.; Roohinejad, S.; Režek Jambrak, A.; Granato, D.; Montesano, D.; Bursać Kovačević, D. Novel food processing and extraction technologies of high-added value compounds from plant materials. *Foods* **2018**, *7*, 106. [CrossRef] [PubMed]
2. Giacometti, J.; Bursać Kovačević, D.; Putnik, P.; Gabrić, D.; Bilušić, T.; Krešić, G.; Stulić, V.; Barba, F.J.; Chemat, F.; Barbosa-Cánovas, G.; et al. Extraction of bioactive compounds and essential oils from mediterranean herbs by conventional and green innovative techniques: A review. *Food Res. Int.* **2018**, *113*, 245–262. [CrossRef] [PubMed]
3. Şahin, S.; Bilgin, M. Olive tree (*olea europaea* L.) leaf as a waste by-product of table olive and olive oil industry: A review. *J. Sci. Food Agric.* **2018**, *98*, 1271–1279. [CrossRef] [PubMed]
4. Chemat, F.; Vian, M.A.; Cravotto, G. Green extraction of natural products: Concept and principles. *Int. J. Mol. Sci.* **2012**, *13*, 8615–8627. [CrossRef] [PubMed]
5. Watson, W.J.W. How do the fine chemical, pharmaceutical, and related industries approach green chemistry and sustainability? *Green Chem.* **2012**, *14*, 251–259. [CrossRef]
6. Soni, M.G.; Burdock, G.A.; Christian, M.S.; Bitler, C.M.; Crea, R. Safety assessment of aqueous olive pulp extract as an antioxidant or antimicrobial agent in foods. *Food Chem. Toxicol.* **2006**, *44*, 903–915. [CrossRef] [PubMed]
7. Japón-Luján, R.; Luque-Rodríguez, J.M.; Luque de Castro, M.D. Dynamic ultrasound-assisted extraction of oleuropein and related biophenols from olive leaves. *J. Chromatogr. A* **2006**, *1108*, 76–82. [CrossRef]
8. Benavente-García, O.; Castillo, J.; Lorente, J.; Ortuño, A.; Del Rio, J.A. Antioxidant activity of phenolics extracted from olea europaea l. Leaves. *Food Chem.* **2000**, *68*, 457–462. [CrossRef]
9. Ferreira, I.C.F.R.; Barros, L.; Soares, M.E.; Bastos, M.L.; Pereira, J.A. Antioxidant activity and phenolic contents of *Olea europaea* l. leaves sprayed with different copper formulations. *Food Chem.* **2007**, *103*, 188–195. [CrossRef]
10. Lockyer, S.; Rowland, I.; Spencer, J.P.E.; Yaqoob, P.; Stonehouse, W. Impact of phenolic-rich olive leaf extract on blood pressure, plasma lipids and inflammatory markers: A randomised controlled trial. *Eur. J. Nutr.* **2016**, *56*, 1421–1432. [CrossRef]

11. Wainstein, J.; Ganz, T.; Boaz, M.; Bar Dayan, Y.; Dolev, E.; Kerem, Z.; Madar, Z. Olive leaf extract as a hypoglycemic agent in both human diabetic subjects and in rats. *J. Med. Food* **2012**, *15*, 605–610. [CrossRef] [PubMed]
12. Vizza, D.; Lupinacci, S.; Toteda, G.; Puoci, F.; Ortensia, I.P.; De Bartolo, A.; Lofaro, D.; Scrivano, L.; Bonofiglio, R.; Russa, A.L.; et al. An olive leaf extract rich in polyphenols promotes apoptosis in cervical cancer cells by upregulating p21$^{cip/waf1}$ gene expression. *Nutr. Cancer* **2019**, *71*, 320–333. [CrossRef] [PubMed]
13. Jung, Y.-C.; Kim, H.W.; Min, B.K.; Cho, J.Y.; Son, H.J.; Lee, J.Y.; Kim, J.-Y.; Kwon, S.-B.; Li, Q.; Lee, H.-W. Inhibitory effect of olive leaf extract on obesity in high-fat diet-induced mice. *In Vivo* **2019**, *33*, 707–715. [CrossRef] [PubMed]
14. Kaneko, Y.; Sano, M.; Seno, K.; Oogaki, Y.; Takahashi, H.; Ohkuchi, A.; Yokozawa, M.; Yamauchi, K.; Iwata, H.; Kuwayama, T.; et al. Olive leaf extract (*Olea Vita*) suppresses inflammatory cytokine production and NLRP3 inflammasomes in human placenta. *Nutrients* **2019**, *11*, 970. [CrossRef] [PubMed]
15. Sarbishegi, M. Antioxidant effects of olive leaf extract in prevention of Alzheimer's disease and Parkinson's disease. *Gene. Cell Tissue* **2018**, *5*, e79847. [CrossRef]
16. Ganjalikhan Hakemi, S.; Sharififar, F.; Haghpanah, T.; Babaee, A.; Eftekhar-Vaghefi, S.H. The effects of olive leaf extract on the testis, sperm quality and testicular germ cell apoptosis in male rats exposed to busulfan. *Int. J. Fertil. Steril.* **2019**, *13*, 57–65. [PubMed]
17. Vidičević, S.; Tošić, J.; Stanojević, Ž.; Isaković, A.; Mitić, D.; Ristić, D.; Dekanski, D. Standardized Olea europaea L. leaf extract exhibits protective activity in carbon tetrachloride-induced acute liver injury in rats: The insight into potential mechanisms. *Arch. Physiol. Biochem.* **2019**, *11*, 1–9.
18. Ranieri, M.; Di Mise, A.; Difonzo, G.; Centrone, M.; Venneri, M.; Pellegrino, T.; Russo, A.; Mastrodonato, M.; Caponio, F.; Valenti, G.; et al. Green olive leaf extract (OLE) provides cytoprotection in renal cells exposed to low doses of cadmium. *PLoS ONE* **2019**, *14*, e0214159. [CrossRef] [PubMed]
19. Wang, Y.; Wang, S.; Cui, W.; He, J.; Wang, Z.; Yang, X. Olive leaf extract inhibits lead poisoning-induced brain injury. *Neural. Regen. Res.* **2013**, *8*, 2021–2029.
20. Caponio, F.; Difonzo, G.; Calasso, M.; Cosmai, L.; De Angelis, M. Effects of olive leaf extract addition on fermentative and oxidative processes of table olives and their nutritional properties. *Food Res. Int.* **2019**, *116*, 1306–1317. [CrossRef] [PubMed]
21. Martínez, L.; Castillo, J.; Ros, G.; Nieto, G. Antioxidant and antimicrobial activity of rosemary, pomegranate and olive extracts in fish patties. *Antioxidants* **2019**, *8*, 86. [CrossRef] [PubMed]
22. Khemakhem, I.; Fuentes, A.; Lerma-García, M.J.; Ayadi, M.A.; Bouaziz, M.; Barat, J.M. Olive leaf extracts for shelf life extension of salmon burgers. *Food Sci. Technol. Int.* **2018**, *25*, 91–100. [CrossRef] [PubMed]
23. Putnik, P.; Barba, F.J.; Španić, I.; Zorić, Z.; Dragović-Uzelac, V.; Bursać Kovačević, D. Green extraction approach for the recovery of polyphenols from Croatian olive leaves (*Olea europea*). *Food Bioprod. Process.* **2017**, *106*, 19–28. [CrossRef]
24. Žugčić, T.; Abdelkebir, R.; Alcantara, C.; Collado, M.C.; García-Pérez, J.V.; Meléndez-Martínez, A.J.; Režek Jambrak, A.; Lorenzo, J.M.; Barba, F.J. From extraction of valuable compounds to health promoting benefits of olive leaves through bioaccessibility, bioavailability and impact on gut microbiota. *Trends Food Sci. Technol.* **2019**, *83*, 63–77. [CrossRef]
25. Chemat, F.; Rombaut, N.; Meullemiestre, A.; Turk, M.; Perino, S.; Fabiano-Tixier, A.-S.; Abert-Vian, M. Review of green food processing techniques. Preservation, transformation, and extraction. *Innov. Food Sci. Emerg.* **2017**, *41*, 357–377. [CrossRef]
26. Madej, K.; Kalenik, T.K.; Piekoszewski, W. Sample preparation and determination of pesticides in fat-containing foods. *Food Chem.* **2018**, *269*, 527–541. [CrossRef] [PubMed]
27. Zaanouni, N.; Gharssallaoui, M.; Eloussaief, M.; Gabsi, S. Heavy metals transfer in the olive tree and assessment of food contamination risk. *Environ. Sci. Pollut. Res.* **2018**, *25*, 18320–18331. [CrossRef]
28. Alcázar Román, R.; Amorós, J.A.; Pérez de los Reyes, C.; García Navarro, F.J.; Bravo, S. Major and trace element content of olive leaves. *OLIVAE* **2014**, *119*, 1–7.
29. Batista Napotnik, T.; Miklavčič, D. In vitro electroporation detection methods—An overview. *Bioelectrochemistry* **2018**, *120*, 166–182. [CrossRef]
30. Režek Jambrak, A.; Donsì, F.; Paniwnyk, L.; Djekic, I. Impact of novel nonthermal processing on food quality: Sustainability, modelling, and negative aspects. *J. Food Qual.* **2019**, *2019*, 2. [CrossRef]
31. Režek Jambrak, A.; Vukušić, T.; Donsi, F.; Paniwnyk, L.; Djekic, I. Three pillars of novel nonthermal food technologies: Food safety, quality, and environment. *J. Food Qual.* **2018**, *2018*, 18. [CrossRef]

32. Chemat, F.; Zill, E.H.; Khan, M.K. Applications of ultrasound in food technology: Processing, preservation and extraction. *Ultrason. Sonochem.* **2011**, *18*, 813–835. [CrossRef] [PubMed]
33. Attri, P.; Arora, B.; Choi, E.H. Retracted article: Utility of plasma: A new road from physics to chemistry. *Rsc Adv.* **2013**, *3*, 12540–12567. [CrossRef]
34. Li, Z.; Fan, Y.; Xi, J. Recent advances in high voltage electric discharge extraction of bioactive ingredients from plant materials. *Food Chem.* **2019**, *277*, 246–260. [CrossRef] [PubMed]
35. Shortle, E.; O'Grady, M.N.; Gilroy, D.; Furey, A.; Quinn, N.; Kerry, J.P. Influence of extraction technique on the anti-oxidative potential of hawthorn (*Crataegus monogyna*) extracts in bovine muscle homogenates. *Meat. Sci.* **2014**, *98*, 828–834. [CrossRef] [PubMed]
36. Achat, S.; Tomao, V.; Madani, K.; Chibane, M.; Elmaataoui, M.; Dangles, O.; Chemat, F. Direct enrichment of olive oil in oleuropein by ultrasound-assisted maceration at laboratory and pilot plant scale. *Ultrason Sonochem.* **2012**, *19*, 777–786. [CrossRef] [PubMed]
37. Benzie, I.F.F.; Strain, J.J. The ferric reducing ability of plasma (FRAP) as a measure of "antioxidant power": The FRAP assay. *Anal. Biochem.* **1996**, *239*, 70–76. [CrossRef] [PubMed]
38. Williams, P.J.; Geladi, P.; Britz, T.J.; Manley, M. Near-infrared (NIR) hyperspectral imaging and multivariate image analysis to study growth characteristics and differences between species and strains of members of the genus Fusarium. *Anal. Bioanal. Chem.* **2012**, *404*, 1759–1769. [CrossRef]
39. Bursać Kovačević, D.; Putnik, P.; Dragović-Uzelac, V.; Pedisić, S.; Režek Jambrak, A.; Herceg, Z. Effects of cold atmospheric gas phase plasma on anthocyanins and color in pomegranate juice. *Food Chem.* **2016**, *190*, 317–323. [CrossRef]
40. Xi, J.; He, L.; Yan, L.-G. Continuous extraction of phenolic compounds from pomegranate peel using high voltage electrical discharge. *Food Chem.* **2017**, *230*, 354–361. [CrossRef]
41. Garcia-Castello, E.M.; Rodriguez-Lopez, A.D.; Mayor, L.; Ballesteros, R.; Conidi, C.; Cassano, A. Optimization of conventional and ultrasound assisted extraction of flavonoids from grapefruit (*Citrus paradisi* L.) solid wastes. *LWT-Food Sci. Technol.* **2015**, *64*, 1114–1122. [CrossRef]
42. Bellumori, M.; Innocenti, M.; Binello, A.; Boffa, L.; Mulinacci, N.; Cravotto, G. Selective recovery of rosmarinic and carnosic acids from rosemary leaves under ultrasound- and microwave-assisted extraction procedures. *C. R. Chim.* **2016**, *19*, 699–706. [CrossRef]
43. Boussetta, N.; Vorobiev, E. Extraction of valuable biocompounds assisted by high voltage electrical discharges: A review. *C. R. Chim.* **2014**, *17*, 197–203. [CrossRef]
44. El Kantar, S.; Boussetta, N.; Rajha, H.N.; Maroun, R.G.; Louka, N.; Vorobiev, E. High voltage electrical discharges combined with enzymatic hydrolysis for extraction of polyphenols and fermentable sugars from orange peels. *Food Res. Int.* **2018**, *107*, 755–762. [CrossRef] [PubMed]
45. Barba, F.J.; Brianceau, S.; Turk, M.; Boussetta, N.; Vorobiev, E. Effect of alternative physical treatments (ultrasounds, pulsed electric fields, and high-voltage electrical discharges) on selective recovery of bio-compounds from fermented grape pomace. *Food Bioprocess. Technol.* **2015**, *8*, 1139–1148. [CrossRef]
46. Almohammed, F.; Koubaa, M.; Khelfa, A.; Nakaya, M.; Mhemdi, H.; Vorobiev, E. Pectin recovery from sugar beet pulp enhanced by high-voltage electrical discharges. *Food Bioprod. Process.* **2017**, *103*, 95–103. [CrossRef]
47. Roselló-Soto, E.; Barba, F.J.; Parniakov, O.; Galanakis, C.M.; Lebovka, N.; Grimi, N.; Vorobiev, E. High voltage electrical discharges, pulsed electric field, and ultrasound assisted extraction of protein and phenolic compounds from olive kernel. *Food Bioprocess. Technol.* **2015**, *8*, 885–894. [CrossRef]
48. Sarkis, J.R.; Boussetta, N.; Tessaro, I.C.; Marczak, L.D.F.; Vorobiev, E. Application of pulsed electric fields and high voltage electrical discharges for oil extraction from sesame seeds. *J. Food Eng.* **2015**, *153*, 20–27. [CrossRef]
49. Nyström, M.; Aimar, P.; Luque, S.; Kulovaara, M.; Metsämuuronen, S. Fractionation of model proteins using their physiochemical properties. *Colloids Surf. A Phys. Eng. Asp.* **1998**, *138*, 185–205. [CrossRef]
50. Anastas, P.T.; Warner, J.C. *Green Chemistry: Theory and Practice*; Oxford University Press: Oxford, UK, 2000.

© 2019 by the authors. Licensee MDPI, Basel, Switzerland. This article is an open access article distributed under the terms and conditions of the Creative Commons Attribution (CC BY) license (http://creativecommons.org/licenses/by/4.0/).

Article

The Impact of Pulsed Electric Field on the Extraction of Bioactive Compounds from Beetroot

Malgorzata Nowacka [1], Silvia Tappi [2], Artur Wiktor [1,*], Katarzyna Rybak [1], Agnieszka Miszczykowska [1], Jakub Czyzewski [1], Kinga Drozdzal [1], Dorota Witrowa-Rajchert [1] and Urszula Tylewicz [2,3]

1. Department of Food Engineering and Process Management, Faculty of Food Sciences, Warsaw University of Life Sciences, Nowoursynowska 159c, 02-776 Warsaw, Poland
2. Interdepartmental Centre for Agri-Food Industrial Research, University of Bologna, Via Quinto Bucci 336, 47521 Cesena, Italy
3. Department of Agricultural and Food Sciences, University of Bologna, Piazza Goidanich 60, 47521 Cesena, Italy
* Correspondence: artur_wiktor@sggw.pl; Tel.: +48-22-593-75-60

Received: 2 June 2019; Accepted: 2 July 2019; Published: 5 July 2019

Abstract: Beetroot is a root vegetable rich in different bioactive components, such as vitamins, minerals, phenolics, carotenoids, nitrate, ascorbic acids, and betalains, that can have a positive effect on human health. The aim of this work was to study the influence of the pulsed electric field (PEF) at different electric field strengths (4.38 and 6.25 kV/cm), pulse number 10–30, and energy input 0–12.5 kJ/kg as a pretreatment method on the extraction of betalains from beetroot. The obtained results showed that the application of PEF pre-treatment significantly ($p < 0.05$) influenced the efficiency of extraction of bioactive compounds from beetroot. The highest increase in the content of betalain compounds in the red beet's extract (betanin by 329%, vulgaxanthin by 244%, compared to the control sample), was noted for 20 pulses of electric field at 4.38 kV/cm of strength. Treatment of the plant material with a PEF also resulted in an increase in the electrical conductivity compared to the non-treated sample due to the increase in cell membrane permeability, which was associated with leakage of substances able to conduct electricity, including mineral salts, into the intercellular space.

Keywords: pulsed electric field; extraction; bioactive compounds; red beet

1. Introduction

Beetroot (*Beta vulgaris* L.) is part of the Chenopodiaceae family and has originated in Asia and Europe. The red beetroot variety, a cultivated form of *Beta vulgaris* subsp. *vulgaris* (conditiva), is widely used all over the world to produce pickles, salad, or juice [1]. Beetroot contains many functional components, such as vitamins, minerals (potassium, sodium, phosphorous, calcium, magnesium, copper, iron, zinc, manganese), phenolics, carotenoids, nitrate, and ascorbic acids that promote health benefits [2]. In fact, polyphenols, carotenoids, and vitamins present in beetroot have been recognized to have antioxidant, anti-inflammatory, anticarcinogenic, and hepato-protective activities, which can help in the prevention of many diseases, such as cardiovascular disease or hypertension and diabetes [3].

The characteristic red purple color of beetroot derives from betalain, water soluble pigments found in plants of the Caryophyllales order [4]. Depending on their chemical structure, betalains can be divided into red-purple and violet betacyanins (betanin, isobetanin, probetanin, and neobetanin) or yellow betaxanthins (vulgaxanthin, miraxanthin, portulaxanthin, and indicaxanthin) [5]; and the redness of beetroot depends on the ratio between the two classes [6]. For example, it was observed that the intact beetroot plant extracts contain about 40 mg/g dm of betalains, from which 20.75 mg/g dm are betacyanins and 19.01 mg/g dm are betaxanthins [2]. Slavov et al. [3] studied the individual betalain compounds content in pressed juice from beetroot, showing that the most abundant were betanin, followed by vulgaxanthin and isobetanin.

Betalain pigments can generally exhibit health benefits as antioxidants, anti-cancer, anti-lipidemic, and antimicrobial agents [5,7]. They are mostly used as food dyes due to non-precarious, non-toxic, non-carcinogenic, and non-poisonous nature [2]. Therefore, there is an increasing interest of food industries in the extraction of this natural food colorant. In order to extract the pigments from plant material, the disruption of membranes is necessary, and this is usually obtained through the application of detergents, solvents, or thermal treatments [8]. The latter is the preferred method for color concentrates production; however, betalain is highly sensitive to heat. Therefore, alternative methods are necessary in order to prevent the discoloration of the pigments [9]. In recent years, the use of novel and mild methods has been investigated, including, among others, the microwave [9,10] and ultrasound [11,12] assisted extraction. Another promising technique could be application of pulsed electric field (PEF) that, applied as a pre-treatment for the extraction, allows the selective recovery of bioactive compounds, at the same time reducing the energy and time required in the process.

PEF is the application of short time pulsed with high voltage into the food product placed between two electrodes [13,14], thus promoting the modification of membrane permeability and the increase of the extraction yield [15,16].

Recently, PEF treatment has been applied in order to recover pigments from beetroot. Fincan et al. [17] observed a release of about 90% of total red coloring from beetroot subjected to 270 rectangular pulses of 10 µs at 1 kV/cm field strength and total energy of 7 kJ/kg. López et al. [18] and Luengo et al. [19] studied the influence of different PEF parameters, such as electric field strength, pulse number, pulse duration, and specific energy applied on the extractability of betanin from the beetroot cylinders. Similarly, Chalermchat et al. [20] evaluated the impact of PEF treatment on the red pigment in beetroot discs. However, in the cited works, PEF was applied only on a single geometry (cylinder or disc), while we assumed that this parameter could also have an influence on the extraction. Moreover, in most cases, the efficiency of extraction was assessed only by measuring the difference in the color of the extracts. Hence, our goals in the present research were to compare (i) the impact of mechanical preparation (cylinders or pulp) and (ii) the impact of PEF pre-treatments on different geometric forms on extractability of beetroot pigments. For these aims, beetroot was subjected to cutting into cylinders or pulping and different PEF conditions (electric field strength between 4.38 and 6.25 kV/cm, pulse number in the range of 10–30, energy input in the range of 0–12.5 kJ/kg) as pre-treatments for the extraction of selected red (betanin) and yellow (vulgaxanthin) pigments.

2. Materials and Methods

2.1. Preparation of Samples

Beetroot (*Beta vulgaris* L.) was purchased from a local market (Warsaw, Poland). Products with a similar degree of maturity and characterized by similar dimensions were selected for the study. The raw materials were stored at 4 °C until the start of the experiments (not longer than one week). Before each experiment, the vegetables were removed from the cooling chamber, allowed to reach room temperature, and then washed. The beetroot was cut into 7 mm height slices, and then cylinders with a diameter of d = 15 mm were obtained using a cork borer. An electric mill was used to crush the cylinders to obtain a pulp.

2.2. PEF Treatement

A prototype PEF generator ERTEC-RI-1B (ERTEC, Wroclaw, Poland) with output high-voltage impulse up to 30 kV and capacitance of 0.25 µF reactor was used. The generator provided monopolar, exponential shaped pulses with an average width of 10 µs and an interval between pulses equal to 2 s, which allows minimization of the temperature increase during the electric field application. The treatment chamber (diameter of 40 mm, height of 16 mm), made of dielectric material (Corian) and two stainless-steel lids, consists of a spark gap controlling the flow of electrical impulses and a set of electrodes (stationary and mobile). Both stationary and mobile electrodes were connected to the PEF generator.

Beetroot samples (8 cylinders) were placed in the treatment chamber. Next, to ensure a good contact between the electrodes and the tested material, the chamber was filled with a phosphate buffer at pH = 6.5 and closed using a mobile electrode. In the case of pulp, the entire volume of the electrical processing chamber (20 cm^3) was carefully filled with the sample without additional solvent. PEF application was carried out using two different electric field intensities (4.38 and 6.25 kV/cm) and three pulse numbers (10, 20, 30). Each combination of the experiment was performed in three repetitions. The specific energy intake was calculated according to the following equation [14]:

$$W_s = (V^2 C n)/(2m), \tag{1}$$

where V (V) is the voltage, C (F) is a capacitance of the energy storage capacitor, n is number of pulses, and m (kg) is mass of the sample in the treatment chamber.

The energy values (W_s) supplied to the sample are shown in Table 1. After PEF treatment, cylinder samples were dried on filter paper. Directly after PEF application, the temperature increase of the samples did not exceed 9.2 °C.

Table 1. Parameters of pulsed electric field (PEF) treatment: electric field intensity, pulse number, and energy delivered to the sample used in the investigation.

Sample Code	Electric Field Intensity (kV/cm)	Pulse Number (-)	Energy (kJ/kg)
0_0	0	0	0
4.38_10	4.38	10	2.43
4.38_20	4.38	20	4.86
4.38_30	4.38	30	7.28
6.25_10	6.25	10	4.96
6.25_20	6.25	20	9.92
6.25_30	6.25	30	14.88

2.3. Electrical Conductivity

The electrical conductivity of the sample was measured using a conductometer (CPC-505, Elmetron, Gliwice, Poland) with a platinum dual-needle probe, and it was used to evaluate the effectiveness of electroporation. The electrical conductivity was measured continuously for at least 5 min at room temperature. The measurements were conducted in five repetitions [21].

2.4. Color Determination

The color of the beetroot cylinders and extracts were measured in the CIE L*a*b* color system using a Konica Minolta CR-5 Chroma Meter (Osaka, Japan) with D65 light source and standard 2° observer. The L* parameter defines the lightness of the sample, the a* chromatic coordinate determines the green (−) and red (+) component, and the b* coordinate describes the blue (−) and yellow (+) component [22]. The measurements were conducted at sixteen repetitions. Moreover, the supernatant (solution remaining after extraction) was subjected to the color measurements in the transmittance mode.

2.5. Betalain Content

The betalain content in raw and PEF-treated material was determined by the chemical method, which is based on the extraction of dyes using phosphate buffer (pH 6.5) and the simultaneous determination of red and yellow dyes measured spectrophotometrically [23]. In the case of raw (untreated) material and beetroot pulp, 0.5 g of sample was weighted and then quantitatively transferred to a falcon tube, added a 50 cc phosphate buffer, and placed on a Vortex shaker for 20 min at 2000 rpm. After shaking, the obtained supernatant was filtered and the absorbance of the solution was measured in a spectrophotometer HeΛios ThermoSpectronic γ (Thermo Electron Corporation, USA) at 476, 538, and 600 nm using a phosphate buffer solution as standard. In the case of PEF-treated material, a single beetroot cylinder along with the phosphate buffer was quantitatively transferred to the falcon. Each time, the PEF chamber was rinsed three times with buffer and added to the falcon in order to avoid betalain losses. The falcon was filled with a buffer to the 50 cc volume and then analyzed as described above. The assay was performed three times.

The absorbance value for red dyes (B)] at λ = 538 nm including light absorption due to the presence of impurities was calculated as:

$$E_B = 1.095 \times (E_{538} - E_{600}), \qquad (2)$$

where E_{538} absorbance at λ = 538 nm, E_{600} absorbance at λ = 600 nm, and 1.095 is a absorption coefficient at λ = 538 nm resulting from impurities present.

The content of red dyes (B) (mg betanin/100 g dm) was calculated as:

$$B = (c \times E_B)/(1120 \times a \times b), \qquad (3)$$

where c (mg) mass of the sample with buffer, 1120 (-) value resulting from the absorbance of a 1% betanin solution measured at 538 nm in a 1 cm cuvette, a (g) sample weight, and b (g/g dm) is dry matter content.

The absorbance value for yellow dyes (W) at λ = 476 nm including light absorption due to the presence of impurities and red dyes and was calculated as:

$$E_W = E_{476} - E_{358} + 0.667 \times E_B, \qquad (4)$$

where E_{476} absorbance at λ = 476 nm, E_{538} absorbance at λ = 538 nm, and E_B is the absorbance value for red pigments.

The content of yellow dyes (W) (mg vulgaxanthin /100 g dm) was calculated as:

$$W = (c \times E_W)/(750 \times a \times b), \qquad (5)$$

where c (mg) mass of the sample with buffer, 750 (-) value resulting from the absorbance of a 1% betanin solution measured at 476 nm in a 1 cm cuvette, a (g) sample weight, and b (g/g dm) is dry matter content.

The content of red and yellow dyes ware quantified as mg of betanin and vulgaxanthin, respectively, in 100 g dry material.

2.6. Statitical Analysis—PCA and Pearson's Correlation

The ANOVA and the Tukey test (α = 0.05) were applied to evaluate significant differences between investigated samples. The two-way ANOVA and comparison of obtained partial η^2 values have been used to assess the effect size of the analyzed variables on the electrical conductivity. The Pearson's correlation coefficient was calculated in order to evaluate the dependence between selected parameters of analyzed tissue. Hierarchical Cluster Analysis (HCA) with Ward's agglomeration method was done for complex assessment of obtained data. Euclidian distance was used to express obtained results by the means of relative distance. All calculations were performed using STATISTICA 13 (StatSoft Inc., Washington, DC, USA) and Excel (Microsoft, New York, NY, USA) software.

3. Results and Discussion

3.1. Electical Conductivity

Electrical conductivity measurement is used to evaluate the effectiveness of PEF treatment in biological tissue [24]. Following electroporation and intracellular content leakage, electrical conductivity of plant materials rises, indicating cell membrane rupture [25]. Figure 1 shows the results of both electrical conductivity and energy input delivered to the beetroot samples. As expected, all PEF pre-treated samples exhibited higher electrical conductivity in comparison to intact tissue. For instance, material treated by 10 pulses at 4.38 kV/cm was characterized by electrical conductivity equal to 9.05×10^{-4} S/m. When the number of pulses increased to 30 (hence increasing the energy input) at the same electric field intensity, the electrical conductivity rose significantly ($p < 0.05$) to 14.2×10^{-4} S/m. The application of the higher electric field intensity (6.25 kV/cm) lead to a further increase of electrical conductivity. However, in this case, the difference between samples pre-treated by 10, 20, and 30 pulses were statistically insignificant ($p > 0.05$). In other words, despite the energy input increase, the electroporation efficiency did not grow because of the saturation of the electroporation phenomenon that was previously observed in the literature for materials other than beetroot [14,26,27].

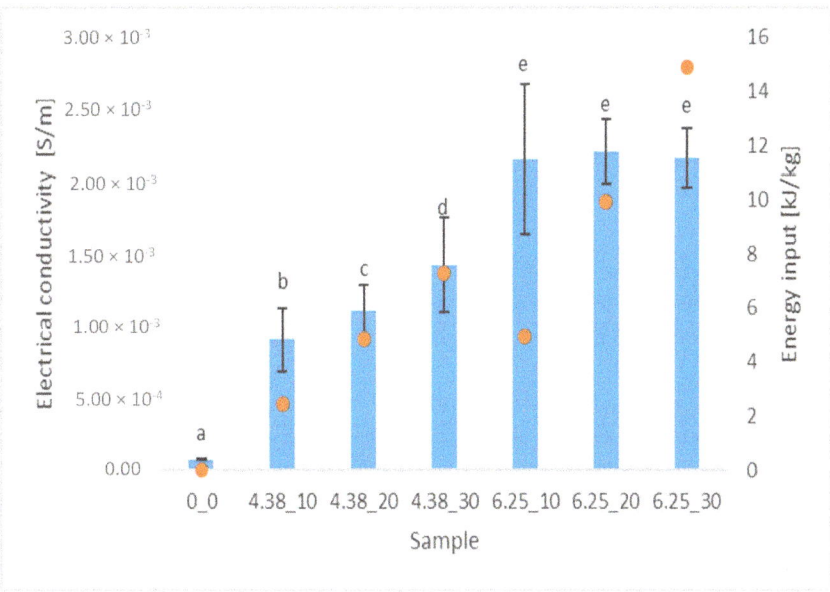

Figure 1. Electrical conductivity S/m (bars) and total energy input kJ/kg (dots) of differently treated beetroot samples. The same letter on the same column means no significant difference between electrical conductivities by the Tukey test ($p < 0.05$).

The performed statistical analysis showed that electric field intensity and number of pulses significantly ($p < 0.05$) influenced the electrical conductivity of the beetroot samples. Moreover, the interaction of these parameters, which could be interpreted as the influence of energy input, also had a significant ($p < 0.05$) impact on the electroporation efficiency measured by electrical conductivity measurement. However, comparing the partial η^2 values, it can be stated that the biggest size effect was found for electric field intensity ($\eta^2 = 0.753$). For comparison, partial η^2 was equal to 0.0114 and 0.113, for number of pulses and the interaction between electric field intensity and number of pulses, respectively. Similar results were obtained previously by Wiktor et al. [21], suggesting that the cell disintegration index could be influenced not only by electric field strength and total energy input but also by the pulses number applied during the treatment, simply because it increases the chance of electroporation to occur.

The PEF parameters that have been used in further analysis were selected in order to allow to study the extraction of pigments from (i) samples that have been treated with similar energy input but manifested different electrical conductivity (4.38_20 and 6.25_10) and (ii) samples that exhibited similar electroporation efficiency (expressed as similar electrical conductivity) but treated with different energy input (6.25_10 and 6.25_30).

3.2. Color Determination

Color is a major quality index for fresh and processed fruit and vegetables, being one of the main influencing parameters for consumers. Moreover, color could be also indicator of the presence and the quantity of bioactive compounds, e.g., betanins which are responsible for the characteristic red-purple color of the beetroot [4].

Table 2 reports the color parameters of untreated and PEF-treated beetroot tissue before and after extraction. It can be observed that the application of PEF decreased significantly all the color parameters, and this decrease was more accentuated in the samples 4.38_20 and 6.25_30. The decrease of a* and b* values indicates the loss of the pigments (red betanin and yellow vulgaxanthin) from the beetroot to the extracting solution.

Table 2. Color parameters (L* = lightness, a* = red/green index, b* = yellow/blue index) of untreated and PEF-treated beetroot cylinders before and after extraction.

Sample	L*		a*		b*	
	Before	After	Before	After	Before	After
0_0	21.1 ± 0.8 [a]	21 ± 1 [AB]	27 ± 2 [a]	29 ± 2 [A]	7.74 ± 0.6 [a]	8.8 ± 0.9 [A]
4.38_20	18.9 ± 0.9 [c]	20.7 ± 0.7 [A]	18 ± 3 [c]	20 ± 2 [B]	4.6 ± 0.9 [c]	4.9 ± 0.7 [C]
6.25_10	20 ± 1 [b]	24.5 ± 0.8 [C]	23 ± 3 [b]	31 ± 2 [A]	6 ± 1 [b]	6.1 ± 0.9 [BC]
6.25_30	18 ± 1 [c]	23 ± 1 [BC]	18 ± 2 [c]	29 ± 3 [A]	4.6 ± 0.8 [c]	7 ± 1 [B]

The same letter on the same column means no significant difference by the Tukey test ($p < 0.05$).

The application of 4.38 kV/cm caused a significant decrease in both parameters in the beetroot tissue after extraction, while for other PEF conditions these values were mostly unchanged. When comparing the color parameters before and after extraction, it is visible that cylinders after extraction exhibited higher values of L*, a* and b*. Such a situation can be linked to the diffusion of pigments from the inner parts of the tissue towards its surface during extraction and by lower water content after extraction which influenced the light reflection during measurement.

Table 3 shows the color parameters of beetroot extracts obtained from untreated and PEF-treated beetroot cylinders and pulp. It is possible to observe a significant ($p < 0.05$) decrease in the lightness and a significant increase of the red color component of the beetroot cylinders subjected to all PEF treatments, without significant differences due to treatment parameters (Figure A1). The color intensity is usually directly related to the content of betalains extracted from beets. The use of PEF clearly increases the color intensity, and therefore it increases the concentration of these compounds [28]. Concerning the yellow index, a decrease was observed only in samples treated at 6.25 kV/cm, probably due to the degradation of the yellow dye—vulgaxanthin.

Concerning the extract obtained from the pulped beetroot (Table 3), it was possible to observe that the pulping process itself caused a significant reduction in the brightness of the extract, as well as an increase in the share of red color. The application of PEF did not result in significant differences between the values of the L* and a* parameters. On the other side, the b* coordinate was characterized by a lower value in samples treated at 6.25 kV/cm, further reduced increasing pulses from 10 to 30, probably indicating a lower amount of vulgaxanthin in the extract.

Table 3. Color parameters (L* = Lightness, a* = red/green index, b* =yellow/blue index) of extracts obtained from untreated and PEF-treated beetroot cylinders and pulp.

Sample	L*		a*		b*	
	Cylinders	Pulp	Cylinders	Pulp	Cylinders	Pulp
0_0	96.2 ± 0.2 [a]	72 ± 1 [A]	7.2 ± 0.4 [a]	52 ± 2 [A]	−0.6 ± 0.1 [a]	−1.7 ± 0.4 [A]
4.38_20	76 ± 3 [b]	70 ± 2 [A]	43 ± 5 [b]	55 ± 4 [A]	1.5 ± 1.6 [a]	−1.4 ± 0.5 [A]
6.25_10	81 ± 2 [b]	70 ± 1 [A]	36 ± 4 [b]	57 ± 2 [A]	−5 ± 2 [b]	−6.16 ± 0.06 [B]
6.25_30	79 ± 2 [b]	70 ± 2 [A]	41 ± 3 [b]	58 ± 4 [A]	−5.1 ± 0.7 [b]	−9.3 ± 06 [C]

The same letter on the same column means no significant difference by the Tukey test ($p < 0.05$).

3.3. Betalain Content

Betanin and vulgaxanthin are the most abundant betalains in beetroot [3]. Figure 2 shows the betanin content, while Figure 3 shows the vulgaxanthin content of differently treated beetroot samples. The application of PEF to beetroot cylinders allowed to increase the extraction of betanin in a similar way (of about 3-fold) for all the PEF conditions applied. The use of PEF at the intensity of 4.38 kV/cm significantly ($p < 0.05$) increased the yield of betanin and vulgaxanthin extraction from beetroot cylinders by 329% and 244%, respectively, compared to the control samples. This increase indicates that PEF facilitates their extraction by increasing cell membrane permeability, which is also confirmed by the color parameters. While the values of the red and yellow color components decreased in the cylinders after extraction, an increase was recorded in the post-extraction solution. In addition, the electrical conductivity of beet tissue after application of 20 pulses at 4.38 kV/cm was significantly higher compared to the untreated sample (Figure 1).

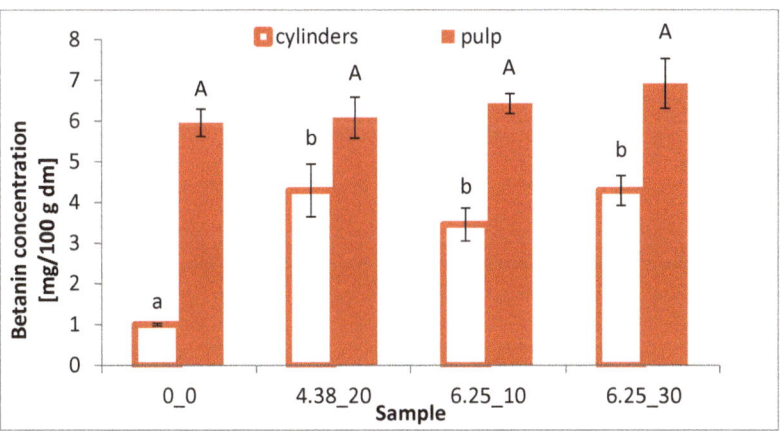

Figure 2. Betanin concentration of differently treated beetroot cylinders and pulp. The same letter (and the same letter size) on the same column means no significant difference by the Tukey test ($p < 0.05$).

In the case of beetroot pulp, an increase in betanin content (Figure 2), even if not statistically significant, can be observed, as well as a decrease in vulgaxanthin extractability (statistically significant) with increasing intensity of the electric field (Figure 3). In addition, the fragmentation of the raw material tissue significantly increased the content of both betanin and vulgaxanthin in the extract, compared to their content in the beetroot cylinders. The highest content of vulgaxanthin was observed in the extract of untreated pulp (control), while the lowest was after the treatment with the electric field of 6.25 kV/cm and 30 pulses, which caused statistically significant ($p < 0.05$) drop in vulgaxanthin content by 27%. In turn, in the case of betanin, an inverse relationship was demonstrated, in the PEF application with the parameters E = 6.25 kV/cm, n = 30 promoted the higher extraction efficiency, increased by 16% in relation to the reference sample; however, this was not statistically significant.

Information on the influence of PEF on the vulgaxanthin content in the literature is really scarce. Nevertheless, the decrease in its content in the beetroot pulp, along with the increasing intensity of the electric field, could be related to the physicochemical properties of this pigment. Vulgaxanthin is less stable in an acid environment and at room temperature (and higher) compared to betanin. The solvent used in the test was a slightly acidic buffer (pH = 6.5), and in addition, although PEF treatment is a non-thermal method, the temperature was slightly risen during its application. In addition, the fragmentation of the material facilitated the extraction of pigment from the tissue, which resulted in longer exposure of vulgaxanthin to a temperature higher than room temperature, thus it could cause its degradation [29].

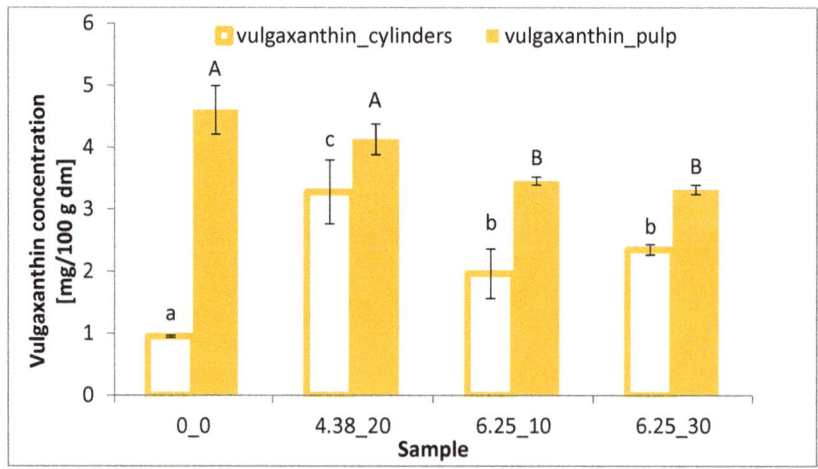

Figure 3. Vulgaxanthin concentration of differently treated beetroot cylinders and pulp. The same letter (and the same letter size) on the same column means no significant difference by the Tukey test ($p < 0.05$).

The use of electrical treatment allows achievement of a high degree of disintegration of cells with only small temperature increases (up to 9.1 °C). Two examples in the literature deal with the identification of the optimal amount of energy to maximize pigment extraction from beetroot tissue (between 2.5 and 7 kJ/kg obtained using a different electric field intensity). For example, a PEF treatment at 1 kV/cm (270 pulses, 10 µs, 7 kJ/kg) allowed to release about 90% of the total red pigment after 1 h of extraction compared to less than 5% in the untreated samples [17,30]. López et al. [18] instead observed that the application of 7 kV/cm (5 pulses of 2 µs, 2.5 kJ/kg) caused a release of about 90% of total betanin from beetroot samples in 300 min. This release was 5-fold quicker than the control samples, while the application of higher electric field strength (9 kV/cm) decreased the extraction yield of the betanin. In this case, the energy supplied to the beetroot was lower than in the present study and in the work of Fincan et al. [17].

3.4. HCA and Pearson's Correlation

Figure 4 presents the results of HCA, which was carried out using normalized color data (L*, a* and b*) and concentration of betanin and vulgaxanthin. Performed analysis allowed the samples to be grouped in two big clusters. The first group consisted of most of the samples in cylindrical form (C_0_0; C_6.25_10; C_6.25_30), whereas the second group included all samples in pulp form and one cylindrical sample—the one treated with 20 pulses at 4.38 kV/cm. This sample formed a smaller aggregation that included untreated pulp form beetroot (P_0_0) and the pulp form material treated by the same parameters. Hence, it can be stated that PEF treatment can lead to similar changes in the tissue brought about by PEF influencing extraction are similar to the changes due to the mechanical disintegration obtained by pulping. Beetroot samples treated with 6.25 kV/cm were grouped in two separate aggregates depending on their physical form, cylinder or pulp. This information is very valuable because it indicates that increasing the number of pulses does not bring additional benefits. Such a behavior in literature is defined as 'saturation level of electroporation', meaning that no more changes can be achieved by PEF treatment despite of utilization of higher energy input [31].

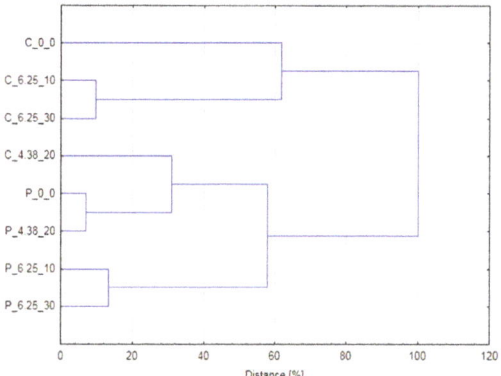

Figure 4. The results of hierarchical cluster analysis (HCA) performed using color parameters (L*, a*, b*) and concentration of betanin and vulgaxanthin. (C_0_0 = untreated beetroot cylinders; C_4.38_20 = beetroot cylinders treated by 20 pulses at 4.38 kV/cm; C_6.25_10 = beetroot cylinders treated by 10 pulses at 6.25 kV/cm; C_6.25_30 = beetroot cylinders treated by 30 pulses at 6.25 kV/cm; P_0_0 = untreated beetroot pulp; P_4.38_20 = beetroot pulp treated by 20 pulses at 4.38 kV/cm; P_6.25_10 = beetroot pulp treated by 10 pulses at 6.25 kV/cm; P_6.25_30 = beetroot pulp treated by 30 pulses at 6.25 kV/cm).

The results of Pearson's correlation analysis of selected variables are presented in Table 4. Significant correlation has been found between lightness (L*) and a* coordinate. What is more interesting is that both L* and a* values presented a significant relationship with betanin and vulgaxanthin concentration. However, higher values of Pearson's correlation coefficient were found between color parameters and betanin concentration. Similar observations were also reported by Antigo et al. [32], who examined the betanin content of microcapsules of beetroot extract obtained by spray drying and freeze-drying.

Table 4. The results of Pearson's correlation analysis.

Variable	L*	a*	b*	BC	VC
L*	-	r = −0.9977 p < 0.001	r = 0.3593 p = 0.382	r = −0.9760 p < 0.001	r = −0.8853 p = 0.003
a*	r = −0.9977 p < 0.001	-	r = −0.4214 p = 0.298	r = 0.9821 p < 0.001	r = 0.8594 p = 0.006
b*	r = 0.3593 p = 0.382	r = −0.4214 p = 0.298	-	r = −0.4692 p = 0.241	r = 0.0077 p = 0.986
BC	r = −0.9760 p < 0.001	r = 0.9821 p < 0.001	r = −0.4692 p = 0.241	-	r = 0.8623 p = 0.006
VC	r = −0.8853 p = 0.003	r = 0.8594 p = 0.006	r = 0.0077 p = 0.986	r = 0.8623 p = 0.006	-

p-Values lower than 0.05 indicate the significant character of the correlation between normalized variables. L*, a*, b* = color parameters of the extracts; BC = betanin concentration in the sample; VC = vulgaxanthin concentration in the sample.

4. Conclusions

The use of PEF led to an increase of the electrical conductivity of beetroot tissue, thus leading to an increase of the cell membrane permeability. When lower electric field intensity (4.38 kV/cm) was applied, the electrical conductivity increased proportionally to the pulse number applied, whereas with the higher intensity (6.25 kV/cm) this effect was negligible, showing an effect of saturation level of electroporation. PEF treatment caused changes in the color of red beetroot tissue that were associated with a better extraction of pigments as a result of the electroporation. The use of PEF with an intensity of 4.38 kV/cm and 20 impulses and energy consumption of 4.10 kJ/kg, allowed an increase in the yield of betanin and vulgaxanthin extraction from beetroot cylinders by 329% and 244%, respectively, compared to the control. Furthermore, the fragmentation of the tissue (pulping) resulted in an increased extraction of pigments from the plant tissue; however, the use of PEF did not bring additional advantages.

Author Contributions: Conceptualization, A.W. and D.W.R.; methodology, A.W., M.N., K.R.; validation, K.R., A.M. and J.C.; formal analysis, U.T., S.T., M.N. and A.W.; investigation, K.R., A.M., A.W., J.C.; writing—original draft preparation, U.T., S.T., M.N., K.D., and A.W.; writing—review and editing, U.T., S.T., M.N. and A.W.; visualization, U.T., S.T., M.N. and A.W.; supervision, A.W., D.W.R.

Funding: This research received no external funding.

Conflicts of Interest: The authors declare no conflict of interest.

Appendix A

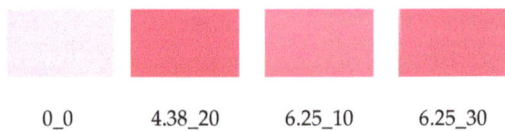

0_0   4.38_20   6.25_10   6.25_30

Figure A1. Changes in the color of the extract from the beetroot slices depending on the parameters of the applied pulsed electric field treatment.

References

1. Wruss, J.; Waldenberger, G.; Huemer, S.; Uygun, P.; Lanzerstorfer, P.; Müller, U.; Höglinger, O.; Weghuber, J. Compositional characteristics of commercial beetroot products and beetroot juice prepared from seven beetroot varieties grown in Upper Austria. *J. Food Compos. Anal.* **2015**, *42*, 46–55. [CrossRef]
2. Chhikara, N.; Kushwaha, K.; Sharma, P.; Gat, Y.; Panghal, A. Bioactive compounds of beetroot and utilization in food processing industry: A critical review. *Food Chem.* **2019**, *272*, 192–200. [CrossRef]
3. Slavov, A.; Karagyozov, V.; Denev, P.; Kratchanova, M.; Kratchanov, C. Antioxidant activity of red beet juices obtained after microwave and thermal pretreatments. *Czech J. Food Sci.* **2013**, *31*, 139–147. [CrossRef]

4. Strack, D.; Vogt, T.; Schliemann, W. Recent advances in betalain research. *Phytochemistry* **2003**, *62*, 247–269. [CrossRef]
5. Gengatharan, A.; Dykes, G.A.; Choo, W.S. Betalains: Natural plant pigments with potential application in functional foods. *LWT—Food Sci. Technol.* **2015**, *64*, 645–649. [CrossRef]
6. Szopinska, A.A.; Gawęda, M. Comparison of yield and quality of red beet roots cultivated using conventional, integrated and organic method. *J. Hort. Res.* **2013**, *21*, 107–114. [CrossRef]
7. Tanabtabzadeh, M.S.; Javanbakht, V.; Golshirazi, A.H. Extraction of Betacyanin and Betaxanthin Pigments from Red Beetroots by Chitosan Extracted from Shrimp Wastes. *Waste Biomass Valoriz.* **2019**, *10*, 641–653. [CrossRef]
8. Celli, G.B.; Brooks, M.S.-L. Impact of extraction and processing conditions on betalains and comparison of properties with anthocyanins—A current review. *Food Res. Int.* **2017**, *100*, 501–509. [CrossRef]
9. Singh, A.; Ganesapillai, M.; Gnanasundaram, N. Optimizaton of extraction of betalain pigments from *beta vulgaris* peels by microwave pretreatment. In *IOP Conference Series: Materials Science and Engineering*; IOP Publishing: Bristol, UK, 2017.
10. Cardoso-Ugarte, G.A.; Sosa-Morales, M.E.; Ballard, T.; Liceaga, A.; San Martín-González, M.F. Microwave-assisted extraction of betalains from red beet (*Beta vulgaris*). *LWT—Food Sci. Technol.* **2014**, *59*, 276–282. [CrossRef]
11. Ramli, N.S.; Ismail, P.; Rahmat, A. Influence of conventional and ultrasonic-assisted extraction on phenolic contents, betacyanin contents, and antioxidant capacity of red dragon fruit (Hylocereus polyrhizus). *The Sci. World J.* **2014**, *2014*, 964731. [CrossRef]
12. Laqui-Vilca, C.; Aguilar-Tuesta, S.; Mamani-Navarro, W.; Montaño-Bustamante, J.; Condezo-Hoyos, L. Ultrasound-assisted optimal extraction and thermal stability of betalains from colored quinoa (Chenopodium quinoa Willd) hulls. *Ind. Crop. Prod.* **2018**, *111*, 606–614. [CrossRef]
13. Tylewicz, U.; Tappi, S.; Mannozzi, C.; Romani, S.; Dellarosa, N.; Laghi, L.; Ragni, L.; Rocculi, P.; Dalla Rosa, M. Effect of pulsed electric field (PEF) pre-treatment coupled with osmotic dehydration on physico-chemical characteristics of organic strawberries. *J. Food Eng.* **2017**, *213*, 2–9. [CrossRef]
14. Wiktor, A.; Sledz, M.; Nowacka, M.; Rybak, K.; Chudoba, T.; Lojkowski, W.; Witrowa-Rajchert, D. The impact of pulsed electric field treatment on selected bioactive compound content and color of plant tissue. *Innov. Food Sci. Emerg. Technol.* **2015**, *30*, 69–78. [CrossRef]
15. Vorobiev, E.; Lebovka, N.I. Enhanced extraction from solid foods and biosuspensions by pulsed electrical energy. *Food Eng. Rev.* **2010**, *2*, 95–108. [CrossRef]
16. Barba, F.J.; Parniakov, O.; Pereira, S.A.; Wiktor, A.; Grimi, N.; Boussetta, N.; Saraiva, J.A.; Raso, J.; Martin-Belloso, O.; Witrowa-Rajchert, D.; et al. Current applications and new opportunities for the use of pulsed electric fields in food science and industry. *Food Res. Int.* **2015**, *77*, 773–798. [CrossRef]
17. Fincan, M.; DeVito, F.; Dejmek, P. Pulsed electric field treatment for solid-liquid extraction of red beetroot pigment. *J. Food Eng.* **2004**, *64*, 381–388. [CrossRef]
18. López, N.; Puértolas, E.; Condón, S.; Raso, J.; Alvarez, I. Enhancement of the extraction of betanine from red beetroot by pulsed electric fields. *J. Food Eng.* **2009**, *90*, 60–66. [CrossRef]
19. Luengo, E.; Martinez, J.M.; Álvarez, I.; Raso, J. Comparison of the efficacy of pulsed electric fields treatments in the millisecond and microsecond range for the extraction of betanine from red beetroot. In Proceedings of the 1st World Congress on Electroporation and Pulsed Electric Fields in Biology, Medicine and Food & Environmental Technologies (WC 2015), Portorož, Slovenia, 6–10 September 2015; pp. 375–378.
20. Chalermchat, Y.; Fincan, M.; Dejmek, P. Pulsed electric field treatment for solid-liquid extraction of red beetroot pigment: mathematical modelling of mass transfer. *J. Food Eng.* **2004**, *64*, 229–236. [CrossRef]
21. Wiktor, A.; Iwaniuk, M.; Śledź, M.; Nowacka, M.; Chudoba, T.; Witrowa-Rajchert, D. Drying kinetics of apple tissue treated by pulsed electric field. *Dry Technol.* **2013**, *31*, 112–119. [CrossRef]
22. Fijalkowska, A.; Nowacka, M.; Witrowa-Rajchert, D. The physical, optical and reconstitution properties of apples subjectes to ultrasound before drying. *Ital. J. Food Sci.* **2017**, *29*, 343–356.
23. Fijalkowska, A.; Nowacka, M.; Witrowa-Rajchert, D. The influence of ultrasound pre-treatment on drying kinetics and the colour and betalains content in beetroot. *Zeszyty Problemowe Postępów Nauk Rolniczych* **2015**, *581*, 11–20.
24. Lebovka, N.; Bazhal, M.; Vorobiev, E. Estimation of characteristic damage time of food materials in pulsed electric fields. *J. Food Eng.* **2002**, *54*, 337–346. [CrossRef]

25. Tylewicz, U.; Aganovic, K.; Vannini, M.; Toepfl, S.; Bortolotti, V.; Dalla Rosa, M.; Oey, I.; Heinz, V. Effect of pulsed electric field treatment on water distribution of freeze-dried apple tissue evaluated with DSC and TD-NMR techniques. *Innov. Food Sci. Emerg. Technol.* **2016**, *37*, 352–358. [CrossRef]
26. Lebovka, N.I.; Bazhal, M.I.; Vorobiev, E. Simulation and experimental investigation of food material breakage using pulsed electric field treatment. *J. Food Eng.* **2000**, *44*, 213–223. [CrossRef]
27. Bazhal, M.; Lebovka, N.; Vorobiev, E. Optimisation of pulsed electric field strength for electroplasmolysis of vegetable tissues. *Biosyst. Eng.* **2003**, *86*, 339–345. [CrossRef]
28. Puértolas, E.; Saldaña, G.; Raso, J. Pulsed electric field treatment for fruit and vegetable processing. In *Handbook of Electroporation*; Miklavčič, D., Ed.; Springer International Publishing AG: Basel, Switzerland, 2017.
29. Manchali, S.; Murthy, K.N.C.; Nagaraju, S.; Neelwarne, B. Stability of Betalain Pigments of Red Beet. In *Red Beet Biotechnology: Food and Pharmaceutical Applications*; Neelwarne, B., Ed.; Springer Science and Business Media: New York, NY, USA, 2013.
30. Donsì, F.; Ferrari, G.; Pataro, G. Applications of pulsed electric field treatments for the enhancement of mass transfer from vegetable tissue. *Food Eng. Rev.* **2010**, *2*, 109–130. [CrossRef]
31. Canatella, P.J.; Karr, J.F.; Petros, J.A.; Prausnitz, M.R. Quantitative study of electroporation-mediated molecular uptake and cell viability. *Biophys. J.* **2001**, *80*, 755–764. [CrossRef]
32. Antigo, J.L.D.; Bergamasco, R.D.C.; Madrona, G.S. Effect of pH on the stability of red beet extract (*Beta vulgaris* L.) microcapsules produced by spray drying or freeze drying. *Food Sci. Technol.* **2018**, *38*, 72–77. [CrossRef]

© 2019 by the authors. Licensee MDPI, Basel, Switzerland. This article is an open access article distributed under the terms and conditions of the Creative Commons Attribution (CC BY) license (http://creativecommons.org/licenses/by/4.0/).

Article

Valorization of Tomato Surplus and Waste Fractions: A Case Study Using Norway, Belgium, Poland, and Turkey as Examples

Trond Løvdal [1,*], Bart Van Droogenbroeck [2], Evren Caglar Eroglu [3], Stanislaw Kaniszewski [4], Giovanni Agati [5], Michel Verheul [6] and Dagbjørn Skipnes [1]

1. Department of Process Technology, Nofima—Norwegian Institute of Food, Fisheries and Aquaculture Research, N-4068 Stavanger, Norway
2. ILVO—Institute for Agricultural and Fisheries Research, Technology and Food Science Unit, 9090 Melle, Belgium
3. Department of Food Technology, Alata Horticultural Research Institute, 33740 Mersin, Turkey
4. Department of Soil Science and Vegetable Cultivation, InHort—Research Institute of Horticulture, 96-100 Skierniewice, Poland
5. Consiglio Nazionale delle Ricerche, Istituto di Fisica Applicata 'Nello Carrara', 50019 Sesto Fiorentino, Italy
6. NIBIO—Norwegian Institute of Bioeconomy Research, N-4353 Klepp Stasjon, Norway
* Correspondence: trond.lovdal@nofima.no

Received: 1 June 2019; Accepted: 24 June 2019; Published: 27 June 2019

Abstract: There is a large potential in Europe for valorization in the vegetable food supply chain. For example, there is occasionally overproduction of tomatoes for fresh consumption, and a fraction of the production is unsuited for fresh consumption sale (unacceptable color, shape, maturity, lesions, etc.). In countries where the facilities and infrastructure for tomato processing is lacking, these tomatoes are normally destroyed, used as landfilling or animal feed, and represent an economic loss for producers and negative environmental impact. Likewise, there is also a potential in the tomato processing industry to valorize side streams and reduce waste. The present paper provides an overview of tomato production in Europe and the strategies employed for processing and valorization of tomato side streams and waste fractions. Special emphasis is put on the four tomato-producing countries Norway, Belgium, Poland, and Turkey. These countries are very different regards for example their climatic preconditions for tomato production and volumes produced, and represent the extremes among European tomato producing countries. Postharvest treatments and applications for optimized harvest time and improved storage for premium raw material quality are discussed, as well as novel, sustainable processing technologies for minimum waste and side stream valorization. Preservation and enrichment of lycopene, the primary health promoting agent and sales argument, is reviewed in detail. The European volume of tomato postharvest wastage is estimated at >3 million metric tons per year. Together, the optimization of harvesting time and preprocessing storage conditions and sustainable food processing technologies, coupled with stabilization and valorization of processing by-products and side streams, can significantly contribute to the valorization of this underutilized biomass.

Keywords: tomato; valorization; sustainable production; processing; lycopene; waste reduction; vegetables; postharvest physiology; healthy food

1. Introduction

The tomato (*Solanum lycopersicum* (L.)), which is neither a vegetable nor a fruit but botanically speaking a berry, is currently spread across the world and is a key element in most cultures cuisines. The tomato originated in South America, from where it was imported to Mexico. Tomato came to Europe from the Spanish colonies in the 1500s along with several other "new" plants such as maize, potato, and tobacco. The tomato plant was immediately cultivated in the Mediterranean countries, but was initially poorly received further north in Europe. The skepticism of the tomato was due to that it was long suspected to be poisonous. As a curiosity, the tomato was not found in Norwegian grocery shelves until well into the 1950s, and was thus more exotic than oranges and bananas. Nowadays, however, tomatoes have definitely become an essential ingredient also in the North European and the Nordic cuisine. For example, in Norway, tomatoes have in recent years been the most sold product in the fresh vegetable segment, with a total turnover of approximately 15 million € and an annual consumption of 7.3 kg per capita, of which 1/3 is produced in Norway [1]. Including also processed tomato products, annual per capita consumption in Norway increases to 16.3 kg, whereas it is 23.5 and 27.5 kg in Poland and Belgium, respectively. These are however still low values compared to the Mediterranean diet; annual per capita consumption in Turkey, Armenia, and Greece is 94, 85, and 77 kg, respectively, whereas in Italy, Spain, Portugal, and Ukraine it is approximately 40 kg [2]. For a detailed list of European tomato production and consumption, see Supplementary Table S1.

A joint FAO/WHO Expert Consultation report on diet, nutrition and the prevention of chronic diseases, recommended several years ago the intake of a minimum of 400 g of fruit and vegetables per day (excluding potatoes and other starchy tubers), for the prevention of chronic diseases such as cardiovascular diseases (CVD), diabetes, and obesity, as well as for the prevention and alleviation of several micronutrient deficiencies [3]. However, in the western world, the consumption of vegetables is still far less than recommended. There is thus a socioeconomic gain if one succeeds in stimulating increased consumption of tomato-based products high in lycopene and β-carotene that may lead to reduced incidents of cancers and CVD. To increase the consumption of vegetables, it is important to provide raw materials with high quality, diversity, and availability. Fruit and vegetables are important components of a healthy diet, and their sufficient daily consumption could help prevent major diseases such as CVD and certain cancers. According to the World Health Report 2002, low fruit and vegetable intake is estimated to cause ~31% of ischemic heart disease and 11% of stroke worldwide [4]. Overall, it is estimated that up to 2.7 million lives could potentially be saved each year if fruit and vegetable consumption were sufficiently increased.

There is a large potential in Europe for optimization in valorization of crop biomass in the vegetable food supply chain. For example, there is occasionally overproduction of tomatoes for fresh consumption, and a fraction of the production is unsuited for fresh consumption sale (unacceptable color, shape, maturity, lesions, etc.). These tomatoes are normally destroyed and used as landfilling or animal feed, which represents an economic loss for producers and negative environmental impact. In Norway and Belgium, this surplus/waste fraction amounts to about 200 (Unpublished data from Rennesøy Tomat & Fruktpakkeri AS (2012), the biggest tomato packaging station in Norway) and 500 tons per year [5], respectively. A conservative estimate of € 4 per kg price increase for this raw material will thus yield a potential of 0.8 and 2.0 million € per year, respectively, for Norway and Belgium alone. Besides overproduction, part of the tomatoes produced in the greenhouse might not reach the market because they do not reach the local market standards. This can be due to cosmetic defects such as color, shape, size, etc.

Additionally, there is loss of fresh tomato at the retailer's level. In Norway, the general retailer loss is 10% for cluster tomatoes, 3–6% for single retail ("ordinary round") tomatoes, and approximately 1% for cherry tomatoes, but it can be substantially higher in the peak of the growing season [6]. In the case of Norway, assuming a mean loss at retailers level of 5%, and that this loss can be halved by improved market regulation, there is a value increasing potential of a further 0.8 million € per year, again with a conservative price estimate of € 4 per kg. It is assumed that the total loss fraction is approximately the same in the other European countries, implying a much higher value potential in countries producing more tomatoes than Norway, considering that Norway is indeed a small tomato producer by European standards (Supplementary Table S1). An estimate for the total European market, based on a total waste fraction assuming a 15% waste in the tomato processing industry (including what is left in the field), and 5% waste in fresh market tomato, the waste fraction amounts to 3 million metric tons per year (Supplementary Table S1). The tomato processing waste quantities worldwide in 2010 was estimated to be between 4.3 and 10.2 million metric tons [7]. Based on this, it is beneficial to develop processing technology for the best possible utilization of this resource to improve economic sustainability of tomato production.

2. Tomato Production in Norway, Belgium, Poland and Turkey

2.1. Norway

Tomato production in Norway has been stable for the last 15 years with volumes between 9000 and 12,000 tons per year. Ninety percent of the production takes place in the county of Rogaland, on the southwest coast. Practically all tomatoes produced in Norway are destined for the domestic fresh market. The production is costly because the Norwegian climate necessitates the use of heated glass houses and artificial light for year-round production. Because of the high production costs, Norwegian tomatoes have traditionally not been subject to processing. Recently, some tomato farmers have found ways to alleviate the energy expenses by innovative solutions. One example of this is the 'Miljøgartneriet' which can be translated as 'the Environmental plant nursery' [8]. This glass house was built in 2010, covers 77,000 m^2, and employs 70–85 workers in the high season. The innovations consist of amongst others the use of surplus CO_2 and warm wastewater from a nearby dairy plant for plant feed and heating, respectively. Combined with other energy-efficient solutions in the construction of the glass house and recirculation of water, the production becomes more sustainable and with a low carbon footprint. An optimized year-round cultivation system achieved a yield of over 100 kg m^{-2} in commercial production, with an estimated maximum potential of 125–140 kg m^{-2} [9]. In Norway, the surplus fraction resulting from high-season overproduction amounts to 200 tons per year, corresponding to approximately 2% of total production. Total waste, i.e., combined with the waste at the retailer level and in the greenhouses, is estimated at ~6%. One of the main reasons for waste at the retailer level was found to be due to packaging. Comparing packaged cluster tomatoes to loose, unpacked single tomatoes revealed, contrary to expectation, that the former had significantly more wastage [6]. It was speculated that this was because packaging may lead to condensation and subsequent growth of molds. Therefore, packaging of warm tomatoes should be avoided [6]. Temperature abuse during transport and in the stores was also considered as main factors leading to waste [6].

Efforts have been made to produce tomato sauce of the surplus tomatoes. However, due to small volumes and high production costs, this turned out not to be economically viable and production stopped. At present, the fraction is primarily used as cattle feed, so that costs for disposal can be minimized. In order to overcome the seasonality problem, tomato surplus and waste fractions may be sorted and stored frozen in order to collect volumes for subsequent processing and perhaps to add this to batches of imported tomatoes for processing. A project is now starting up in Norway to look into this possibility for valorization and to identify and overcome the challenges related to this strategy.

2.2. Belgium

Belgian tomato production takes predominantly place in Flanders, where some 250 growers produced 220 to 260,000 tons per year on about 500 hectares between 2006 and 2016. Tomato is the second biggest crop under glass after lettuce, but generates the biggest economic return, ~180 million € per year, leaving the second and third place to strawberry and bell pepper, respectively. Belgian tomato growers deliver tomatoes for the internal fresh market during a period of about nine months each year. There are three areas in Flanders where the tomato growers cluster together, that is around

Mechelen, Hoogstraten, and Roeselare. Roughly, half of the production is on the vine and the other half are loose tomatoes [10]. Also in Belgium, practically all tomatoes are produced in greenhouses and for fresh consumption. The average price the growers get for their tomatoes (loose and on the vine) is about € 0.75 per kg. As opposed to Norway, Belgian tomato export volumes are considerable. Today, approximately 70% of all tomatoes produced in Belgium are exported [10]. According to Lava, the cooperative of all Belgian fruit and vegetable auctions, the main exporting partners are France, Germany, the Netherlands, the UK, and the Czech Republic [11]. The tomato surplus fraction in Belgium makes up about 500 tons per year, corresponding to approximately 2% [5], as in Norway, and total waste (retailers, etc.) is estimated at approximately 5%. A recent study in Belgium estimated the losses of tomato that cannot be marketed due to cosmetic reasons to be ~1–2% only. Similar numbers were recorded for bell pepper and cucumber. This is very low compared to the percentages of other crops (e.g., zucchini 11.5% and lettuce 9.1%) [12].

2.3. Poland

Polish tomato production is different from Norwegian and Belgian production in several ways. First, the polish production volume is 3 times the Belgian and >60 times the Norwegian with a yearly production amounting to ~920,000 tons (~250,000 tons in the field and ~670,000 tons in greenhouses) [13], placing them among the top eight in Europe (Supplementary Table S1). Moreover, the production is carried out both in open field and under cover. The cultivation area under cover is approximately 27% of the total, but it is increasing [14]. Approximately 70% of the production takes place in Greater Poland, Kuyaivan-Pomeranian, Mazovian, and Switokrzyskie Provinces. Since Poland's entry into the EU in 2004, fresh tomato exports have doubled, and accounts now for about 11% of production [14]. Approximately 1/3 of the total production is processed domestically, mainly into tomato paste and canned tomatoes (approximately 40,000 tons per year), and ketchup and tomato sauce (approximately 135,000 tons per year), whereof ~50% are exported [13]. The production of greenhouse tomatoes is intended for the fresh market and nearly 80% of production is sold on the internal market. The remaining 20% is export, and the main recipients are Ukraine, Belarus, the Czech Republic, Germany and the United Kingdom. The processing waste value can be assumed to be up to approximately 8.5%. This value consists of 1–3% seed waste, 2.8–3.5% skin, and up to 2% as whole fruit waste. Measures to reduce losses like choosing correct harvest time, avoiding damage during harvest, storage of crops protected from sunlight and immediate cool storage, the removal of damaged fruit, and using clean packaging material and proper transport are also important in Poland. The results obtained in open field tomato production in Poland depend very much on weather conditions. In some years, maturation is delayed and the quality of the fruits is poor, and it is very important to protect plants from diseases. These detrimental effects can be increased through improper nitrogen fertilization, which can delay the maturity of the fruits. In addition, the growing season in some years may be shorter due to the occurrence of early autumn frosts. In these conditions, unripen tomato fruits remain in the field and is lost. To reduce losses, proper nitrogen fertilization, early varieties with concentrated fruiting, and the use of ethylene to accelerate ripening are proposed.

2.4. Turkey

Turkey is the fourth largest tomato producer after China, India and the United States, yielding more than 7.2% of the world tomato production. The production amount was ~11.8 million tons in 2014 and 12.7 million tons in 2017. Sixty-seven percent of total production was evaluated as table tomato and 33% were industrially processed. More than 25% of the total production and 40% of table tomato is cultivated in greenhouses. Three-and-a-half-million tons (~28%) of the tomato production is being processed into paste, while 500,000 tons (4%) is used as sun-dried and canned (whole peeled, cubic chopped, puree, etc.). Due to the climate advantage, sun-dried tomatoes have great potential and almost all (97%) of them are exported. Tomato is the undisputed and clear leader product of the vegetable industry in Turkey. Tomato export is almost 40% of total fresh vegetable exportation of Turkey. From 10 to 18% of the total processing raw material is gone to waste [15]. Skin, seeds, fiber, etc. make up ~7% of this fraction, and the rest is mainly due to bad transportation in the tomato paste industry. Between 2010 and 2017, the average losses during harvesting were 3.5% and loss after harvesting was 10 to 15%. In 2017, pre- and postharvest losses were more than 2.1 million metric tons, corresponding to 16.5% of the total production [16]. Occurrences of exceptional high tomato wastage of up to 28% in specific regions have been reported [17]. Measures to reduce losses are summarized as choosing correct harvest time, avoiding damage during harvest, storage of crops protected from sunlight and immediate cool storage, the removal of damaged fruit, and using clean packaging material and proper transport [18]. The following precautions were proposed in order to reduce loss between harvesting and processing or wholesale: Choosing an earlier harvest time, using better packaging material at the farm stage instead of only traditional wooden or plastic cases, and refrigerated transport to the packaging or processing facilities [19].

3. The Significance of Lycopene in Tomato

Consumers are increasingly demanding naturally nutritious and healthy products that are produced without the use of genetic modification or additives and pesticide residues. It is therefore a large potential for developing processed products based on the part of the tomato production that does not go to fresh consumption. It turns out that the willingness to pay among the modern consumer increases when positive health effects attributed to the products can be documented. Most of the adult consumers are aware of the health benefits attributed to lycopene and other phytonutrients found in tomato, and thus lycopene is the second most important driver for consumer preferences, after price [20].

Lycopene is a member of the carotenoid family of compounds and is a key intermediate in the biosynthesis of many carotenoids. Lycopene is a pigment found in small amounts in many fruits and vegetables, and which, like carotene, gives rise to red color. Tomatoes are the main source of lycopene, while chili peppers may contain comparable amounts, and watermelon, red bell pepper, carrot, spinach, guava, papaya, and grapefruit contain relatively moderate amounts [21,22]. Lycopene occurs in several forms (isomers), some of which are taken up more easily by the human body than others [22–24]. The all-*trans* form is predominating in fresh tomato (~90%) [25], whereas it is the *cis* isomer that is most easily bioavailable to the human body [26–28]. Besides being an important nutrient, lycopene is also a very potent and sought after natural colorant with many applications in industrial food processing [29].

Research has shown that by means of processing it is possible to increase the proportion of the most bioavailable forms and stabilize these to thereby provide an increased health benefit. There is evidence that heat treatment and the addition of vegetable oils in tomato products increases the body's absorption of lycopene compared with corresponding consumption of fresh tomato [25,27,30]. For lycopene to be absorbed in the duodenum, it must be dissolved in fat. The fat should not contain components which compete with lycopene for absorbing, such as vitamin E and K [25,31]. Although the biochemical mechanisms that make lycopene so beneficial to health are largely unknown, there is much to suggest that antioxidant and provitamin A properties can be crucial.

3.1. Lycopene Content in Tomato

The lycopene range (0.03–20.2 mg/100 g) as reviewed in Table 1 is comparable to original results presented by Adalid et al. [32] where 49 diverse accessions of tomato from 24 countries on four continents displayed a span from 0.04 to 27.0 mg lycopene/100 g. In some wild species of tomato (*S. pimpinellifolium*), the lycopene concentration can be as high as 40 mg/100 g [33]. As shown in Table 1, the type and variety of tomato is also crucial for lycopene. Even the origin and the geographic location of their cultivation appears to play a major role [34,35]. This is probably due to different growing conditions and the degree of maturity [36,37], storage and transport conditions, etc.

Table 1. Lycopene content in tomato varieties (converted to mg/100 g fresh weight (FW)). Literature review. Values in italics are obtained by spectrophotometry, otherwise high-performance liquid chromatography (HPLC).

Variety	Total Lycopene	Type	Origin	Growth Conditions	Reference
Ministar	3.11	plum			
Juanita	10.51	cherry	SW Norway	Greenhouse, soil free	
Dometica	4.08	salad			This study
Volna	8.15	salad	Skierniewice, Poland	Field	
Calista	10.75	processing			
Pearson	*10.77*	N/A	California, USA	Field	[38]
DX-54	*~12*	N/A	Utah, USA	Field	[39]
Unknown	15.8	N/A	Florida, USA	Unknown	[40]
Amico	7.73				
Casper	6.61				
Góbé	5.92				
Ispana	6.22				
Pollux	5.14				
Soprano	8.65				
Tenger	7.66				
Uno	7.09	processing			
Zaphyre	6.95				
Draco	6.87				
Jovanna	11.61				
K-541	9.95				
Nivo	8.46				
Simeone	9.88			Field	[41]
Sixtina	10.51				
Monika	7.22		Gödöllö, Hungary		
Delfine	6.51				
Marlyn	5.53				
Fanny	5.26				
Tiffany	6.23				
Alambra	5.40	salad			
Regulus	6.59				
Petula	6.68				
Diamina	6.48				
Brillante	8.47				
Furone	5.18				
Linda	5.69				
Early Fire	*10.1–14.0*				
Bonus	*8.5–12.7*				
Falcorosso	*8.0–11.1*	processing		Field	[42]
Korall	*8.1–11.3*				
Nívó	*9.7–15.5*				
Strombolino	*5.3–10.3*	cherry, processing	Gödöllö, Hungary	Field	[43]
12 unnamed local varieties	*5.04–13.46*	N/A			
ACE 55 VF	6.38	Flattened globe	SE Spain	Field	[44]
Marglobe	*8.46*	Round			
Marmande	*7.01*	Flattened globe			
CIDA-62	6.23	cherry			
CIDA-44A	2.95				
CIDA-59A	2.70	round	Spain	Organic, field	[45]
BGV-004123	5.50				
BGV-001020	3.66	flattened and ribbed			
Baghera	4.64	round			
CXD277	15.33				
H9661	12.21				
H9997	14.96	processing	Spain	Field	[46]
H9036	11.36				
ISI-24424	17.01				
Kalvert	16.71				
Kalvert	20.2				
Hly18	19.5	processing	Lecce, Italy	Field	[47]
Donald	9.5				
Incas	9.3				

Table 1. Cont.

Variety	Total Lycopene	Type	Origin	Growth Conditions	Reference
143	9.47				
Stevens	10.2				
Poly20	16.0				
Ontario	6.54				
Sel6	9.73				
Poly56	14.2				
1447	5.61				
977	10.0				
1513	9.21	N/A	San Marzano, Italy	Field	[48]
988	14.1				
Cayambe	13.5				
Heline	9.46				
1512	3.25				
1438	6.35				
Motelle	16.9				
Momor	13.3				
981	2.33				
Poly27	11.0				
Shasta	6.7–7.7				
H9888	8.9–9.7	Early-season varieties			
Apt410	9.1–10.0				
CXD179	9.2–10.4				
CXD254	10.5–12.0	Mid-season varieties	California, USA	Field	[49]
H8892	8.7–10.1				
CXD222	9.8–13.2				
H9665	8.7–12.2	Late-season varieties			
H9780	9.2–13.0				
Bos3155	14.92				
CXD510	15.37	Red varieties			
CXD514	11.80				
CXD276	2.47	Light color Tangerine variety	California, USA	Field	[50]
CX8400	0.08	Yellow variety			
CX8401	0.68	Orange variety			
CX8402	0.03	Green variety			
SEL-7	3.23	N/A	Haryana, India		[51]
ARTH-3	4.03				
Laura	12.20	N/A	New Jersey, USA	Greenhouse	[52]
Brigade	12.9	Processing	Salerno, Italy	N/A	[53]
PC 30956	18.7	High lycopene experimental hybrid,			
Cheers	3.7	N/A	Southern France	Greenhouse	[54]
Lemance	3.7–6.9	N/A	N/A	Greenhouse	[36]
Ohio-8245	9.93				
92-7136	7.76				
92-7025	6.46	Tomato pulp fraction	Ontario, Canada	N/A	[55]
H-9035	10.19				
CC-164	10.70				
Dasher	3.98				
Iride	4.45				
Navidad	4.89	Plum	Italy	Greenhouse	[56]
Sabor	5.22				
292	4.57				
738	4.77				
Cherubino	3.43	Cherry			
Crimson, green,	0.52	Salad	Ohio, USA	Purchased from local market	[57]
Crimson, breaker	3.84				
Crimson, red	5.09				
Unknown	10.14	Cherry	California, USA	Purchased from local supermarket	[58]
Unknown	5.98	On-the-vine			
Roma	8.98	Processing			
Jennita	1.60–5.54	Cherry	SW Norway	Greenhouse, soil free	[59] [a]
Naomi	7.1–12.0	Cherry	Sicily, Italy	Cold greenhouse	[60] [b]
Naomi	12.4–13.3	Cherry			
Ikram	8.5–8.9	Cluster	Italy	Cold greenhouse	[34]
Eroe	2.1–2.8	Salad			
Corbarino	6.8–14.6	Cherry	Battipaglia, Italy	Field grown	[61] [c]

(a) Harvested twice monthly from May to October. (b) Harvested at six different times throughout the year. (c) As an effect of N and P fertilization load.

It is well known that tomato lycopene is concentrated in the skin and the water-insoluble fraction directly beneath the skin [53]. Table 2 demonstrates this partitioning and underpins that the tomato skin waste fractions is a good source of lycopene. Since lycopene and other carotenoids are most concentrated in and just inside the skin, lycopene is often higher per volume in small tomatoes of cherry type, because they have a relatively high peel to volume ratio.

Table 2. Lycopene content in peel versus pulp in some tomato varieties (converted to mg/100 g FW). Literature review. Values in italics are obtained by spectrophotometry, otherwise HPLC.

Cultivar	Total Lycopene (Converted to mg/100 g FW)		Comment	Reference
	Peel	Pulp		
HLT-F61	89.3	28.0		
HLT-F62	50.8	16.7	Field grown, Northern Tunisia	[62]
Rio Grande	42.4	10.1		
8-2-1-2-5	14.3	6.7		
Castle Rock	13.1	6.2		
IPA3	10.2	4.0	Harvested at mature green stage	
Pb Chhuhra	8.6	4.6	(Ludhiana, India) and stored at	[37]
UC-828	6.5	3.7	20 °C until ripe	
WIR 4285	6.5	3.1		
WIR-4329	8.1	4.3		
818 cherry	14.1	6.9		
DT-2	8.1	5.2		
BR-124 cherry	10.2	4.9		
5656	10.7	4.5		
7711	9.0	4.4		
Rasmi	10.8	4.3		
Pusa Gaurav	10.2	4.0	Field grown, New Delhi, India	[63]
T56 cherry	12.0	3.8		
DTH-7	4.8	2.7		
FA-180	7.6	2.5		
FA-574	6.1	2.2		
R-144	6.2	2.0		
Grapolo	6.0	1.2		
Italian cherry tomato	7.2	2.0	Purchased in supermarket or	
Croatian cherry tomato	5.3	1.6	open-air market, Zagreb, Croatia,	[64]
Croatian large size tomato	3.5	1.3		
Turkish large size tomato	3.3	1.2		

FW: fresh weight.

Since lycopene is the pigment responsible of the red hue of tomatoes, it can be derived that unripe tomatoes and light color tangerine varieties and green, orange and yellow varieties are lower in lycopene than mature red tomatoes [47,62]. Tomatoes with lower lycopene can be stored under special light and temperature so that they may accumulate lycopene before processing. Figure 1 illustrates the correlation between maturity stage, color, and lycopene content.

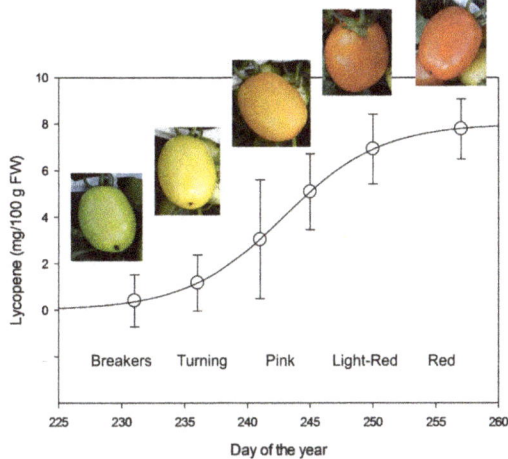

Figure 1. Lycopene evolution in processing tomatoes, cv. Calista, as measured during ripening in the field by a nondestructive optical method as previously described in Ciaccheri et al., 2019 [65]. FW: fresh weight.

Sikorska-Zimny et al., 2019 [66] proposed that, although tomatoes harvested at the full-ripe stage maintained 90% of their lycopene content for three weeks of storage, a compromise between firmness and storability may be found by harvesting at an earlier stage in order to balance the organoleptic and nutraceutical quality of the fruit. This means, at least for fresh-market tomatoes, that they can be harvested un-ripe in order to obtain storability without compromising neither on sensory or nutraceutical qualities, as long as proper storage conditions are obtained.

3.2. Effect of Processing

Processing strategies for tomatoes range from the very simple, as for fresh-market tomatoes, to complicated, as for the production of, e.g., tomato paste which includes multiple steps and several heat treatments such as drying, hot-break, and pasteurization [67]. Conclusions in relation to lycopene are that it is only slowly broken down by boiling (100 °C), and therefore constitutes no restriction for the heat treatment (Table 3) [67–71]. On the contrary, boiling for around two hours results in breakdown of carotenoid-associated protein structures so that lycopene is released, isomerization occurs and bioavailability increases [69,72]. Interestingly, Seybold et al., 2004 [70] found that lycopene isomerization occurred readily as an effect of thermal treatment in a standard lycopene solution, but this was not the case in tomatoes treated at the similar time/temperature conditions. Nevertheless, in freeze-dried lycopene powder, it was found that high temperatures (120 °C) and relatively short exposure time resulted in profound isomerization in both water and oil medium, but that loss of lycopene was significantly less in oil medium, presumably because oxidation was avoided [73,74]. Effects of thermal treatment on a range of health-beneficial antioxidants in tomato are reviewed in Capanoglu et al., 2010 [67], and it may seem that most of the other antioxidants (e.g., vitamin C and tocopherols, phenolics and flavonoids) are less heat-stable than lycopene. Mechanical and thermal treatment have significant effects on the consistency of tomatoes, the former mainly due to the release of pectin [75]. Mechanical treatment does not seem to affect the content of lycopene to any significant degree, but it may enhance bioavailability, especially when combined with thermal treatment [75]. Factors such as light, pH, and temperature is very critical to the stability of lycopene and carotenes [76]. Wrong processing or storage (i.e., exposure to light and oxygen) may, therefore, affect the ratio between isomers or totally degrade the beneficial compounds. However, when optimal storage criteria are met, lycopene is a very stable molecule [77]. Traditional processing methods have only little effect on the level of lycopene or isomerization [25]. In fact, thermal processing may generally increase the bioavailability of lycopene despite decrease of the total concentration of lycopene [27,57]. Studies that have followed the evolution of lycopene through the different processing steps of commercial tomato paste production are inconclusive, either reporting a small increase [41] or a significant decrease [78] as the tomatoes are processed into paste. Comparing rapid industrial scale continuous flow microwave pasteurization to conventional thermal processing of tomato juice, revealed that this novel energy-efficient technology resulted in a product with a higher antioxidant capacity and similar organoleptic, physiochemical and microbiological qualities [79]. High pressure processing (HPP) may increase lycopene extractability compared to conventional processing and result in higher carotenoid content, including lycopene, in tomato purées [80,81]. HPP also results in less lycopene cis-isomers compared to thermal processing [82].

Table 3. Effects of heat treatment on tomato products.

Processing	Heat Treatment	Effect	Texture	Taste	Lycopene	Color
Chopping raw	Mild < 80 °C	Enzymes are released, pectin degraded and hexanal/hexanol formed	Thick before heating, then soup	Vivid Green	Unchanged	Poor, controlled by pH
	Strong > 80 °C			Moderate green	Increased	Acceptable
Chopping raw, waiting for thickening before cooking, for example 2 h	Instantly to 100 °C, medium shortly, to thicken		Slightly thick and thickens with increased cooking time	Vivid Green	Somewhat increased	Acceptable
Chopping cooked	Mild < 80 °C	Enzymes inactivated only partially	Thin	Moderate green	Unchanged	Poor, controlled by pH
	Strong > 80 °C	Enzymes inactivated	Thick	Green aroma	Unknown	Acceptable
Puree, unpeeled	2 h 100 °C	Carotenoid content maximum after 2 h.	Unknown	A little green	Most after 2 h *	Most after 2 h
Puree, peeled		Carotenoid content low and stable unaffected by time			Less than unpeeled	Less than unpeeled

* Carotenoid associated protein structures are broken down so that lycopene is released and isomerization occurs so that the bioavailability increases.

Regarding lycopene and processing, the challenge is to limit the breakdown and stimulate the desired isomerization. In order to optimize the contents of isomerized lycopene, the kinetics of both isomerization and breakdown have to be known for the specific process. Experiments including a high number of time/temperature combinations should be done for a number of situations, e.g., aerobe vs. anaerobe processing and at different pH. It is only highly concentrated (e.g., dried powder) and, to a certain extent, concentrated and sterilized (canned) products that generally exhibit enlarged lycopene concentration compared to fresh tomato (Tables 4 and 5). However, a heavy heat treatment is very energy intensive, and often leads to undesirable sensory properties.

Table 4. Lycopene content in tomato products (converted to mg/100 g FW). Literature review. Values in italics are obtained by spectrophotometry, otherwise HPLC.

Product	Total Lycopene	Comment	Reference
Pulp	10.6–18.7	Commercial products, Salerno, Italy	[53]
Purée	12.7–19.6		
Paste	57.87	Commercial products, California, USA	[58]
Purée	23.46		
Juice	10.33		
Ketchup	12.26–14.69		
Juice, heat concentrated	2.34	Experimentally processed from Crimson-type tomatoes purchased from local markets, Ohio, USA	[57]
Paste, heat concentrated	9.93		
Soup, retorted	10.72		
Sauce, retorted	10.22		

Table 4. *Cont.*

Product	Total Lycopene	Comment	Reference
Juice	7.83	Experimentally processed from tomatoes purchased from local markets and heat treated according to standardized industrial food processing requirements	[25]
Soup, condensed	7.99		
Canned whole tomato	11.21		
Canned pizza sauce	12.71		
Paste	30.07		
Powder, spray dried	126.49		
Powder, sun dried	112.63		
Sun dried in oil	46.50		
Ketchup	13.44		
Tangerine tomato sauce	4.86	Experimentally processed from tomatoes grown at the Ohio State University, USA	[30]
Tangerine tomato juice	2.19		
Red tomato juice	7.63		
Regular salad tomatoes, Gran Canaria, Spain	1.15	Fresh	[69]
	1.09	Boiled 10 min	
	0.99–1.18	LTLT, 60 °C 40 min	
	1.07–1.23	HTST, 90 °C 4 min	
Bella Donna on the vine, Netherlands	3.80	Fresh	
	3.06	Boiled 20 min	
	3.91–4.31	LTLT, 60 °C 40 min	
	3.43–4.15	HTST, 90 °C 10 min	
Daniella, Spain	2.37	Fresh purée	[81]
Daniella, Spain	0.99	Fresh	[80]
	1.48	HP (400 MPa, 25 °C, 15 min)	
	0.86	Pasteurization (70 °C, 30 s)	
	0.95	Pasteurization (90 °C, 60 s)	
Torrito, Spain	39.67	Fresh	This study
	26.39	HTST (90 °C, 15 min)	
	23.77	HP (400 MPa, 90 °C, 15 min)	
Torrito, the Netherlands	11.44	Fresh	
	7.57	HTST (90 °C, 15 min)	
	10.10	HP (400 MPa, 90 °C, 15 min)	
	10.00	HP (400 MPa, 20 °C, 15 min)	
	5.41	HP (600 MPa, 90 °C, 15 min)	
	4.08	HP (600 MPa, 20 °C, 15 min)	
Heinz purée, USA	6.62	Puré, fresh	[83]
	6.61	Boiled 5 min	
	6.57	Boiled 10 min	
	6.48	Boiled 30 min	
	6.39	Boiled 60 min	
Double concentrated commercial canned tomato purée, Netherlands	39	Unheated	[68]
	31	Autoclaved 100 °C, 20 min	
	29	Autoclaved 100 °C, 60 min	
	29	Autoclaved 100 °C, 120 min	
	28	Autoclaved 120 °C, 20 min	
	30	Autoclaved 120 °C, 60 min	
	29	Autoclaved 120 °C, 120 min	
	31	Autoclaved 135 °C, 20 min	
	33	Autoclaved 135 °C, 60 min	
	32	Autoclaved 135 °C, 120 min	
Experimental purée	3.79	Unheated	[84]
	5.93	Steam retorted, 90 °C, 110 min	
	5.20	Steam retorted, 100 °C, 11 min	
	4.74	Steam retorted, 110 °C, 1.1 min	
	3.37	Steam retorted, 120 °C, 0.11 min	
FG99-218, USA	16.04	Juice, fresh	[85]
	16.05	Juice, hot break	
	17.95	Juice, HP (700 Mpa/45 °C/10 min)	
	17.12	Juice, HP (600 Mpa/100 °C/10 min)	
	15.50	Juice, TP (100 °C/35 min)	
OX325, USA	9.84	Juice, fresh	
	10.22	Juice, hot break	
	10.88	Juice, HP (700 Mpa/45 °C/10 min)	
	10.29	Juice, HP (600 Mpa/100 °C/10 min)	
	8.49	Juice, TP (100 °C/35 min)	

LTLT: Low Temperature, Long Time; HTST: High Temperature, Short Time; HP: High Pressure processing.

Table 5. Lycopene content (mg/100 g FW) in tomato and tomato products including the fractions of *trans*- and *cis*-isomers.

Produkt	Total Lycopene	All *Trans* Lycopene (% of Total)	*Cis* Lycopene (% of Total)	Reference
Conesa tomato paste, Spain, 0.16% fat Batch 1 (2014)	32.1	29.2 (91.0)	2.9 (9.0)	
Conesa tomato paste Spain, 0.16% fat Batch 2 (2015)	26.6	23.6 (88.7)	3.0 (11.3)	
Conesa tomato paste Spain, 0.16% fat Batch 2 (2015)—Autoclaved	22.8	19.9 (87.3)	2.9 (12.7)	This study
Conesa tomato paste Spain, 0.16% fat Batch 2 (2015)—Microwaved	22.9	20.1 (87.8)	2.8 (12.2)	
Conesa tomato fine chopped, Spain, 0.04% fat	6.5	5.9 (90.8)	0.6 (9.2)	
Heinz ketchup 0.1% fat	11.0	9.4 (85.5)	1.6 (14.5)	
Eldorado tomato puree, Italy, 1% fat	32.1	29.4 (91.6)	2.7 (8.4)	
Cherry tomatoes	10.14	8.91 (87.9)	1.23 (12.1)	
On-the-vine tomatoes	5.98	5.00 (83.6)	0.98 (16.4)	
Roma tomatoes	8.98	7.88 (87.7)	1.10 (12.3)	
Tomato paste	57.87	45.94 (79.4)	11.93 (20.6)	[58]
Tomato purée	23.46	17.85 (76.1)	5.61 (23.9)	
Tomato juice	10.33	8.47 (82.0)	1.86 (18.0)	
Tomato ketchup	12.26–14.69	9.40–9.47 (64.4–76.7)	2.86–5.22(23.3–35.6)	

4. Utilization of Tomato Side Streams and By-Products

The valorization strategies for tomato waste biomass may be different depending on whether the primary production is originally intended for the fresh market or for industrial processing. For the former case, the biomass may mainly consist of surplus tomato due to seasonal overproduction or fractions perceived as unmarketable for cosmetics reasons, and in the latter of side streams and byproducts from the processing. Thus, the remainder of this chapter is divided into 'Fresh tomato' and 'Processing'. However, the strategies described are not understood to be necessarily fixed in these categories, and can be interchanged (i.e., postharvest ripening can also be applied for processing tomatoes). Nevertheless, chosen strategies will depend on the available technologies and the volumes of the available biomass, and type of by-product/side stream fraction, which varies considerably in the countries subject to this case study.

4.1. Fresh Tomato

In Norway and Belgium, as mentioned above, domestically grown tomatoes are at present predominantly meant for the fresh market. Hence, processing by-products and side streams is not a big issue. However, mainly due to seasonal overproduction and, to a lesser extent, that a fraction of the tomatoes is not suitable for fresh market sale (wrong color, maturity level, shape, and injuries), there have been attempts to develop processing technology for this fraction. The valorization of this biomass is currently mainly impeded by the high moisture content and corresponding fast decay. The small volumes, geographical dispersity, and the seasonality make it even more challenging to process by conventional processing technologies. Alternatively, flexible and mobile processing technologies may be looked upon to valorize the underutilized tomato biomass. An example is the proposed novel spiral-filter press technology to refine horticultural by-products including tomato [86]. This technology alleviates the need of stabilizing the biomass by using expensive drying technology, and besides, it is flexible and may be used to produce a range of volumes as well as handle a multitude of different textures [87]. This implies that it may be used for, e.g., apple, berries, and carrot processing after the high-season tomato processing is over.

Regards the surplus tomato fraction that is predominantly made up of unripe or underpigmented tomatoes, research has shown that these tomatoes can be turned into marketable tomatoes very effectively by simple means. In the SusFood1 era-net project 'SUNNIVA' [88], a range of elicitor treatments were tested in postharvest trials to identify efficient elicitor treatments as tools to influence the content of health-beneficial phytochemicals (HBPC) in tomato raw material and waste fractions.

Products both for industry use and fresh market use were targeted. Results showed that the waste fractions of tomato could be utilized as valuable sources of HBPC, and also provide better raw material utilization when subjected to efficient postharvest elicitor treatments. Among the most promising elicitor treatments for tomatoes were ethylene treatments for pink and waste fractions (Figure 2). An important point of attention to maximize health benefits of industrial tomato products as well as tomatoes for fresh consumption is that different types or cultivars of tomatoes reach their maximum level of the HBPC at different maturity stages.

Figure 2. Example of ethylene treatment. Examples of Calista (**a**) and Volna (**c**) varieties that were not ripe at the time of harvesting and the respective varieties after six days of storage under ethylene atmosphere (**b**,**d**). Adapted from Grzegorzewska et al., 2017 [89], with permission.

Studies have shown that hormic dosage of ultraviolet radiation (UV-C) can be applied to delay the senescence of fruit and vegetables, suggesting that photochemical treatment may have the potential for postharvest preservation of tomato [90]. The effects of UV-C and temperature on postharvest preservation of tomato are summarized in Tjøstheim, 2011 [91], and long-term controlled atmosphere and temperature storage in Batu, 2003 [92] and Dominguez et al., 2016 [93]. In short, temperatures from 12.8 to 15 °C appear to be optimal, but there are large variations between different cultivars. An example of postharvest lycopene evolution in pink tomatoes at different storage temperatures is shown in Figure 3.

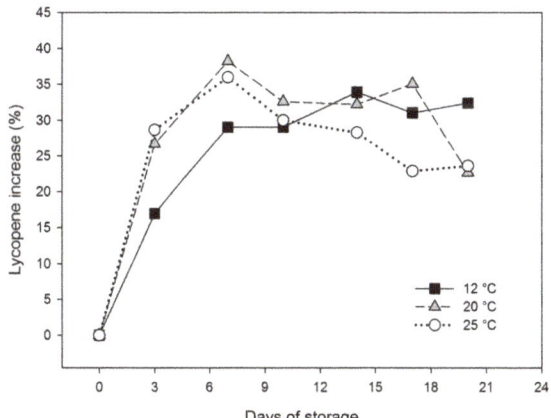

Figure 3. Lycopene increase (%) in pink harvested processing tomatoes, cv. Calista, during storage in the dark at 3 different temperatures (12, 20, and 25 °C) and 80% relative humidity. The initial level of lycopene was about 6.5 mg/100g fresh weight (FW). Rearranged from data previously published in Sikorska-Zimny et al., 2019 [66], with permission.

A completely different way of valorizing the fraction of tomatoes in the sub-optimal food (SOF) category, i.e., tomatoes with a color or shape that may be considered undesirable, is to target the consumers and try to get them more aware of the consequences of food waste. Consumers appear receptive to discounts on vegetables with imperfections [94]. Since October 2013, under its own brand "Wunderlinge" (translated as 'odds') such fruit and vegetables have been offered in Austria, and similar actions rapidly spread to neighboring countries [95]. Depending on season, and what is available, these fruits and vegetables, which, despite their idiosyncratic appearance is flawless in taste, are offered at a cheaper price. Similar campaigns and the establishments of 'food-banks' is becoming more common throughout Europe, but many actions are still at an experimental stage.

Then there will still be left fractions that are not suitable for recycling into the food chain. Upon extraction, both tomato fruit waste and vegetative by-products may be utilized as sources of compounds with pharmaceutical and therapeutic benefits (e.g., phenolic compounds like quercetins, kaempferol, and apigenin) or cosmetics ingredients (e.g., lactic acids) [96]. Side-flows and waste from vegetable processing can also be recirculated back to the field in the form of compost and used as growth substrates. Tomato waste compost may be used to replace partially peat-based substrate used for vegetable transplants production in nurseries [97]. Tomato side streams may also be used as raw material for the production of organic fertilizer or soil amendment. However, more research is needed to document the bio-stimulating effect of tomato waste streams for its potential use as an input source for such products [88].

4.2. Processing

During tomato processing, three to seven percent of the raw material is lost as waste [7,98]. The press cake resulting from tomato juice and sauce production consists of skin and seeds [99]. The seeds constitute approximately 10% of the fruit and 60% of the total waste, and is a source of protein and fat [100].

The chemical composition of tomato processing waste fractions was characterized by Al-Wandawi et al., 1985 [101]. The seed fraction was rich in oleic and palmitic acids, a high protein content with threonine and lysine as the dominating amino acids, and K, Mg, Na, and Ca as the dominating elements. Whereas the skin fraction was also rich in proteins with lysine, valine, and leucine as the predominating essential amino acids, and Ca, K, Na, and Mg as the major elements [101].

Pure lycopene has traditionally been extracted from tomatoes through processes using chemical solvents. Innovative supercritical fluid extraction (SFE) methods do not leave behind the chemical residues associated with other forms of lycopene extraction and were demonstrated by researchers at the University of Florida to be very efficient and with a greater yield than conventional methods [102]. Supercritical CO_2 extraction using ethanol as a solvent is an efficient method to recover lycopene and β-carotene from tomato skin by-products [103]. Lenucci et al., 2015 [104] performed studies on enzymatic treatment of tomato biomass prior to supercritical CO_2 extraction of lycopene, and the results showed that the enzymatic pretreatment could increase the yield of lycopene extraction by 153% as compared to solvent extraction. Besides enzymatic pretreatment, ultrasound and microwave-assisted extraction methods, on their own or combined, have been developed for the extraction of lycopene, resulting in higher extraction yield. Lianfu & Zelong, 2008 [105] compared combined ultrasound/microwave-assisted extraction (UMEAE) and ultrasonic assisted extraction of lycopene from tomato paste and achieved a yield of 97.4% and 89.4% for UMAE and UAE, respectively. UMEAE has thus shown to be highly effective and may also provide rapid extraction (367 s in the mentioned study [105]). The use of UAE was reviewed by Chemat et al., 2017 [106], who concluded that the process can produce extracts in concentrate form, free from any residual solvents, contaminants, or artifacts, and one of the most promising hybrid techniques is UMAE. Supercritical CO_2 extraction has recently been optimized by modelling and resulted in a lycopene yield of 1.32 mg of extract per kg of raw material obtained by a peel/seed ratio of 70/30 [107], opening for a very promising future. Similarly, the use of pulsed electric fields (PEF) to improve carotenoid extraction from tomato was demonstrated [108]. The recent developments in carotenoid extraction methods was recently reviewed by Saini & Keum, 2018 [109], comparing enzyme-assisted extraction to the methods mentioned above and Soxhlet extraction.

Lycopene is a high-value compound, costing on the order of 2000 €/kg in its pure form. Nevertheless, at a 10 mg lycopene per 100 g FW basis, it takes 10,000 kg of tomato fruit to produce 1 kg of pure lycopene even at 100% yield. Hence, in order to be economically sustainable, large volumes of the tomato raw material are needed. Consequently, this would hardly be feasible in Norway and Belgium, but may be proposed as a viable valorization strategy in countries like Poland and Turkey. Some strategies for valorization of tomato side streams and by-products are exemplified in Table 6.

Table 6. Proposed utilization of tomato side streams and by-products from food processing.

Product	Active Ingredients	Fraction	Reference
Color pigments, Antioxidants	Lycopene	Skin, pomace, whole fruit	[72,102,110]
Tomato seed oil	Unsaturated fatty acids (linoleic acid)	Seeds	[111,112]
Thickening agent	Pectin	Dried Pomace	[113,114]
Comminuted and vegetarian sausages	Dried and bleached tomato pomace	Dried Pomace	[115]
Tomato seed meals	Protein, polyphenols, etc.	Seeds, pomace	[116]
Nutrient supplements	Vitamin B12	Pomace	[117]
Cosmetics	Phenolic compounds, antioxidants, lactic acid, etc.	Whole plant	[96]
Compost, growth substrates, fertilizer	Phytochemicals	Whole plant	[97]

5. Conclusions

Tomato side streams, by-products and surplus fractions are underutilized resources estimated to amount to in excess of 3 million metric tons per year in Europe. The ratios of this biomass that is consisting of whole fruit versus the processing side-streams are largely unknown. However, in regions where production is largely dependent on greenhouse production for fresh market sale due to climatic preconditions (e.g., Norway and Belgium), the fraction is predominantly whole fruit, and the opposite is the case where tomato processing constitutes a larger industry (e.g., Poland and Turkey). For the former case, strategies to prolong the postharvest storability of the fruit by, e.g., controlled atmosphere, elicitor, light, and temperature to overcome surplus tomatoes due to seasonal over production for the fresh market may be proposed, combined with novel, sustainable, low-energy, flexible processing technologies. For the latter case, where volumes of the side fractions make more sophisticated and targeted technologies economically and environmentally sustainable, several options for bioeconomical valorization exists, including utilization of by-products in comminuted hybrid and vegetarian food items, and the extraction of valuable health-beneficial compounds for the production of functional ingredients, protein-dense meals, and nutrient supplements. Strategies for utilization of inedible fractions, including the vegetative parts of the tomato plants may be found in the production of organic fertilizers, biobased materials such as paper, fiberboard, or extracts used in the cosmetics industries.

The notion that modern consumers are becoming more aware of the health beneficial properties of tomato and tomato products, and lycopene in particular, should not go unnoticed. Cultivation and processing practices may be further designed to meet consumer demands and preferences related to health and nutritional issues, and consequently add value to the tomato supply chain, also through the fabrication of functional and nutraceutical ingredients from biomass traditionally considered as waste.

Supplementary Materials: The following are available online at http://www.mdpi.com/2304-8158/8/7/229/s1, Table S1: Tomato production Europe.

Author Contributions: T.L. performed the literature review and wrote the original draft. B.V.D., E.C.E., S.K., G.A., M.V., and D.S. provided critical feedback on the draft, edited the text, and helped compile figures and tables. All authors contributed to the final version of the manuscript, reviewed the final version, and approved the submission.

Funding: This research was funded by the Research Council of Norway (RCN), grant numbers 238207, 284235, and 255613. Nofima funded the APC.

Acknowledgments: This study was financed through the two SusFood era-net projects SUNNIVA and InProVe and RCN project BIOFRESH. T.L. and D.S. acknowledge funding from RCN (project no. 238207 and 284235). B.V.D. acknowledges funding from the Agency for Innovation by Science and Technology (IWT)—Belgium. E.C.E. acknowledges funding from the General Directory of Agricultural Research and Policies (GDAR)—Republic of

Turkey. S.K. acknowledges funding from the National Centre for Research and Development (NCBiR)—Poland. G.A. acknowledges funding from the Daniel and Nina Carasso Foundation (DNCF) and the Ministry of Agricultural, Food and Forestry Policies (MiPAAF)—Italy. M.V. acknowledges funding from RCN (project No. 255613). The authors wish to thank Rune Slimestad (NIBIO) and Romain Larbat (INRA, Nancy) for conducting HPLC analysis of lycopene, and the remaining SUNNIVA consortium for fruitful collaboration.

Conflicts of Interest: The authors declare no conflict of interest.

References

1. *Total Overview 2004–2014; Fresh Fruit, Berries, Vegetables, and Potatoes*; Opplysningskontoret for Frukt og Grønt: OSLO, Norway, 2015.
2. Tomato Production and Consumption by Country. Available online: Http://Chartsbin.Com/View/32687 (accessed on 30 May 2019).
3. World Health Organization. Diet, nutrition and the prevention of chronic diseases. Report of a joint FAO/WHO expert consultation. In *WHO Technical Report Series*; World Health Organization: Geneva, Switzerland, 2003.
4. World Health Organization. *The World Health Report 2002; Reducing Risks, Promoting Healthy Life*; World Health Organization: Geneva, Switzerland, 2002.
5. Kips, L.; Van Droogenbroeck, B. *Valorisation of Vegetable and Fruit by-Products and Waste Fractions: Bottlenecks and Opportunities*; Instituut voor Landbouw, Visserij-en Voedingsonderzoek: Melle, Belgium, 2014.
6. Vold, M.; Møller, H.; Hanssen, O.J.; Langvik, T.Å. *Value-Chain-Anchored Expertise in Packaging/Distribution of Food Products*; Stiftelsen Østfoldforskning: Fredrikstad, Norway, 2006.
7. Van Dyk, S.J.; Gama, R.; Morrison, D.; Swart, S.; Pletschke, B.I. Food processing waste: Problems, current management and prospects for utilisation of the lignocellulose component through enzyme synergistic degradation. *Renew. Sustain. Energy Rev.* **2013**, *26*, 521–531. [CrossRef]
8. Miljøgartneriet. Available online: http://miljogartneriet.no/hjem (accessed on 30 May 2019).
9. Verheul, M.; Maessen, H.; Grimstad, O.S. Optimizing a year round cultivation system of tomato under artificial light. *Acta Hortic.* **2012**, *956*, 389–394. [CrossRef]
10. Belgian Tomatoes' Strong Focus on Exports. Eurofresh. 2016. Available online: https://www.eurofresh-distribution.com/news/belgian-tomatoes%E2%80%99-strong-focus-exports (accessed on 30 May 2019).
11. Lava. Available online: https://www.lava.be/ (accessed on 30 May 2019).
12. Gellynck, X.; Lambrecht, E.; de Pelsmaeker, S.; Vandenhaute, H. *The Impact of Cosmetic Quality Standards on Food Loss—Case Study of the Flemish Fruit and Vegetable Sector*; Department of Agriculture and Fisheries: Flandern, Belgium, 2017.
13. Statistics Poland. *Statistical Yearbook of Agriculture-2018*; Rozkrut, D., Ed.; Agricultural Department: Warsaw, Poland, 2018.
14. Greenhouse Tomatoes: A Greater Role on the Polish Scene. *Eurofresh*. 2016. Available online: https://www.eurofresh-distribution.com/news/greenhouse-tomatoes-greater-role-polish-scene (accessed on 30 May 2019).
15. Sarisacli, E. *Tomato Paste*; The Prime Ministry Foreign Trade Undersecretariat of the Turkey Republic: Ankara, Turkey, 2007.
16. TUIK. Crop Products Balance Sheet, Vegetables. 2000/01–2017/18. Available online: www.tuik.gov.tr (accessed on 29 May 2019).
17. Tatlidil, F.; Kiral, T.; Gunes, A.; Demir, K.; Erdemir, G.; Fidan, H.; Demirci, F.; Erdogan, C.; Akturk, D. *Economic Analysis of Crop Losses during Pre-Harvest and Harvest Periods in Tomato Production in the Ayas and Nallihan Districts of the Ankara Province*; TUBITAK-TARP: Ankara, Turkey, 2003.
18. Celikel, F. Maintaining the postharvest quality of organic horticultural crops. *Turk. J. Agric. Food Sci. Technol.* **2018**, *6*, 175–182.
19. Buyukbay, O.E.; Uzunoz, M.; Bal, H.S.G. Post-Harvest Losses in Tomato and Fresh Bean Production in Tokat Province of Turkey. *Sci. Res. Essays* **2011**, *6*, 1656–1666.
20. Simonne, H.A.; Behe, B.K.; Marshall, M.M. Consumers Prefer Low-Priced and High-Lycopene-Content Fresh-Market Tomatoes. *HortTechnology* **2006**, *16*, 674–681. [CrossRef]
21. Fadupin, T.G.; Osadola, O.T.; Atinmo, T. Lycopene Content of Selected Tomato Based Products, Fruits and Vegetables Consumed in South Western Nigeria. *Afr. J. Biomed. Res.* **2012**, *15*, 187–191.
22. Nguyen, M.; Schwartz, S.J. Lycopene: Chemical and Biological Properties. *Food Technol.* **1999**, *53*, 38–45.

23. Kin-Weng, K.W.; Khoo, H.E.; Prasad, K.N.; Ismail, A.; Tan, C.P.; Rajab, N.F. Revealing the Power of the Natural Red Pigment Lycopene. *Molecules* **2010**, *15*, 959–987.
24. Agarwal, S.; Rao, A.V. Tomato Lycopene and Low Density Lipoprotein Oxidation: A Human Dietary Intervention Study. *Lipids* **1998**, *33*, 981–984. [CrossRef]
25. Nguyen, M.; Schwartz, S.J. Lycopene Stability During Food Processing. *Proc. Soc. Exp. Biol. Med.* **1998**, *218*, 101–105. [CrossRef]
26. Story, N.E.; Kopec, R.E.; Schwartz, S.J.; Harris, G.K. An Update on the Health Effects of Tomato Lycopene. *Annu. Rev. Food Sci. Technol.* **2010**, *1*, 187–210. [CrossRef] [PubMed]
27. Stahl, W.; Sies, H. Uptake of Lycopene and Its Geometrical-Isomers Is Greater from Heat-Processed Than from Unprocessed Tomato Juice in Humans. *J. Nutr.* **1992**, *122*, 2161–2166. [CrossRef] [PubMed]
28. Cooperstone, L.J.; Ralston, R.A.; Riedl, K.M.; Haufe, T.C.; Schweiggert, R.M.; King, S.A.; Timmers, C.D.; Francis, D.M.; Lesinski, G.B.; Schwartz, S.J. Enhanced Bioavailability of Lycopene When Consumed as Cis-Isomers from Tangerine Compared to Red Tomato Juice, a Randomized, Cross-over Clinical Trial. *Mol. Nutr. Food Res.* **2015**, *59*, 658–669. [CrossRef] [PubMed]
29. Papaioannou, H.E.; Liakopoulou-Kyriakides, M.; Karabelas, A.J. Natural Origin Lycopene and Its Green Downstream Processing. *Crit. Rev. Food Sci. Nutr.* **2016**, *56*, 686–709. [CrossRef]
30. Cooperstone, L.J.; Francis, D.M.; Schwartz, S.J. Thermal Processing Differentially Affects Lycopene and Other Carotenoids in Cis-Lycopene Containing, Tangerine Tomatoes. *Food Chem.* **2016**, *210*, 466–472. [CrossRef]
31. Graziani, G.; Pernice, R.; Lanzuise, S.; Vitaglione, P.; Anese, M.; Fogliano, V. Effect of Peeling and Heating on Carotenoid Content and Antioxidant Activity of Tomato and Tomato-Virgin Olive Oil Systems. *Eur. Food Res. Technol.* **2003**, *216*, 116–121. [CrossRef]
32. Adalid, M.A.; Rosello, S.; Nuez, F. Evaluation and Selection of Tomato Accessions (*Solanum* Section *Lycopersicon*) for Content of Lycopene, Beta-Carotene and Ascorbic Acid. *J. Food Compos. Anal.* **2010**, *23*, 613–618. [CrossRef]
33. Porter, W.J.; Lincoln, R.E. *Lycopersicon* Selections Containing a High Content of Carotenes and Colorless Polyenes. II. The Mechanism of Carotene Biosynthesis. *Arch. Biochem.* **1950**, *27*, 390–403.
34. Zanfini, A.; Dreassi, E.; la Rosa, C.; D'Addario, C.; Corti, P. Quantitative Variations of the Main Carotenoids in Italian Tomatoes in Relation to Geographic Location, Harvest Time, Varieties and Ripening Stage. *Ital. J. Food Sci.* **2007**, *19*, 181–190.
35. Aherne, S.A.; Jiwan, M.A.; Daly, T.; O'Brien, N.M. Geographical Location has Greater Impact on Carotenoid Content and Bioaccessibility from Tomatoes than Variety. *Plant Foods Hum. Nutr.* **2009**, *64*, 250–256. [CrossRef]
36. Brandt, S.; Pék, Z.; Barna, É.; Lugasi, A.; Helyes, L. Lycopene content and colour of ripening tomatoes as affected by environmental conditions. *J. Sci. Food Agric.* **2006**, *86*, 568–572. [CrossRef]
37. Kaur, D.; Sharma, R.; Wani, A.A.; Gill, B.S.; Sogi, D. Physicochemical Changes in Seven Tomato (*Lycopersicon esculentum*) Cultivars During Ripening. *Int. J. Food Prop.* **2006**, *9*, 747–757. [CrossRef]
38. Yamaguchi, M.; Howard, F.D.; Luh, B.S.; Leonard, S.J. Effect of Ripeness and Harvest Dates on the Quality and Composition of Fresh Canning Tomatoes. *Proc. Am. Soc. Horticult. Sci.* **1960**, *76*, 560–567.
39. Wu, M.T.; Jadhav, S.J.; Salunkhe, O.K. Effects of Sub-atmospheric pressure storage on ripening of tomato fruits. *J. Food Sci.* **1972**, *37*, 952–956. [CrossRef]
40. Sadler, G.; Davis, J.; Dezman, D. Rapid Extraction of Lycopene and B-Carotene from Reconstituted Tomato Paste and Pink Grapefruit Homogenates. *J. Food Sci.* **1990**, *55*, 1460–1461. [CrossRef]
41. Abushita, A.A.; Daood, H.G.; Biacs, P.A. Change in Carotenoids and Antioxidant Vitamins in Tomato as a Function of Varietal and Technological Factors. *J. Agric. Food Chem.* **2000**, *48*, 2075–2081. [CrossRef] [PubMed]
42. Helyes, L.; Pék, Z.; Brandt, S.; Lugasi, A. Analysis of Antioxidant Compounds and Hydroxymethylfurfural in Processing Tomato Cultivars. *HortTechnology* **2006**, *16*, 615–619. [CrossRef]
43. Pék, Z.; Szuvandzsiev, P.; Daood, H.; Neményi, A.; Helyes, L. Effect of irrigation on yield parameters and antioxidant profiles of processing cherry tomato. *Open Life Sci.* **2014**, *9*, 383–395. [CrossRef]
44. Gómez, R.; Costa, J.; Amo, M.; Alvarruiz, A.; Picazo, M.; Pardo, J.E. Physicochemical and Colorimetric Evaluation of Local Varieties of Tomato Grown in SE Spain. *J. Sci. Food Agric.* **2001**, *81*, 1101–1105.

45. Gonzalez-Cebrino, F.; Lozano, M.; Ayuso, M.C.; Bernalte, M.J.; Vidal-Aragon, M.C.; González-Gómez, D. Characterization of traditional tomato varieties grown in organic conditions. *Span. J. Agric. Res.* **2011**, *9*, 444. [CrossRef]
46. Lahoz, I.; Leiva-Brondo, M.; Martí, R.; Macua, J.I.; Campillo, C.; Roselló, S.; Cebolla-Cornejo, J. Influence of high lycopene varieties and organic farming on the production and quality of processing tomato. *Sci. Hortic.* **2016**, *204*, 128–137. [CrossRef]
47. Lenucci, S.M.; Caccioppola, A.; Durante, M.; Serrone, L.; de Caroli, M.; Piro, G.; Dalessandro, G. Carotenoid Content During Tomato (*Solanum Lycopersicum* L.) Fruit Ripening in Traditional and High-Pigment Cultivars. *Ital. J. Food Sci.* **2009**, *21*, 461–472.
48. Frusciante, L.; Carli, P.; Ercolano, M.R.; Pernice, R.; Di Matteo, A.; Fogliano, V.; Pellegrini, N. Antioxidant nutritional quality of tomato. *Mol. Nutr. Food Res.* **2007**, *51*, 609–617. [CrossRef] [PubMed]
49. Akbudak, B. Effects of harvest time on the quality attributes of processed and non-processed tomato varieties. *Int. J. Food Sci. Technol.* **2010**, *45*, 334–343. [CrossRef]
50. Akbudak, B. Comparison of Quality Characteristics of Fresh and Processing Tomato. *J. Food Agric. Environ.* **2012**, *10*, 133–136.
51. Gupta, A.; Kawatra, A.; Sehgal, S. Physical-Chemical Properties and Nutritional Evaluation of Newly Developed Tomato Genotypes. *Afr. J. Food Sci. Technol.* **2010**, *2*, 167–172.
52. Arias, R.; Lee, T.-C.; Logendra, L.; Janes, H. Correlation of Lycopene Measured by HPLC with the L*, a*, b* Color Readings of a Hydroponic Tomato and the Relationship of Maturity with Color and Lycopene Content. *J. Agric. Food Chem.* **2000**, *48*, 1697–1702. [CrossRef] [PubMed]
53. Giovanelli, G.; Pagliarini, E. Antioxidant Composition of Tomato Products Typically Consumed in Italy. *Ital. J. Food Sci.* **2009**, *21*, 305–316.
54. Georgé, S.; Tourniaire, F.; Gautier, H.; Goupy, P.; Rock, E.; Caris-Veyrat, C. Changes in the Contents of Carotenoids, Phenolic Compounds and Vitamin C During Technical Processing and Lyophilisation of Red and Yellow Tomatoes. *Food Chem.* **2011**, *124*, 1603–1611. [CrossRef]
55. Sharma, K.S.; LeMaguer, M. Lycopene in Tomatoes and Tomato Pulp Fraction. *Ital. J. Food Sci.* **1996**, *8*, 107–113.
56. Muratore, G.; Licciardello, F.; Maccarone, E. Evaluation of the Chemical Quality of a New Type of Small-Sized Tomato Cultivar, the Plum Tomato (*Lycopersicon lycopersicum*). *Ital. J. Food Sci.* **2005**, *17*, 75–81.
57. Nguyen, M.; Francis, D.; Schwartz, S. Thermal isomerisation susceptibility of carotenoids in different tomato varieties. *J. Sci. Food Agric.* **2001**, *81*, 910–917. [CrossRef]
58. Toma, R.B.; Frank, G.C.; Nakayama, K.; Tawfik, E. Lycopene content in raw tomato varieties and tomato products. *J. Foodserv.* **2008**, *19*, 127–132. [CrossRef]
59. Slimestad, R.; Verheul, M.J. Seasonal Variations in the Level of Plant Constituents in Greenhouse Production of Cherry Tomatoes. *J. Agric. Food Chem.* **2005**, *53*, 3114–3119. [CrossRef] [PubMed]
60. Raffo, A.; La Malfa, G.; Fogliano, V.; Maiani, G.; Quaglia, G. Seasonal variations in antioxidant components of cherry tomatoes (*Lycopersicon esculentum* cv. Naomi F1). *J. Food Compos. Anal.* **2006**, *19*, 11–19. [CrossRef]
61. Di Cesare, L.F.; Migliori, C.; Viscardi, D.; Parisi, M. Quality of Tomato Fertilized with Nitrogen and Phosphorous. *Ital. J. Food Sci.* **2010**, *22*, 186–191.
62. Ilahy, R.; Tlili, I.; Riahi, A.; Sihem, R.; Ouerghi, I.; Hdider, C.; Piro, G.; Lenucci, M.S. Fractionate analysis of the phytochemical composition and antioxidant activities in advanced breeding lines of high-lycopene tomatoes. *Food Funct.* **2016**, *7*, 574–583. [CrossRef] [PubMed]
63. George, B.; Kaur, C.; Khurdiya, D.; Kapoor, H. Antioxidants in tomato (*Lycopersium esculentum*) as a function of genotype. *Food Chem.* **2004**, *84*, 45–51. [CrossRef]
64. Markovic, K.; Krbavčić, I.P.; Krpan, M.; Bicanic, D.; Vahčić, N. The lycopene content in pulp and peel of five fresh tomato cultivars. *Acta Aliment.* **2010**, *39*, 90–98. [CrossRef]
65. Ciaccheri, L.; Tuccio, L.; Mencaglia, A.A.; Mignani, A.G.; Hallmann, E.; Sikorska-Zimny, K.; Kaniszewski, S.; Verheul, M.J.; Agati, G. Directional versus total reflectance spectroscopy for the in situ determination of lycopene in tomato fruits. *J. Food Compos. Anal.* **2018**, *71*, 65–71. [CrossRef]
66. Sikorska-Zimny, K.; Badelek, E.; Grzegorzewska, M.; Ciecierska, A.; Kowalski, A.; Kosson, R.; Tuccio, L.; Mencaglia, A.A.; Ciaccheri, L.; Mignani, A.G.; et al. Comparison of Lycopene Changes between Open-Field Processing and Fresh Market Tomatoes During Ripening and Post-Harvest Storage by Using a Non-Destructive Reflectance Sensor. *J. Sci. Food Agric.* **2019**, *99*, 2763–2774. [CrossRef] [PubMed]

67. Capanoglu, E.; Beekwilder, J.; Boyacioglu, D.; De Vos, R.C.; Hall, R.D. The Effect of Industrial Food Processing on Potentially Health-Beneficial Tomato Antioxidants. *Crit. Rev. Food Sci. Nutr.* **2010**, *50*, 919–930. [CrossRef] [PubMed]
68. Luterotti, S.; Bicanic, D.; Marković, K.; Franko, M. Carotenes in processed tomato after thermal treatment. *Food Control* **2015**, *48*, 67–74. [CrossRef]
69. Svelander, C.A.; Tibäck, E.A.; Ahrné, L.M.; Langton, M.I.; Svanberg, U.S.; Alminger, M.A. Processing of tomato: Impact on in vitro bioaccessibility of lycopene and textural properties. *J. Sci. Food Agric.* **2010**, *90*, 1665–1672. [CrossRef] [PubMed]
70. Seybold, C.; Fröhlich, K.; Bitsch, R.; Otto, K.; Böhm, V. Changes in Contents of Carotenoids and Vitamin E during Tomato Processing. *J. Agric. Food Chem.* **2004**, *52*, 7005–7010. [CrossRef] [PubMed]
71. Zanoni, B.; Pagliarini, E.; Giovanelli, G.; Lavelli, V. Modelling the effects of thermal sterilization on the quality of tomato puree. *J. Food Eng.* **2003**, *56*, 203–206. [CrossRef]
72. Kaur, D.; Sogi, D.; Wani, A.A. Degradation Kinetics of Lycopene and Visual Color in Tomato Peel Isolated from Pomace. *Int. J. Food Prop.* **2006**, *9*, 781–789. [CrossRef]
73. Kessy, H.H.; Zhang, H.W.; Zhang, L.F. A Study on Thermal Stability of Lycopene in Tomato in Water and Oil Food Systems Using Response Surface Methodology. *Int. J. Food Sci. Technol.* **2011**, *46*, 209–215. [CrossRef]
74. Chen, C.J.; Shi, J.; Xue, S.J.; Ma, Y. Comparison of Lycopene Stability in Water-And Oil-Based Food Model Systems under Thermal-and Light-Irradiation Treatments. *Food Sci. Technol.* **2009**, *42*, 740–747. [CrossRef]
75. Tibäck, E.A.; Svelander, C.A.; Colle, I.J.; Altskär, A.I.; Alminger, M.A.; Hendrickx, M.E.; Ahrné, L.M.; Langton, M.I. Mechanical and Thermal Pretreatments of Crushed Tomatoes: Effects on Consistency and In Vitro Accessibility of Lycopene. *J. Food Sci.* **2009**, *74*, E386–E395. [CrossRef]
76. Kaur, C.; George, B.; Deepa, N.; Singh, B.; Kapoor, H.C. Antioxidant Status of Fresh and Processed Tomato—A Review. *J. Food Sci. Technol.-Mysore* **2004**, *41*, 479–486.
77. Spigno, G.; Maggi, L.; Amendola, D.; Ramoscelli, J.; Marcello, S.; De Faveri, D.M. Bioactive Compounds in Industrial Tomato Sauce after Processing and Storage. *Ital. J. Food Sci.* **2014**, *26*, 252–260.
78. Takeoka, G.R.; Dao, L.; Flessa, S.; Gillespie, D.M.; Jewell, W.T.; Huebner, B.; Bertow, D.; Ebeler, S.E. Processing Effects on Lycopene Content and Antioxidant Activity of Tomatoes. *J. Agric. Food Chem.* **2001**, *49*, 3713–3717. [CrossRef] [PubMed]
79. Stratakos, A.C.; Delgado-Pando, G.; Linton, M.; Patterson, M.F.; Koidis, A. Industrial scale microwave processing of tomato juice using a novel continuous microwave system. *Food Chem.* **2016**, *190*, 622–628. [CrossRef] [PubMed]
80. Sanchez-Moreno, C.; Plaza, L.; de Ancos, B.; Cano, M.P. Impact of High-Pressure and Traditional Thermal Processing of Tomato Puree on Carotenoids, Vitamin C and Antioxidant Activity. *J. Sci. Food Agric.* **2006**, *86*, 171–179. [CrossRef]
81. Sánchez-Moreno, C.; Plaza, L.; de Ancos, B.; Cano, M.P. Effect of Combined Treatments of High-Pressure and Natural Additives on Carotenoid Extractability and Antioxidant Activity of Tomato Puree (*Lycopersicum esculentum* Mill). *Eur. Food Res. Technol.* **2004**, *219*, 151–160. [CrossRef]
82. Knockaert, G.; Pulissery, S.K.; Colle, I.; Van Buggenhout, S.; Hendrickx, M.; Van Loey, A. Lycopene degradation, isomerization and in vitro bioaccessibility in high pressure homogenized tomato puree containing oil: Effect of additional thermal and high pressure processing. *Food Chem.* **2012**, *135*, 1290–1297. [CrossRef] [PubMed]
83. Shi, J.; Maguer, M.; Bryan, M.; Kakuda, Y. Kinetics of lycopene degradation in tomato puree by heat and light irradiation. *J. Food Process. Eng.* **2003**, *25*, 485–498. [CrossRef]
84. Anese, M.; Falcone, P.; Fogliano, V.; Nicoli, M.; Massini, R. Effect of Equivalent Thermal Treatments on the Color and the Antioxidant Activity of Tomato Puree. *J. Food Sci.* **2002**, *67*, 3442–3446. [CrossRef]
85. Gupta, R.; Balasubramaniam, V.M.; Schwartz, S.J.; Francis, D.M. Storage Stability of Lycopene in Tomato Juice Subjected to Combined Pressure–Heat Treatments. *J. Agric. Food Chem.* **2010**, *58*, 8305–8313. [CrossRef]
86. Kips, L.; De Paepe, D.; Bernaert, N.; Van Pamel, E.; De Loose, M.; Raes, K.; Van Droogenbroeck, B. Using a novel spiral-filter press technology to biorefine horticultural by-products: The case of tomato. Part I: Process optimization and evaluation of the process impact on the antioxidative capacity. *Innov. Food Sci. Emerg. Technol.* **2016**, *38*, 198–205. [CrossRef]
87. Siewert, N. Vaculiq—The System for Juice, Smoothie and Puree? *Fruit Process.* **2013**, *2*, 54–57.
88. Løvdal, T.; Vågen, I.; Agati, G.; Tuccio, L.; Kaniszewski, S.; Grzegorzewska, M.; Kosson, R.; Bartoszek, A.; Erdogdu, F.; Tutar, M.; et al. *Sunniva-Final Report*; Nofima: Tromsø, Norway, 2018.

89. Grzegorzewska, M.; Badelek, E.; Sikorska-Zimny, K.; Kaniszewski, S.; Kosson, R. Determination of the Effect of Ethylene Treatment on Storage Ability of Tomatoes. In Proceedings of the XII International Controlled & Modified Atmosphere Research Conference, Warsaw, Poland, 18 June 2017.
90. Maharaj, R.; Arul, J.; Nadeau, P. Effect of photochemical treatment in the preservation of fresh tomato (*Lycopersicon esculentum* cv. Capello) by delaying senescence. *Postharvest Biol. Technol.* **1999**, *15*, 13–23. [CrossRef]
91. Tjøstheim, I.H. Effekten Av Forskjellige Lagringsforhold På Tomatenes Kvalitet (*Lycopersicon esculentum* Mill.) Fra Høsting Til Konsum. Master's Thesis, Universitetet for Miljø-og Biovitenskap, Ås, Norway, 2011.
92. Batu, A. Effect of Long-Term Controlled Atmosphere Storage on the Sensory Quality of Tomatoes. *Ital. J. Food Sci.* **2003**, *15*, 69–77.
93. Domínguez, I.; Lafuente, M.T.; Hernández-Muñoz, P.; Gavara, R. Influence of modified atmosphere and ethylene levels on quality attributes of fresh tomatoes (*Lycopersicon esculentum* Mill.). *Food Chem.* **2016**, *209*, 211–219.
94. Aschemann-Witzel, J. Waste Not, Want Not, Emit Less. *Science* **2016**, *352*, 408–409. [CrossRef]
95. Blanke, M. Challenges of Reducing Fresh Produce Waste in Europe—From Farm to Fork. *Agriculture* **2015**, *5*, 389–399. [CrossRef]
96. Mukherjee, P.K.; Nema, N.K.; Maity, N.; Sarkar, B.K. Phytochemical and therapeutic potential of cucumber. *Fitoterapia* **2013**, *84*, 227–236. [CrossRef]
97. Abdel-Razzak, H.; Alkoaik, F.; Rashwan, M.; Fulleros, R.; Ibrahim, M. Tomato Waste Compost as an Alternative Substrate to Peat Moss for the Production of Vegetable Seedlings. *J. Plant Nutr.* **2019**, *42*, 287–295. [CrossRef]
98. Otto, K.; Sulc, D. Herstellung Von Gemüsesaften. In *Frucht-Und Gemüsesäfte*; Schobinger, U., Ed.; Ulmer: Stuttgart, Germany, 2001.
99. Avelino, A.; Avelino, H.T.; Roseiro, J.C.; Collaço, M.A. Saccharification of tomato pomace for the production of biomass. *Bioresour. Technol.* **1997**, *61*, 159–162. [CrossRef]
100. Schieber, A.; Stintzing, F.; Carle, R. By-products of plant food processing as a source of functional compounds—Recent developments. *Trends Food Sci. Technol.* **2001**, *12*, 401–413. [CrossRef]
101. Al-Wandawi, H.; Abdul-Rahman, M.; Al-Shaikhly, K. Tomato processing wastes as essential raw materials source. *J. Agric. Food Chem.* **1985**, *33*, 804–807. [CrossRef]
102. University of Florida. Lycopene-Extraction Method Could Find Use for Tons of Discarded Tomatoes. Available online: https://www.sciencedaily.com/releases/2004/05/040505064902.htm (accessed on 30 May 2019).
103. Baysal, T.; Ersus, S.; Starmans, D.A.J. Supercritical CO_2 extraction of beta-carotene and lycopene from tomato paste waste. *J. Agric. Food Chem.* **2000**, *48*, 5507–5511. [CrossRef] [PubMed]
104. Lenucci, M.S.; De Caroli, M.; Marrese, P.P.; Iurlaro, A.; Rescio, L.; Böhm, V.; Dalessandro, G.; Piro, G. Enzyme-aided extraction of lycopene from high-pigment tomato cultivars by supercritical carbon dioxide. *Food Chem.* **2015**, *170*, 193–202. [CrossRef] [PubMed]
105. Lianfu, Z.; Zelong, L. Optimization and comparison of ultrasound/microwave assisted extraction (UMAE) and ultrasonic assisted extraction (UAE) of lycopene from tomatoes. *Ultrason. Sonochem.* **2008**, *15*, 731–737. [CrossRef] [PubMed]
106. Chemat, F.; Rombaut, N.; Sicaire, A.-G.; Meullemiestre, A.; Fabiano-Tixier, A.-S.; Abert-Vian, M. Ultrasound Assisted Extraction of Food and Natural Products. Mechanisms, Techniques, Combinations, Protocols and Applications. A Review. *Ultrason. Sonochem.* **2017**, *34*, 540–560. [CrossRef] [PubMed]
107. Hatami, T.; Meireles, M.A.A.; Ciftci, O.N. Supercritical carbon dioxide extraction of lycopene from tomato processing by-products: Mathematical modeling and optimization. *J. Food Eng.* **2019**, *241*, 18–25. [CrossRef]
108. Luengo, E.; Álvarez, I.; Raso, J. Improving Carotenoid Extraction from Tomato Waste by Pulsed Electric Fields. *Front. Nutr.* **2014**, *1*, 12. [CrossRef] [PubMed]
109. Saini, R.K.; Keum, Y.-S. Carotenoid extraction methods: A review of recent developments. *Food Chem.* **2018**, *240*, 90–103. [CrossRef]
110. Østerlie, M.; Lerfall, J. Lycopene from tomato products added minced meat: Effect on storage quality and colour. *Food Res. Int.* **2005**, *38*, 925–929. [CrossRef]
111. Roy, B.C.; Goto, M.; Hirose, T. Temperature and pressure effects on supercritical CO_2 extraction of tomato seed oil. *Int. J. Food Sci. Technol.* **1996**, *31*, 137–141. [CrossRef]

112. Shao, D.; Venkitasamy, C.; Li, X.; Pan, Z.; Shi, J.; Wang, B.; Teh, H.E.; McHugh, T. Thermal and storage characteristics of tomato seed oil. *LWT Food Sci. Technol.* **2015**, *63*, 191–197. [CrossRef]
113. Farahnaky, A.; Abbasi, A.; Jamalian, J.; Mesbahi, G. The use of tomato pulp powder as a thickening agent in the formulation of tomato ketchup. *J. Texture Stud.* **2008**, *39*, 169–182. [CrossRef]
114. Grassino, A.N.; Brnčić, M.; Vikić-Topić, D.; Roca, S.; Dent, M.; Brnčić, S.R. Ultrasound assisted extraction and characterization of pectin from tomato waste. *Food Chem.* **2016**, *198*, 93–100. [CrossRef] [PubMed]
115. Savadkoohi, S.; Hoogenkamp, H.; Shamsi, K.; Farahnaky, A. Color, sensory and textural attributes of beef frankfurter, beef ham and meat-free sausage containing tomato pomace. *Meat Sci.* **2014**, *97*, 410–418. [CrossRef]
116. Sarkar, A.; Kaul, P. Evaluation of Tomato Processing By-Products: A Comparative Study in a Pilot Scale Setup. *J. Food Process. Eng.* **2014**, *37*, 299–307. [CrossRef]
117. Haddadin, M.; Abu-Reesh, I.; Haddadin, F.; Robinson, R. Utilisation of tomato pomace as a substrate for the production of vitamin B12-A preliminary appraisal. *Bioresour. Technol.* **2001**, *78*, 225–230. [CrossRef]

© 2019 by the authors. Licensee MDPI, Basel, Switzerland. This article is an open access article distributed under the terms and conditions of the Creative Commons Attribution (CC BY) license (http://creativecommons.org/licenses/by/4.0/).

Article

Understanding the Properties of Starch in Potatoes (*Solanum tuberosum* var. Agria) after Being Treated with Pulsed Electric Field Processing

Setya B.M. Abduh [1,2,3], Sze Ying Leong [1,3], Dominic Agyei [1] and Indrawati Oey [1,3,*]

1. Department of Food Science, University of Otago, Dunedin 9016, New Zealand; setya.abduh@postgrad.student.ac.nz (S.B.M.A.); sze.leong@otago.ac.nz (S.Y.L.); dominic.agyei@otago.ac.nz (D.A.)
2. Department of Food Technology, Diponegoro University, Semarang 50275, Indonesia
3. Riddet Institute, Palmerston North 4442, New Zealand
* Correspondence: indrawati.oey@otago.ac.nz; Tel.: +64-3-479-8735

Received: 26 April 2019; Accepted: 9 May 2019; Published: 10 May 2019

Abstract: The purpose of this study was to investigate the properties of starch in potatoes (*Solanum tuberosum* cv. Agria) after being treated with pulsed electric fields (PEF). Potatoes were treated at 50 and 150 kJ/kg specific energies with various electric field strengths of 0, 0.5, 0.7, 0.9 and 1.1 kV/cm. Distilled water was used as the processing medium. Starches were isolated from potato tissue and from the PEF processing medium. To assess the starch properties, various methods were used, i.e., the birefringence capability using a polarised light microscopy, gelatinisation behaviour using hot-stage light microscopy and differential scanning calorimetry (DSC), thermal stability using thermogravimetry (TGA), enzyme susceptibility towards α-amylase and the extent of starch hydrolysis under in vitro simulated human digestion conditions. The findings showed that PEF did not change the properties of starch inside the potatoes, but it narrowed the temperature range of gelatinisation and reduced the digestibility of starch collected in the processing medium. Therefore, this study confirms that, when used as a processing aid for potato, PEF does not result in detrimental effects on the properties of potato starch.

Keywords: potato starch; pulsed electric fields; birefringence; thermal properties; enzyme susceptibility

---

1. Introduction

Pulsed electric field (PEF) processing has been reported to have a capability in modifying the microstructure of solid plant foods [1]. This leads to the reduction of the cutting force for potato tuber [2] and sweet potato [3] and the oil uptake of these commodities during frying. Another study by [4] has also found that PEF processing in combination with calcium chloride and trehalose solutions could retain the textural properties of frozen potatoes. Due to these benefits to improve product quality, this technology has been used in the commercial potato French fries or chips industries [5].

Giteru, Oey and Ali [6] have recently reported that PEF could affect either the stability or the functional properties of biomacromolecules such as polysaccharides and proteins. Especially on polysaccharides, PEF could affect their microstructure, conformation, solubility, swelling effect, particle size, viscoelastic properties, structural transition and thermal stability [6]. Moreover, other studies have reported that PEF processing applied at an electric field strength up to 50 kV/cm can change the properties of starches dispersed in water, such as the structural properties and the digestibility of waxy rice starch [7]; the microstructure, thermal properties and viscosity of tapioca starch [8]; the microstructure and thermal properties of maize starch [9]; the thermal properties and microstructure of potato starch [10]; and the thermal properties and microstructure of corn starch [11]. So far, limited studies have been conducted to understand the fate of starch inside the potatoes after the tubers are PEF-treated. Starch is the major component in potato [12] that contributes to its nutritional quality [13,14]. Although food processing techniques [14,15], including boiling, cooling, reheating [16], conventional frying and air frying [17], have been shown to change the digestibility of starch, it is still not known whether PEF processing affects the inherent properties of starch in potatoes. In addition, the adoption of PEF technique in potato processing is fairly recent, compared with, say, the use of PEF in the processing of liquid foods, such as in juice extraction, bacterial inactivation in milk, etc. It is also known that consumer perception regarding the safety of PEF-processed foods is a key to their acceptance [18] and this perception is often influenced by information on the effect of PEF technology on the products themselves [19]. A similar phenomenon would potentially happen in the adoption of PEF in potato processing.

Therefore, the purpose of this study was to investigate the properties of starch in potatoes after being treated by PEF. The native state of starch isolated from potatoes after PEF treatment was examined using polarised light microscopy, combined with hot-stage optical microscopy. The gelatinisation behaviour, thermal stability and thermal properties of starch isolated from potatoes after PEF treatment were studied using thermogravimetric analysis (TGA) and differential scanning calorimetry (DSC). The susceptibility of the isolated starch towards heat stable α-amylase as well as the degree of starch hydrolysis during in vitro simulated human oral-gastro-intestinal digestion were studied. In this study, the effects of electric field strength and specific energy for PEF on these properties were investigated. In addition, the properties of any starch found in the PEF processing medium (SPM) (e.g., leached out of the potato due to cutting during sample preparation) were examined to allow a direct comparison with the properties of starch obtained from the same potatoes after being treated with PEF. To the authors' knowledge, this research is the first work to study the properties of starch relevant to PEF processing of potatoes.

2. Materials and Methods

2.1. Chemicals and Reagents

Potassium iodate (KI), iodine (I_2), potassium hydroxide (KOH), sodium hydroxide (NaOH) and hydrochloric acid (HCl), were purchased from Merck (Darmstadt, Germany). Sodium chloride (NaCl) was purchased from BDH Chemicals (Poole, UK). Potassium chloride (KCl) was purchased from Fisher Scientific (Norcross, GA, USA). Sodium bicarbonate ($NaHCO_3$) was purchased from (Riedel-de Haën, Seelze, Germany). Heat stable α-amylase from *Bacillus licheniformis* (3000 U/mL), amyloglucosidase from *Aspergillus niger* (3300 U/mL), glucose oxidase peroxidase (GOPOD) kit and D-glucose standard (1 mg/mL) were purchased from Megazyme (Wicklow, Ireland). Alpha amylase from *Aspergillus oryzae* (30 U/mg) and pancreatin from porcine pancreas (4 × USP) were purchased from Sigma (St. Louis, MO, USA). Pepsin was purchased from PanReac AppliChem (Barcelona, Spain). Porcine bile extract was purchased from Santa Cruz Biotechnology (Dallas, TX, USA).

2.2. Preparation of Samples

A batch of potato tubers (*Solanum tuberosum* cv. Agria) harvested in August 2017 were obtained from Pyper's Produce (Invercargill, New Zealand). In this study, the Agria cultivar was selected as a model system due to its high starch content. Upon arrival, tubers were sorted according to their weight, size and dimensions while any tubers with cuts, bruises or damage were discarded. Fifty potato tubers with uniform size and dimension were selected and randomly divided into five groups (10 tubers per group) which was later used as replicates. Each group of ten potatoes was peeled and shredded (2.79 mm × 2.79 mm) using an MX 260 food processor (Kenwood, Beijing, China) at medium speed. The shredded potato from 10 tubers were pooled together to attain a homogenous sample set and kept in an ice-water bath for no longer than 30 min. The samples were immediately treated with PEF (see Section 2.3) at different electric field strength and energy combinations (Table 1) following the experimental protocol presented in Figure 1. In total, five independent replicates were conducted for this study.

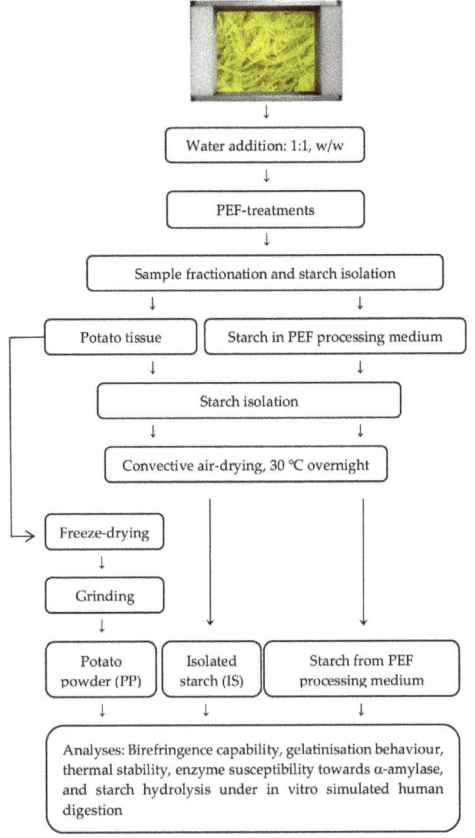

Figure 1. Schematic overview of the experimental design and sample collection followed by preparation.

Table 1. Summary of PEF treatment parameters on potato and the treatment impact on the changes in electrical conductivity, temperature and browning index of the PEF processing medium.

PEF (kV/cm, kJ/kg)	Electric Field Strength (kV/cm)	Specific Energy Input (kJ/kg)	Conductivity Increase (mS/cm) *	Temperature Increase (°C) **	Browning Index
Untreated (No PEF)	0.00	0.00	0.25 ± 0.16 [b]	1.56 ± 2.32 [c]	218.30 ± 33.80 [c]
PEF 1 (0.5, 50)	0.50	50.48 ± 1.10	0.61 ± 0.17 [ab]	6.42 ± 1.25 [b]	279.74 ± 24.23 [ab]
PEF 2 (0.7, 50)	0.70	49.25 ± 0.44	0.44 ± 0.33 [b]	5.90 ± 0.53 [b]	268.36 ± 34.47 [ab]
PEF 3 (0.9, 50)	0.90	49.63 ± 0.31	0.54 ± 0.08 [ab]	6.08 ± 0.99 [b]	294.73 ± 15.97 [a]
PEF 4 (1.1, 50)	1.10	50.10 ± 0.39	0.53 ± 0.07 [ab]	5.86 ± 0.39 [b]	269.13 ± 27.05 [ab]
PEF 5 (0.7, 150)	0.70	151.81 ± 1.72	0.91 ± 0.25 [a]	15.22 ± 1.49 [a]	290.23 ± 32.06 [a]
PEF 6 (0.9, 150)	0.90	153.09 ± 0.77	0.53 ± 0.19 [ab]	14.00 ± 0.60 [a]	252.93 ± 50.83 [bc]
One-way ANOVA result			$F_{(8,39)} = 4.91$ $p = 0.00$	$F_{(8,39)} = 108.47$ $p = 0.00$	$F_{(6,98)} = 9.42$ $p = 0.00$

PEF, pulsed electric fields. * Initial conductivity: 1.61 ± 0.25 mS/cm. ** Initial temperature: 7.22 ± 1.46 °C. All the PEF treatments were carried out at a 20 µs pulse width at 100 Hz. Results are expressed as the mean ± standard deviation of five independent PEF processing experiments. Values in the same column not sharing the same letter are significantly different at $p < 0.05$ analysed with one-way ANOVA and Tukey's post hoc test.

2.3. Pulsed Electric Fields Treatment

PEF processing was performed using an ELCRACK® HVP 5 PEF system (German Institute of Food Technologies, Quakenbruck, Germany) in the batch treatment configuration. The PEF treatment chamber (total volume of 400 mL with dimensions of 100 mm length, 80 mm width and 50 mm depth), consisted of two parallel stainless-steel electrodes (80 mm electrode gap). For each treatment, 125 g of potato samples were placed inside the PEF treatment chamber and then submerged in 125 g of distilled water, which served as the PEF processing medium. The pulse shape generated by the PEF unit was monitored in real-time using an oscilloscope (UTD2042C, Uni-Trend Group Ltd., Dongguan, China). Output parameters, such as electric field strength (kV/cm), pulse voltage (kV), pulse current (A), pulse power (kW), pulse energy (J), total energy (kJ), pulse number (dimensionless) and pulse resistance (ohm), were recorded for each PEF run.

In this study, different field strengths, i.e., 0.5, 0.7, 0.9 and 1.1 kV/cm, accompanied with two specific energy input intensities averaged at 49.87 ± 0.54 and 153 ± 0.91 kJ/kg were used. The specific energy input was calculated using Equation (1):

$$W_{specific}\ (kJ/kg) = \frac{V^2 \cdot (n \cdot m)}{R \cdot W} \tag{1}$$

where V is the pulse voltage (kV), n is the pulse number (dimensionless), m is the pulse width (µs), R is the pulse resistance (ohm) and W is the total weight of the sample and PEF processing medium.

All tested PEF process conditions were achieved by applying a 20 µs pulse width at 100 Hz frequency with the pulse number ranging from 900–6250. Each PEF treatment was carried out in five replicates with each replicate utilising shredded potatoes from 10 tubers. The process of treating potatoes with PEF for each run was standardised and took no more than 3 min. For every PEF run, the untreated sample (i.e., potato samples without PEF treatment) was prepared by soaking the potato samples in distilled water at a ratio of 1:1 for 3 min and afterwards the sample was handled as described in Figure 1.

To test whether PEF causes changes in cell permeability, both conductivity and temperature of the processing medium were first measured just before and immediately after PEF treatment using a CyberScanCON11 (Eutech Instruments, Singapore, Singapore) hand-held conductivity meter to have an indication of increased ion leakage due to PEF treatment. Subsequently, the PEF processing medium was separated from the potato sample using a kitchen sieve and transferred into plastic containers (1000 mL volume). Secondly, the colour of the PEF processing medium was immediately measured on the tristimulus colour combination L*a*b* scale using a MiniScan XEPlus 45/0-L colorimeter (Hunterlab, Reston, VA, USA) in triplicate. The tristimulus colour combination of L*a*b* were then converted into a browning index (BI) using Equations (2) and (3):

$$BI = \frac{100(X - 0.31)}{0.17} \tag{2}$$

with:

$$X = \frac{a + 1.75L}{5.645L + a - 3.012b} \tag{3}$$

2.4. Isolation of Starch after PEF Treatment

Immediately after PEF treatment, both untreated and PEF-treated potato samples were separated from the processing medium using a kitchen sieve. Potato samples (approximately 125 g) and the processing medium were transferred into separate plastic containers for starch isolation as described below.

Potato starch was isolated from 100 g potato samples according to [20] with modifications. Potato samples, either untreated or PEF-treated, were placed into a plastic container containing 300 mL distilled water and, afterwards, the samples were gently macerated by hand for 1 min. The mixture of potato and distilled water was then filtered using a kitchen sieve and the filtrate was collected. The maceration process on the potato samples was repeated with another fresh 300 mL of distilled water. The final filtrate was combined and kept at room temperature for 2 h to allow the starch to settle to the bottom. After 2 h, the water layer on the top of the starch suspension was discarded and replaced with fresh distilled water. This step was repeated twice until clear water was obtained. The starch sediments were then oven dried (Eurotherm 3216, Steridium, Queenstown, Australia) at 30 °C overnight or longer until the starch was completely dry (indicated by no further weight loss). The dried powder, referred as "isolated starch" (later coded as "IS"), was transferred into 1.5 mL tubes and stored in a desiccator filled with silica gel at ambient temperature (17–20 °C) until further analysis.

PEF processing medium (approximately 125 mL) was transferred into a plastic container followed by the addition of 300 mL distilled water. The mixture was kept at room temperature for 2 h to allow the starch to settle. After 2 h, the water layer on the top of starch suspension was discarded and replaced with fresh distilled water. This step was repeated twice until clear water was finally achieved. The starch sediments were then oven dried (Steridium with Eurotherm 3216 controller) at 30 °C overnight or longer until the starch was completely dry (indicated by no further weight loss). The dried powder obtained was referred as "starch from PEF processing medium" (later coded as "SPM"). The sample was kept in 1.5 mL tubes and stored in a desiccator filled with silica gel at ambient temperature (17–20 °C) until analysis.

The remaining potato samples (approximately 25 g) was transferred into plastic bags and kept frozen at −20 °C (Fisher and Paykel, Auckland, New Zealand) for no more than 2 weeks, followed by freeze drying (Labconco Freezone, Kansas City, MO, USA). Subsequently, the freeze-dried samples were homogenised into powder form (thereafter referred as "potato powder or PP" sample) using a laboratory blender (32BL80 Waring, Torrington, CT, USA) for 10 s at high speed. They were sealed tightly inside polypropylene vials under ambient temperature (17–20 °C) until analysis.

2.5. Study on the Birefringence Capability of Starch Granule after PEF Treatment

The native form of starch typically exhibiting birefringence capability is associated with the crystalline structure of starch. The birefringence capability of starch isolated from PEF-treated potatoes was studied by mean of light microscopy and polarised microscopy. Starch suspension was prepared by gentle mixing of 2 mg dried sample and 250 µL deionised water. A drop of starch dispersion was transferred onto a glass slide (LabServ, Waltham, MA, USA) and a small amount of Lugol's iodine dye (5% (w/v) potassium iodate and 0.5% (w/v) iodine at a ratio of 1:1) was added. After the cover slip was placed, the specimen was observed under a BX-50 microscope (Olympus, Tokyo, Japan) with a polariser (U-POT U-P110 model, Olympus, Tokyo, Japan) under a magnification of 400×. The observations under the microscope were captured using a Camedia C4040 Zoom digital camera (Olympus, Tokyo, Japan). The images were saved as TIFF files and later standardised for their brightness using Windows Photos (Microsoft, Redmond, WA, USA).

2.6. Study on the Gelatinisation Behaviour of Starch Using Hot-Stage Microscopy

A light microscope, Motic BA300Pol (OPTIKA SrL, Ponteranica, Italy), complete with polariser, at a magnification of 200× was used. Starch dispersions (2 mg/250 µL deionised water) were prepared and then transferred onto glass slides with cavities, covered with a coverslip and put onto a hot stage (FP82HT model, Mettler Toledo, Columbus, OH, USA) connected with a Mettler Toledo FP90 central processor to control the heating setting of the hot stage. The specimen was heated from 30–80 °C at a rate of 5 °C/min. Live pictures of the specimen were automatically recorded every 15 s which was equal to a temperature increment of 1.25 °C. The pictures were captured using a Nikon Optiphot PFX microscope camera (Nikon, Tokyo, Japan) and the files were saved as BMP files with Image-Pro Plus version 2.7 software (Media Cybernatics Inc., Rockville, MD, USA). The captured images were resized using Windows Photos (Microsoft).

2.7. Study of the Thermal Stability of Starch

Thermogravimetric analysis (TGA) was used to determine the moisture content and the thermal stability of the starch-containing dried samples. Dried sample, about 10 mg, were transferred onto a 100 µL platinum TGA pan (TA Instruments, New Castle, DE, USA). Subsequently, the sample was heated from room temperature to 400 °C inside the TGA 550 unit (TA Instruments). The TGA operation and the data analysis to determine the weight loss and derivative weight loss of the sample during heating were performed using TRIOS software V4.3 (TA Instruments).

2.8. Study on the Gelatinisation Temperature of Starch Using Differential Scanning Calorimetry

Starch-containing samples were weighed closed to 3.0 ± 0.5 mg (in dry basis as predefined using TGA) on Tzero DSC pan (TA Instruments). Adequate amount of distilled water was added to achieve 70% moisture in the sample and the pan was sealed tightly using Tzero hermetic lid with the assistance of a Tzero press with blue die set (TA Instruments). Samples were then allowed to equilibrate to room temperature for 1.5 h prior to analysis in a DSC 250 unit (TA instruments) calibrated with indium (purity >99.9%) and heated from 20–100 °C at a 10 °C/min heating rate. The DSC operation and analysis on the temperatures at which the starch underwent phase transition during heating, i.e., temperature of onset gelatinisation T_o, temperature of peak gelatinisation T_p, temperature of conclusion gelatinisation T_c, range of gelatinisation temperature R and enthalpy change of gelatinisation ΔH were performed using TRIOS software V4.3 (TA Instruments).

2.9. Study on the Susceptibility of Starch towards Enzymes

2.9.1. Starch Susceptibility towards Heat Stable α-Amylase

Twenty milligrams of sample was transferred into a 50 mL falcon tube, added with 980 µL distilled water and equilibrated to room temperature for 10 min. The dispersion was then vortexed and placed on a magnetic stirrer followed by the addition of 1 mL KOH (2M) to dissolve any resistant starch in the sample. The mixture was allowed to stir for 20 min in an ice water bath over the magnetic stirrer. Another 4 mL of sodium acetate buffer (1.2 M; pH 3.8) was then added to neutralise the pH of the mixture. Then, 50 µL of heat stable α-amylase and 50 µL amyloglucosidase were added, followed by incubation at 50 °C in a water bath for 1 h with intermittent vortexing at every 10 min to hydrolyse insoluble starch into soluble branched and dextrin and to hydrolyse the dextrin into D-glucose. After that, the mixture volume was brought up to 40 mL with deionised water, vortexed and centrifuged (Beckman GPR, Beckman, Indianapolis, IN, USA) with an acceleration of 1613× g for 10 min. Fifty microliters of the supernatant was then added with 1.5 mL GOPOD reagent and heated at 50 °C for 20 min. Afterwards, the absorbance was measured at wavelength of λ = 510 nm and temperature of 20 °C using a UV–VIS spectrophotometer (Ultraspec 3300 Pro, Amersham Biosciences, Amersham, UK) with D-glucose (1 mg/mL) solution as the external standard solution. Values were expressed as percent (w/w) hydrolysed starch per sample using a conversion factor of 0.9, which is generally calculated from the molecular weight of starch monomer/molecular weight of glucose (162/180 = 0.9) [21,22].

2.9.2. Starch Hydrolysis under In Vitro Simulated Human Digestion System

In vitro simulated human digestion of starch-containing dried samples consisted of three phases, namely oral, gastric and small intestinal phases, was carried out as outlined in [23] with modifications. Forty milligrams of sample were transferred into glass vials and added with 1960 µL deionised water. Afterwards the mixture was mixed gently with a vortex and equilibrated for 10 min at 20 °C. Subsequently, 2 mL simulated salivary fluid containing NaCl (2 mM), KCl (2 mM) and $NaHCO_3$ (25 mM) and 1 mL α-amylase from *Aspergillus oryzae* (12.5 mg/mL) was added. The mixture was then incubated for 5 min at 37 °C (LabServ, Contherm Scientific Ltd., Wellington, New Zealand) with shaking at 55 strokes per min using a rocking shaker (DLAB Scientific Inc., Beijing, China). Incubation with shaking was carried out for the next 2 h after the pH was adjusted to about 3 with HCl (1 M) and

4 mL simulated gastric juice (1 mM HCl) containing 40 mg/mL pepsin, 151 mM NaCl and 28 mM KCl was added. Upon completion of gastric digestion, the pH was adjusted to pH 7 using NaOH (1 M) and simulated intestinal fluid of NaHCO$_3$ (0.1 M) containing pancreatin from porcine pancreas (10 mg/mL) and bile extract (8.45 mg/mL) was added and incubated with shaking for another 2 h. During the entire course of simulated intestinal digestion, 1 mL digest was withdrawn at time 0, 20, 60, 90 and 120 min and immediately transferred into a 15 mL tube containing 5 mL ethanol (80%). Individual digests were then centrifuged with an acceleration of 1613 g for 10 min at 5 °C and the entire supernatant was used for further analysis.

The supernatant of digested samples was added with 50 µL amyloglucosidase (3300 U/mL) and afterwards incubated at 50 °C for 1 h with an intermittent vortexing for every 10 min. Fifty microliters from the mixture were added with 1.5 mL GOPOD reagent, incubated at 50 °C for 20 min, and the absorbance was measured wavelength of $\lambda = 510$ nm and temperature of 20 °C using a UV–VIS spectrophotometer (Ultraspec 3300 Pro, Amersham Biosciences). D-glucose solution (1 mg/mL) was used as the external standard solution. Values were expressed as mean values with standard deviations of mg glucose per mL digest.

2.10. Statistical Data Analysis

In this study, results are expressed as the mean ± standard deviation of five independent treatment replicates. Statistical analyses were performed with IBM SPSS Statistics version 24 (IBM Corporation, New York, NY, USA) using one-way analysis of variance (ANOVA) followed by the Tukey's HSD post hoc test. Differences between the means were considered significant when $p < 0.05$. Independent samples *t*-test was used to assess the significant differences of the thermal stability, gelatinisation temperature, α-amylase susceptibility and the degree of starch digestibility between untreated and PEF-treated samples.

3. Results and Discussion

3.1. Monitoring the Impact of PEF Treatment on Potatoes

In this study, the changes in the temperature and the conductivity of the product were evaluated to indicate whether all the PEF conditions applied to the potato samples led to cellular damage [1,24,25]. Table 1 clearly showed a considerable increment in the temperature and product conductivity for all the PEF treatment conditions applied. The increase in temperature after PEF treatment at specific energies of 50 and 150 kJ/kg was averaged at 6.07 °C and 14.61 °C, respectively. The temperature increase was due to the external energy generated from the PEF treatment while the increase in product conductivity were at least 0.44 to 0.91 mS/cm higher after PEF treatments than untreated samples (i.e., 0.25 mS/cm, see Table 1) indicating the leaching of ionic species inside the cells from the minerals ions and other soluble solids into the processing medium. The same ionic species were presumably freed during preparation of untreated (No PEF) samples but in the lower concentration which led to an increase in conductivity at a lower value, 0.25 mS/cm.

Browning index in the PEF processing medium was also considered in the present study owing to the possibility of enzymatic reaction occurring between polyphenoloxidase and phenolic compounds released from their cell localisation inside potato tissues [25] into the PEF processing medium after PEF treatment. Result showed that the browning index in the processing medium increased up to 294.73 ± 15.97 after the PEF treatments compared to that of untreated samples (i.e., 218.30 ± 33.80; see Table 1) indicating the leaching of phenolic compounds and polyphenoloxidase. It is important to note that phenolic compounds in potato tissues are localised in the vacuole while polyphenol oxidases are accumulated in the plastids [26]. Therefore, it is likely that the applied PEF treatments had effectively resulted in cellular damage, i.e., microstructural modification due to the formation of cell pores that facilitated the substrate-enzyme interaction in the processing medium. As a consequence, this triggered the enzymatic browning reaction, producing reddish-brown o-quinones compounds that contribute towards the browning of the PEF processing medium.

3.2. Comparison on the Birefringence Capacity of Potato Starch Granules after PEF Treatment

Starch granules have a semi-crystalline structure [27] which exhibits birefringence with a "Maltese cross" feature under a polarised microscope. The starch birefringence has been used as a good indicator to assess the native state of starch [28–30]. Figure 2 presents the microscopy images of starch granules isolated from potatoes after being treated with PEF (i.e., IS samples) and starch found in the PEF processing medium (i.e., SPM samples) using visible and polarised microscopes.

Under polarised microscopy, both starch granules isolated from potatoes (IS) without PEF treatment and from processing medium (SPM) exhibit birefringence with the typical "Maltese cross" feature. In comparison, the same birefringence pattern distinct for native granules was observed for both IS and SPM samples isolated from potatoes PEF-treated at an electric field strength of 0.5 kV/cm with specific energies of 50 kJ/kg (PEF 1) in conjunction with PEF at 0.7 kV/cm with a specific energy of 150 kJ/kg (PEF 5) and also when the electric field strength was increased from 0.5 to 1.1 kV/cm and from 0.7 to 0.9 kV/cm, respectively, at constant specific energies of 50 (PEF 4) and 150 kJ/kg (PEF 5). This study clearly showed that PEF treatment at the processing intensities used in the current study did not influence the molecular crystallinity of starch granules from their native state.

3.3. Understanding the Gelatinisation Behaviour of Starch Isolated from PEF-Treated Potatoes

One of the unique properties of starch is its ability to undergo gelatinisation under sufficient heat and moisture [28]. The starch granules gradually swell with increasing temperature, followed by a loss of crystallinity and the birefringence of the starch [31].

Figure 3 presents the selected microscopy images of starch dispersion being heated at different temperatures. A representative video (Video S1) of gelatinisation behaviour of starch is attached as a supplementary material. Being heated at temperatures of 30, 40, 50 and 55 °C, all starch granules from untreated potato samples continuously showed the existence of birefringence. At 60 °C, most of the starch granules started to lose their birefringence and the birefringence was completely lost at 65 °C. For starch isolated from potato samples treated with PEF at specific energies of 50 (PEF 2) and 150 kJ/kg (PEF 5) with an electric field strength of 0.7 kV/cm, it was found that birefringence of starch was lost extensively at 60 °C, compared with starch from untreated samples that lost birefringence at 65 °C. With respect to PEF treatments on potato samples involving electric field strengths of 1.1 kV/cm and 0.9 kV/cm at specific energies of 50 (PEF 4) and 150 kJ/kg (PEF 6), respectively, there was no obvious indication of the loss of starch birefringence at temperature lower than 60 °C. Likewise, a complete loss of birefringence occurred at a similar temperature as all other starches, i.e., around 65 °C for both untreated and PEF-treated samples. Overall, it was clear that starch isolated from PEF-treated potatoes were retaining the same gelatinisation behaviour as starch from untreated potatoes. It is important to note that it is rather challenging to predict precisely the starting gelatinisation temperatures for the starch under hot-stage microscopy and, hence, the differential scanning calorimetry (DSC) method (Section 3.5), being a more reliable approach, was used to exhibit the onset, peak and conclusion temperatures of the starch gelatinisation process.

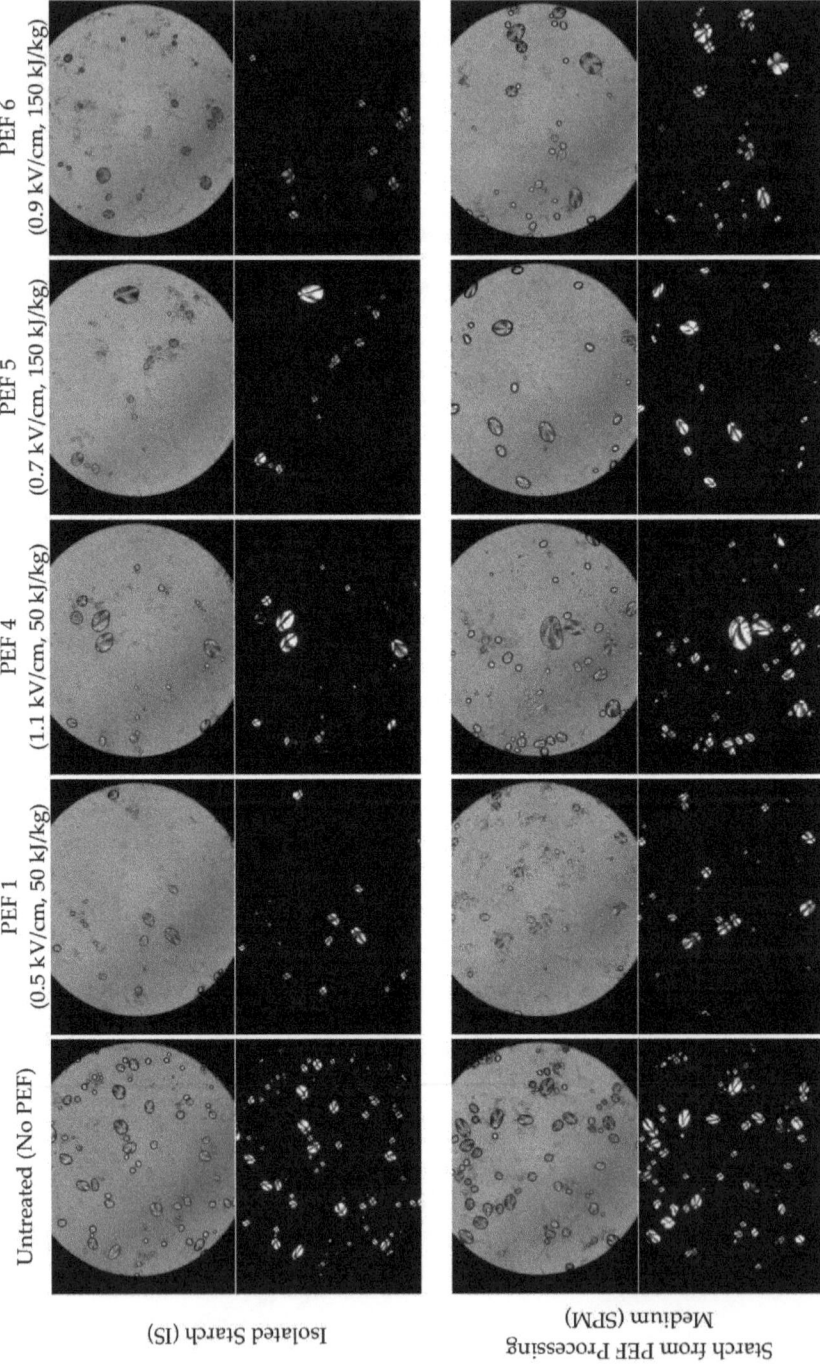

Figure 2. Selected light (**top**) and polarised (**bottom**) micrographs at a magnification of 400× of isolated starch (IS) and starch from PEF processing medium (SPM) of untreated and PEF-treated potatoes.

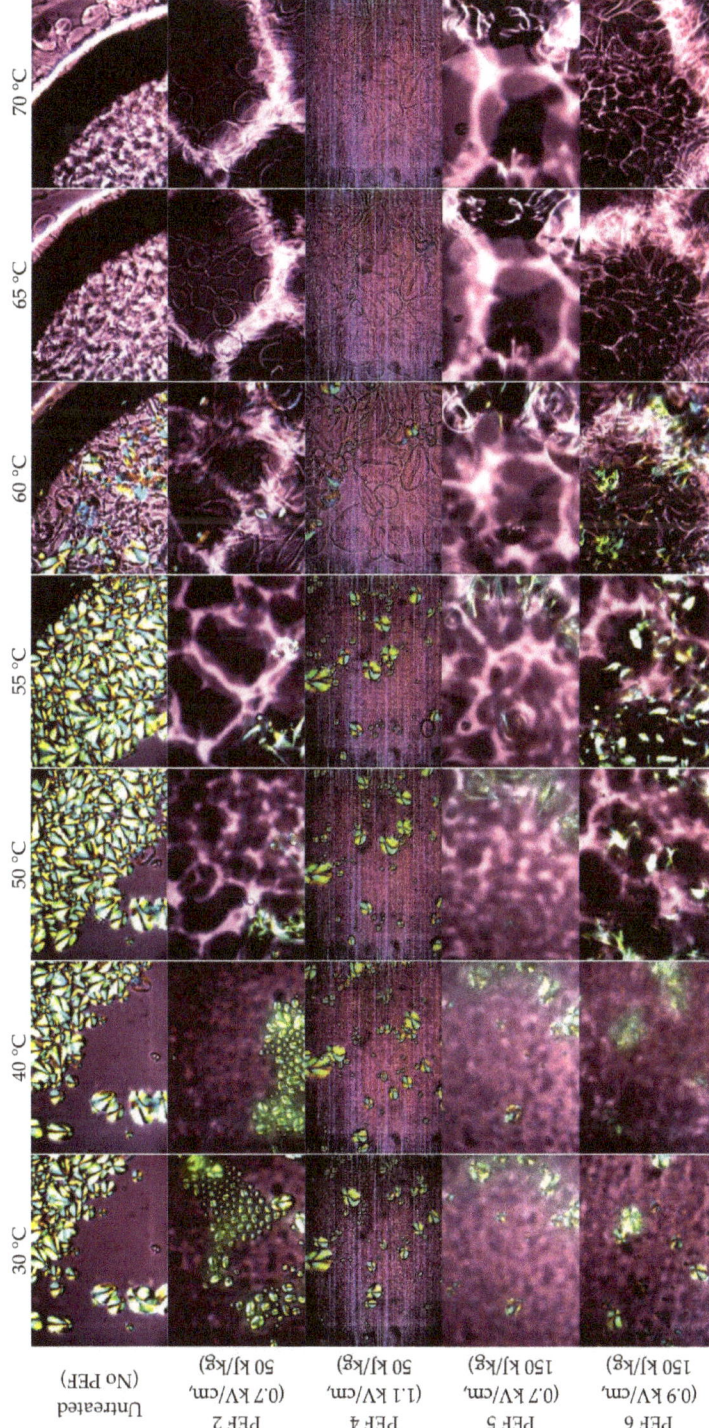

Figure 3. Selected photomicrographs, at a magnification of 200×, of isolated starches from untreated and PEF-treated potatoes, dispersed in water and heated from 30 to 70 °C at a rate of 5 °C/min.

3.4. Effect of PEF on the Thermal Stability of Potato Starch Granules

Figure 4 presents the typical curves of weight (%) and its derivative over the temperature (%/°C) of thermogravimetric analysis (TGA) of starch isolated from potatoes (IS) and starch leached out in the processing medium (SPM). In this study, potato powder (PP) was used as a reference to represent the original sources of the isolated starch.

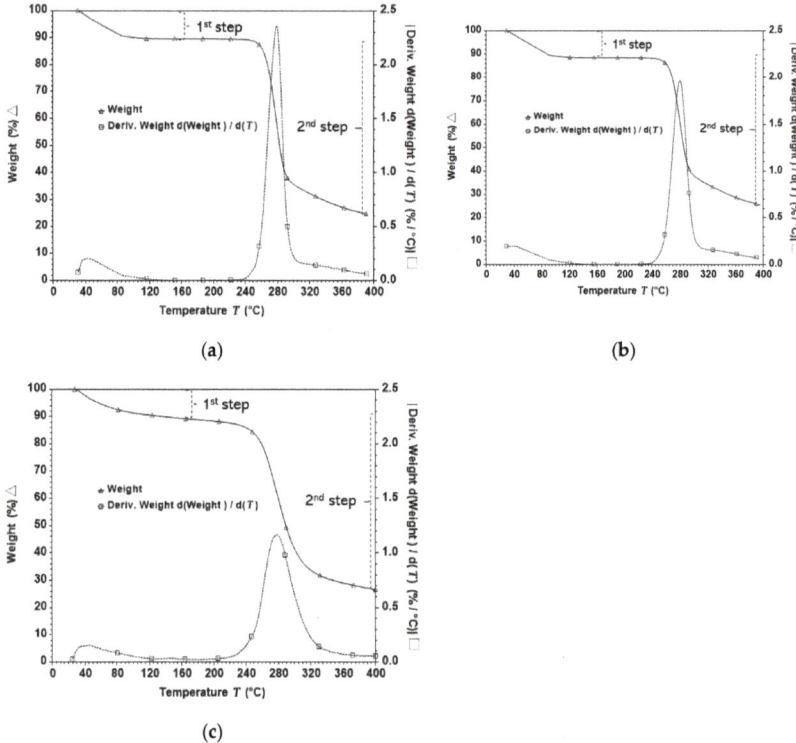

Figure 4. Typical TGA thermograms and their derivative weight of isolated starch (**a**), starch from processing medium (**b**), and potato powder (**c**) from untreated and PEF-treated potatoes. The thermograms are similar for each fraction among PEF treatments analysed in triplicate.

In this study, these three types of samples exhibited similar TGA profiles, which are characterised by two steps of weight loss occurring at similar temperatures. The first step represented the loss of moisture, occurred at about 30 °C and continued until the weight remained constant at about 235 °C. However, compared with those of IS and SPM samples, the initial weight loss on the PP sample occurred slower prior to achieving the constant weight. This indicated that the moisture in the PP sample was bound at a stronger level due to the presence of potato tissue which highly contains water binding compounds, such as pectin, amounting up to 52% of potato cell walls [32].

The second step of weight loss, which occurred at about 260 °C, represents the decomposition of starch polymer resulting in the formation of CO, CO_2 and H_2O due to the degradation C-O and C-C bonds [33]. The most intensive decomposition of polymer was found to occur at a similar temperature across IS, SPM and PP samples, indicating that these three types of samples have polymers with the same properties from where those potato starch granules were extracted. The most intensive decomposition temperatures, indicated as $Tpdw$ in Table 2, were 278.35 ± 1.21 °C, 279.78 ± 1.55 °C, and 276.78 ± 3.77 °C, respectively, for IS, SPM and PP.

Table 2. Thermal stability and gelatinisation temperature of isolated starch (IS), starch from PEF processing medium (SPM), and potato powder (PP) from PEF treated potatoes.

Thermal Properties	Untreated (No PEF)	PEF (kV/cm, kJ/kg)						F Value (6, 14)	p-Value
		PEF 1 (0.5, 50)	PEF 2 (0.7, 50)	PEF 3 (0.9, 50)	PEF 4 (1.1, 50)	PEF 5 (0.7, 150)	PEF 6 (0.9, 150)		
Isolated starch (IS)									
$Tpdtu$ (°C)	278.35 ± 1.21	278.89 ± 1.60	277.13 ± 1.16	279.06 ± 2.20	277.36 ± 1.13	277.82 ± 1.68	274.25 ± 3.62	0.94	0.50
To (°C)	56.67 ± 0.57	56.85 ± 0.79	56.86 ± 0.45	56.52 ± 0.47	56.55 ± 0.37	56.73 ± 0.41	56.67 ± 0.45	0.21	0.97
Tp (°C)	59.81 ± 0.50	60.07 ± 0.62	59.94 ± 0.31	59.56 ± 0.33	59.70 ± 0.24	59.82 ± 0.17	59.88 ± 0.23	0.58	0.74
Tc (°C)	64.83 ± 1.90	66.13 ± 0.81	65.68 ± 0.31	64.88 ± 0.70	65.21 ± 1.46	65.75 ± 0.61	66.02 ± 0.01	0.80	0.59
R (°C)	8.16 ± 1.87	9.28 ± 0.50	8.81 ± 0.45	8.36 ± 0.40	8.67 ± 1.60	9.02 ± 0.88	9.35 ± 0.45	0.55	0.76
ΔH (J/g)	33.37 ± 7.09	38.03 ± 1.15	37.74 ± 2.67	39.37 ± 0.50	35.56 ± 6.36	37.69 ± 1.89	39.43 ± 0.21	0.96	0.49
Starch from PEF processing medium (SPM)									
$Tpdtu$ (°C)	279.78 ± 1.55	280.03 ± 1.35	278.83 ± 1.47	278.02 ± 3.48	276.65 ± 3.00	280.50 ± 0.69	282.30 ± 3.43	1.76	0.18
To (°C)	57.69 ± 0.72	57.59 ± 0.54	57.94 ± 0.42	57.97 ± 0.15	58.23 ± 0.13	58.34 ± 0.07	58.23 ± 0.10	1.67	0.20
Tp (°C)	60.96 ± 0.67	60.83 ± 0.55	61.09 ± 0.20	60.95 ± 0.19	61.21 ± 0.12	61.20 ± 0.03	61.19 ± 0.03	0.56	0.76
Tc (°C)	67.08 ± 0.57	66.56 ± 0.76	66.74 ± 0.25	66.50 ± 0.02	66.57 ± 0.81	66.33 ± 0.07	66.51 ± 0.24	0.72	0.64
R (°C)	9.39 ± 0.20 [a]	8.97 ± 0.23 [ab]	8.80 ± 0.63 [ab]	8.53 ± 0.16 [ab,*]	8.34 ± 0.70 [ab]	7.98 ± 0.05 [b,*]	8.27 ± 0.32 [b,*]	4.33	0.01
ΔH (J/g)	37.32 ± 1.97	38.55 ± 1.58	40.08 ± 1.92	40.98 ± 0.51 *	38.89 ± 2.86	38.50 ± 4.09	39.43 ± 0.41	0.84	0.56
Potato powder (PP)									
$Tpdtu$ (°C)	276.78 ± 3.77	274.40 ± 6.90	273.31 ± 4.63	277.72 ± 1.78	275.21 ± 6.72	276.10 ± 2.69	276.80 ± 4.70	0.31	0.92
To (°C)	57.45 ± 0.53	56.99 ± 0.18	56.57 ± 0.57	57.16 ± 0.34	57.42 ± 0.77	56.69 ± 1.06	57.16 ± 0.49	0.88	0.53
Tp (°C)	62.08 ± 0.10	61.33 ± 0.10 *	61.10 ± 0.25 *	61.44 ± 0.46	61.61 ± 0.58	61.51 ± 0.64	61.34 ± 0.25 *	1.82	0.17
Tc (°C)	67.75 ± 0.42	67.28 ± 0.35	66.64 ± 0.55 *	66.92 ± 0.65	66.86 ± 0.52	66.96 ± 0.91	66.86 ± 0.62	1.13	0.39
R (°C)	10.30 ± 0.47	10.29 ± 0.52	10.07 ± 0.87	09.76 ± 0.55	09.44 ± 1.25	10.26 ± 0.76	09.70 ± 0.34	0.65	0.69
ΔH (J/g)	26.26 ± 0.23 [b]	28.21 ± 0.19 [a,*]	28.31 ± 0.50 [a,*]	27.41 ± 1.01 [ab]	20.68 ± 0.87 [c,*]	28.49 ± 1.09 [a,*]	26.58 ± 0.25 [ab]	46.87	0.00

Result expressed as means ± standard deviation of three independent batches of potato sample (10 potatoes per batch). Means in the same row not sharing the alphabets are significantly different at $p < 0.05$ analysed with one-way ANOVA and Tukey's post hoc test. Means in the same row with asterisk * are significantly different from untreated/No PEF sample (95% interval confidence) analysed with independent t-test. $Tpdtu$: temperature of peak weight loss, To: temperature of onset gelatinisation, Tp: temperature of peak gelatinisation, Tc: temperature of conclusion gelatinisation, R: gelatinisation temperature range, ΔH: enthalpy change of gelatinisation.

The typical TGA curve in the current study was consistent with the work carried out on corn starch with two steps of weight losses [34]. However, on the aforementioned study, the T_{pdw} was found to be about 300 °C, similar to that of maize starch [9]. Therefore, T_{pdw} can be directly and uniquely attributed to the type of crystalline structure of the starch. Starch from cereals is characterised by the A-type crystalline structure while starch granules from tubers are usually characterised by the B-type of crystalline structure [35]. A similar TGA profile was also found in another potato starch study [36]. Furthermore, the slower weight loss occurring at the first step of weight loss on Agria cultivar in the current study is similar with those of Agata and IAPAR Cristina cultivars [37]. However, the latter study showed higher T_{pdw}, i.e., 300 °C and 298 °C, respectively, for Agata and IAPAR Cristina. Thus, thermal stability of potato can also be dependent on the biological origins of samples.

The IS, SPM and PP samples obtained from potatoes treated with PEF at increasing specific energies from 50–150 kJ/kg, as well as at increasing electric field strengths from 0.5–1.1 kV/cm were found to have negligible influence on the T_{pdw}. In other words, the T_{pdw} values for all starch samples obtained from PEF-treated potatoes were not significantly different from their untreated counterparts. This finding is consistent with the work of Han with his co-workers [9] who reported that thermal stability of maize starch remained unchanged (T_{pdw} at about 300 °C) even after PEF treatment at high intensity electric field strengths between 30 and 50 kV/cm.

3.5. Study on the Effect of PEF on the Gelatinisation Temperature of Potato Starch Granules

Temperatures at the onset (T_o), peak (T_p) and concluding stage of gelatinisation (T_c), as well as the temperature range ($R = T_c - T_o$) and enthalpy change during gelatinisation (ΔH), are important parameters in investigating the gelatinisation behaviour of starch. These parameters can be obtained using the method of differential scanning calorimetry (DSC) [20,30,38,39]. Figure 5 presents the DSC thermograms for IS, SPM and PP with the corresponding temperature values are presented in Table 2. In the same manner as TGA assay, potato powder (PP) was used as a reference in this DSC assay to represent the original sources of the isolated starch. The IS, SPM, and PP samples from untreated potatoes gelatinised at T_o of 56.67–57.69 °C followed by T_p between 59.81 and 62.08 °C, and finally reached T_c between 64.83 and 67.75 °C. The corresponding temperature range (R) was narrow for IS samples (8.16 ± 1.87 °C), followed by SPM samples at 9.39 ± 0.20 °C and the widest for PP at 10.30 ± 0.47 °C. Overall, it was demonstrated that IS, SPM and PP samples from untreated potatoes shared some similarities in the gelatinisation temperatures, but DSC thermograms also showed that PP samples experienced the lowest ΔH (26.26 ± 0.23 J/g) compared to IS (33.37 ± 7.09 J/g) and SPM (37.32 ± 1.97 J/g) samples.

The present study found that starches isolated from any PEF-treated potatoes (IS), either at increasing electric field strengths up to 1.1 kV/cm and increasing specific energies up to 150 kJ/kg, were gelatinised in a similar matter as starch isolated from untreated potatoes. It indicated that starch in potato tissue was not prone to PEF treatment probably due to its location in potato tissue. The PEF energy delivered to the potato tissue lead to pore formation on the cell membrane as indicated by the increase in conductivity and browning index, but it was not sufficient to cause the change in the starch structure.

With respect to the starch from PEF processing medium (SPM), those samples derived from PEF-treated potatoes were found to share similar values for gelatinisation temperatures of T_o, T_p and T_c. However, there was a significant difference in the gelatinisation temperature range (R) among the SPM samples owing to the intensity of PEF initially applied to the potato samples. In particular, SPM samples from potatoes after PEF treatment at higher specific energies of 150 kJ/kg (PEF 5 and PEF 6) were found to have a narrower gelatinisation temperature range compared to SPM sample from untreated potatoes, i.e., R decreased from 9.39 to 7.98 °C. Since the range of gelatinisation temperature is inversely proportional to the degree of cohesion between crystallites of starch [40], a narrow range of gelatinisation temperature reflects a stronger cohesion between crystallites, particularly after PEF treatment at high specific energy. Moreover, a narrow gelatinisation temperature range exhibited by

SPM samples at high-energy PEF treatment could be due to no potato tissue was present to protect the starch granules from the PEF energy. Hence, the starch was more prone to the PEF treatment. It is important to note that the starch found in the processing medium is generally represented by starch granules available at the surface of potato tissue which were washed out into the PEF processing medium during the process.

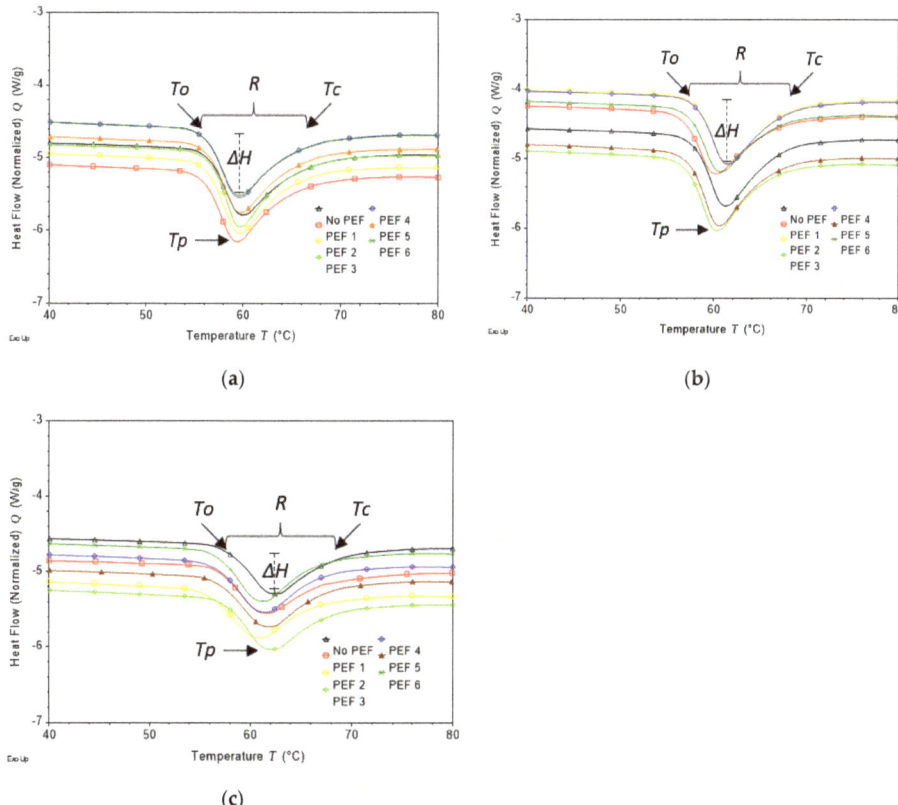

Figure 5. DSC thermograms of isolated starch (a), starch from processing medium (b), and potato powder (c) from untreated (No PEF) and PEF-treated potatoes at 50 kJ/kg specific energy with field strengths (kV/cm) of 0.5 (PEF 1), 0.7 (PEF 2), 0.9 (PEF3), 1.1 (PEF 4) and at 150 kJ specific energy with field strength (kV/cm) of 0.7 (PEF 5) and 0.9 (PEF 6). T_o: temperature of onset gelatinisation, T_p: temperature of peak gelatinisation, T_c: temperature of conclusion gelatinisation, R: gelatinisation temperature range, ΔH: enthalpy change of gelatinisation.

On PP samples, apparent changes were found on PEF-treated samples such as in T_p of PEF 1, PEF 2, PEF 6 and ΔH of all PEF treatments. These changes were presumably associated with the variation in the amount of starch in PP sample during DSC analysis considering that the sample size is limited to about 2.45 mg dry sample whereas, in fact, a lesser amount of starch in the sample matters in lowering the peak height of T_{pdw} in the thermal stability of PP sample compared with those of IS and SPM samples, as shown by the finding in TGA analysis.

The effect of PEF on the narrower temperature range of gelatinisation (R) of SPM sample in the current study was different from the work carried out on 8% (w/w) potato starch in water dispersion and PEF-treated at intensities up to 50 kV/cm [10]. In the Han and co-workers study [10] the temperature range of gelatinisation was slightly broadened with increasing field strength from 30 to

50 kV/cm indicating less structuring of the resulted starch granule after PEF treatment. Furthermore, the previous study did not mention whether specific energy was also important in the change of gelatinisation temperature range. The different phenomenon observed in this study and that of Han and co-workers [10] could also be due to the differences in the potato cultivar and PEF processing parameters used.

3.6. Susceptibility of Starch Granules from PEF-Treated Potatoes towards Heat Stable α-Amylase

The susceptibility of starch to heat stable α-amylase is an important property of starch [41]. This indicates the starch damage [42] occurring in the development of porous starch granule [43] and damage found on the starch due to processing [44] which influence starch functionalities. Unlike corn starch and starch from cereal sources, potato starch is described as a very large and smooth granule [45] with a relatively well-ordered and dense structure. For this reason, potato granules are considered to be relatively resistant to hydrolytic enzymes such as amyloglucosidase and α-amylase [12,46]. As was done for the TGA and DSC assays, potato powder (PP) was used as a reference in the enzyme susceptibility assay to represent the original sources of the isolated starch.

Table 3 presents the susceptibility of IS, SPM, and PP from PEF-treated potatoes and untreated samples expressed as total hydrolysed starch. The total hydrolysed starch found in the current study were on average 68.62% ± 4.08%, 73.09% ± 2.65%, and 62.83% ± 6.16%, respectively, for IS, SPM and PP from untreated potatoes. These values are comparable with the hydrolysed starch reported in other potato cultivars which range from about 68–73% [47]. Moreover, PP samples, regardless of the level of PEF-treatment, consistently showed a lower enzyme susceptibility towards heat stable α-amylase compared to their corresponding IS and SPM samples. This could be due to a lower starch content available in the PP samples. Results from TGA (Table 2) further support this assumption since it was found that the height of T_{pdw} peak of the PP samples was typically lower than IS and SPM samples.

With respect to PP samples, it was clear that any PEF treatments applied on the potatoes did not significantly impact their enzyme susceptibility towards heat stable α-amylase. Likewise, for starch inside the potatoes (IS samples), this study showed that PEF treatments led to no statistically significant effect on their susceptibility towards heat stable α-amylase. However, it was interesting to find that IS samples from PEF-treated potatoes at an electric field strength of 0.7 kV/cm regardless of the specific energy applied (PEF 2 and PEF 5) consistently exhibited higher susceptibility towards heat stable α-amylase compared with those of IS samples from untreated samples and those from potatoes treated at other electric field strengths. Such finding indicated that 0.7 kV/cm could be an optimum electric field strength to be applied on potatoes in order to improve the susceptibility of starch in the PEF-treated potatoes towards heat stable α-amylase leading to a better digestibility.

With respect to the starch found in the processing medium (SPM samples), the impact of PEF treatment on their enzyme susceptibility was also not statistically different. However, it is important to note that SPM samples from potatoes treated at an electric field strength of 0.9 kV/cm combined with a specific energy of 50 (PEF 3) showed a significant improvement in the susceptibility towards heat stable α-amylase compared to SPM sample from untreated potatoes. Another interesting finding was that when applying a higher intensity of specific energy (from 50 to 150 kJ/kg) on potatoes at either electric field strength of 0.7 or 0.9 kV/cm resulted in the corresponding SPM samples (PEF 2 vs. PEF 5, PEF 3 vs. PEF 6) to be more susceptible towards heat stable α-amylase. This phenomenon related to the improved enzyme susceptibility as the result of the application of increasing specific energy for potatoes was only observed for SPM starches. IS or PP starches could not be associated with the disruption of starch granules and the crystalline structure responsible for gelatinisation remained unchanged. An enhancement effect on enzyme susceptibility with increasing specific energy for the SPM was consistent with the structural disruption found for PEF-treated potato starch dispersed in water [10,48].

3.7. Enzyme Susceptibility of Starch from PEF-Treated Potatoes under In Vitro Simulated Human Digestion Condition

The in vitro simulated human digestion assay was used to assess the susceptibility of the starches in potatoes after being treated with PEF towards digestive enzymes. Table 4 summarises the glucose release per volume digest (mg/mL) of simulated human intestine phase at different digestion period i.e., 0, 20, 60, 90 and 120 min from IS and SPM isolated from PEF-treated potatoes and untreated control, compared to its reference, PP samples.

This study found that PEF at all treatments did not significantly change the in vitro simulated digestibility of IS but did pose some major influence on selected SPM and PP samples. After undergoing 120 min of in vitro simulated human digestion during the intestinal phase, it was found that SPM samples from potatoes treated with PEF, particularly PEF 4 and 6, showed a slight reduction in starch digestibility and a prominent reduction was found for that of PEF 4 treatment (i.e., a lower amount of glucose released compared to SPM samples from untreated potatoes). With respect to PP samples, the most distinct difference in the starch digestibility was found after these samples had undergone 90 min of in vitro simulated human digestion during the intestinal phase. PP samples from potatoes treated with PEF, particularly PEF 1 and 6, showed a considerable improvement in starch digestibility (i.e., a higher amount of glucose released compared to PP samples from untreated potatoes).

Some general findings can be seen in the starch digestibility. PEF-treated samples of SPM tend to consistently decrease in starch digestibility. It is presumably due to the disruption in starch structure as indicated by the change in narrowed temperature range of gelatinisation R, as shown by DSC analysis. PEF treated samples of PP tend to be consistently higher in starch digestibility than those of non-PEF treated materials. This indicated that potato tissue was more prone than starch to PEF treatment as already proven by the fact that no change was found in starch properties of IS samples after PEF treatment. Consequently, pores on cell membranes were formed after PEF treatment [1] leading to facilitation of enzyme diffusion to reach the starch. Furthermore, the trend of PEF effect on digestibility of SPM and PP samples was consistently found after the most intensive PEF (PEF 6) which shows the most extreme change in starch digestibility compared with their untreated counterparts.

The finding in starch digestibility of PEF-treated potato gave an indication that concern to nutritional attributes and safety of starch from PEF treated potato is unwarranted, as already shown that starch in potato remained unchanged after PEF treatment. The increase in starch digestibility of potato (powder) is an additional impact of modification in microstructure of solid plant food after PEF treatment [1] from other impacts that has been previously studied, such as the reduction of cutting force for potato [2] and sweet potato [3], the oil uptake during frying and retaining the textural properties [4]. Furthermore, PEF treatment at a proper condition without leading to excessive external energy could potentially be used on starch to decrease its digestibility.

Regarding the nutritional status of starch, a decrease in starch digestibility is considered as healthy for some cohorts of consumers. In this context, starch digestibility [22] is divided into (a) rapidly digestible starch (RDS): starch digested within 20 min; (b) slowly digestible starch (SDS): starch digested between 20 and 120 min; and (c) resistant starch (RS): starch digested after over 120 min. The health benefit of decreased digestibility of starch can be attributed either to SDS [49] or RS [50]. SDS is considered as healthy to lower the risk of drastic increase of postprandial blood sugar while RS is considered as healthy to feed the colon microbiome leading to colonic health of the host [50]. Thus, RS has been considered as prebiotic [51] and adopted as a functional ingredient [52].

Therefore, a decrease in digestibility of starch is an intended outcome of food processing to produce resistant starch [53]. Some processing techniques that have been used to lower starch digestibility are gamma irradiation of corn starch [54], high pressure treatment of starch from wheat, tapioca, potato, corn, and waxy corn [55], heat moisture treatment of mung beans [56] and rice starch [57], dual autoclaving-retrogradation of rice starch [58] and annealing of common buckwheat starch [59]. The outcome of the present study shows that PEF can be considered as a technology that contributes to lowering starch resistance.

Table 3. Susceptibility, expressed as % (w/w) hydrolysed starch, of isolated starch, starch from PEF processing medium, and potato powder from PEF-treated potatoes towards heat-stable α-amylase and amyloglucosidase.

Samples	Untreated (No PEF)	PEF Treatments (kV/cm, kJ/kg)						F Value (6,14)	p-Value
		PEF 1 (0.5, 50)	PEF 2 (0.7, 50)	PEF 3 (0.9, 50)	PEF 4 (1.1, 50)	PEF 5 (0.7, 150)	PEF 6 (0.9, 150)		
Isolated starch (IS)	68.62 ± 4.08	65.45 ± 4.39	74.70 ± 3.65	71.23 ± 6.42	67.95 ± 7.91	70.18 ± 6.46	67.86 ± 5.96	0.80	0.58
Starch from PEF processing medium (SPM)	73.09 ± 2.65	78.66 ± 8.74	68.94 ± 3.56	78.78 ± 1.92 *	75.80 ± 3.79	74.53 ± 6.33	83.25 ± 10.9	1.64	0.21
Potato powder (PP)	62.83 ± 6.16	58.73 ± 10.21	61.10 ± 6.55	60.31 ± 5.33	62.28 ± 4.23	58.37 ± 5.71	62.77 ± 8.39	0.22	0.97

Result expressed as mean ± standard deviation (n = 3). Values in the same row not sharing the same superscript are significantly different at $p < 0.05$ analysed with one-way ANOVA and Tukey's post hoc test. Means in the same row with asterisk * are significantly different from untreated/No PEF sample (95% interval confidence) analysed with independent t-test.

Table 4. Digestibility, expressed as milligram amount of glucose released per mL digest, of isolated starch (IS), starch from PEF processing medium (SPM), and potato powder (PP) from PEF-treated potatoes.

Time (min)	Untreated (No PEF)	PEF Treatments (kV/cm, kJ/kg)						F Value (6,35)	p-Value
		PEF 1 (0.5, 50)	PEF 2 (0.7, 50)	PEF 3 (0.9, 50)	PEF 4 (1.1, 50)	PEF 5 (0.7, 150)	PEF 6 (0.9, 150)		
Isolated starch (IS)									
0	4.53 ± 0.49	4.75 ± 0.69	4.63 ± 0.25	4.57 ± 0.17	4.55 ± 0.21	4.88 ± 0.50	4.69 ± 0.13	0.61	0.72
20	4.67 ± 0.24	4.81 ± 0.48	4.62 ± 0.26	4.75 ± 0.28	4.67 ± 0.14	5.14 ± 0.41*	4.59 ± 0.21	2.24	0.06
60	4.74 ± 0.14	4.83 ± 0.32	4.86 ± 0.40	4.70 ± 0.14	4.92 ± 0.38	5.06 ± 0.29*	4.92 ± 0.31	1.01	0.43
90	4.68 ± 0.40	4.82 ± 0.48	4.57 ± 0.29	4.52 ± 0.33	4.54 ± 0.25	4.71 ± 0.41	4.60 ± 0.12	0.59	0.74
120	4.46 ± 0.22	4.53 ± 0.30	4.65 ± 0.29	4.49 ± 0.24	4.50 ± 0.21	4.64 ± 0.42	4.46 ± 0.20	0.51	0.80
Starch from PEF processing medium (SPM)									
0	4.45 ± 0.23	4.26 ± 0.27	4.36 ± 0.40	4.48 ± 0.19	4.41 ± 0.39	4.09 ± 0.29*	4.24 ± 0.28	1.24	0.31
20	4.47 ± 0.41	4.37 ± 0.23	4.46 ± 0.21	4.43 ± 0.15	4.34 ± 0.33	4.32 ± 0.20	4.33 ± 0.26	0.34	0.91
60	4.61 ± 0.38	4.27 ± 0.14	4.36 ± 0.20	4.45 ± 0.16	4.31 ± 0.26	4.37 ± 0.31	4.43 ± 0.13	1.32	0.28
90	4.55 ± 0.17	4.29 ± 0.30	4.35 ± 0.17	4.33 ± 0.16	4.24 ± 0.15*	4.29 ± 0.06*	4.21 ± 0.25*	2.00	0.09
120	4.35 ± 0.18 [a]	4.16 ± 0.10 [ab]	4.17 ± 0.21 [ab]	4.10 ± 0.23 [ab]	3.92 ± 0.22 [b,*]	4.11 ± 0.20 [ab]	4.06 ± 0.23 [ab,*]	2.60	0.04
Potato powder (PP)									
0	4.71 ± 0.18 [b]	5.04 ± 0.64 [ab]	5.13 ± 0.55 [ab]	5.01 ± 0.30 [ab]	4.94 ± 0.26 [ab]	5.02 ± 0.30 [ab]	5.58 ± 0.39 [a,*]	2.57	0.04
20	4.71 ± 0.57	4.90 ± 0.43	5.05 ± 0.32	4.95 ± 0.54	5.16 ± 0.28	5.13 ± 0.50	5.38 ± 0.50	1.30	0.28
60	5.25 ± 0.26	4.97 ± 0.39	5.09 ± 0.49	5.31 ± 0.48	5.17 ± 0.19	5.23 ± 0.17	5.41 ± 0.24	1.06	0.40
90	4.88 ± 0.29 [b]	5.33 ± 0.09 [a,*]	4.98 ± 0.25 [ab]	5.13 ± 0.26 [ab]	5.15 ± 0.27 [ab]	5.09 ± 0.07 [ab]	5.30 ± 0.23 [a,*]	3.02	0.02
120	4.84 ± 0.36	5.18 ± 0.31*	5.01 ± 0.42	5.06 ± 0.15	5.16 ± 0.21	4.96 ± 0.26	5.28 ± 0.30*	1.49	0.21

Result expressed as the mean ± standard deviation of six tests (triplicate samples with duplicate assay). Means in the same row not sharing the same superscript are significantly different at $p < 0.05$ analysed with one-way ANOVA and Tukey's post hoc test. Means in the same row with asterisk * are significantly different from untreated/No PEF sample (95% interval confidence) analysed with an independent t-test.

4. Conclusions

This study confirmed that the starch inside potatoes after being treated with PEF remained in its native state as indicated by the presence of birefringence properties under a polarised microscope. The thermal stability, gelatinisation behaviour, susceptibility of starch in potato to heat-stable α-amylase and digestive enzymes under in vitro simulated human digestion conditions remained unchanged after PEF treatment. However, starch on the surface of potato (that leached into the medium, as SPM) apparently was found to be more prone to PEF treatment as indicated by a narrow range of the gelatinisation temperature especially after PEF treatment at 150 kJ/kg, leading to less digestible starch. Since PEF processing did not change the properties of starch in potatoes as shown in this study it is suggested that the phenomena previously reported in the literature, such as reduced processing intensities for frying or changes in sensory properties after frying, is driven more by other factors, such as structural changes of the potato tissues, and not by modification of potato starch granules.

Supplementary Materials: The following are available online at http://www.mdpi.com/2304-8158/8/5/159/s1, Video S1: Gelatinisation behaviour of starch from potatoes treated with PEF at an electric field strength of 1.1 kV/cm and energy input of 50.1 kJ/kg (PEF 4) observed under a hot-stage microscopy (at a magnification of 200×) from 30–80 °C at a rate of 5 °C/min.

Author Contributions: I.O. and S.B.M.A. conceived and designed the experiments; S.B.M.A. and S.Y.L. performed the experiments; and S.B.M.A. analysed the data; S.B.M.A. wrote the paper with the contribution from S.Y.L., D.A. and I.O.

Funding: Abduh thanks the Indonesia Endowment Fund for Education (LPDP) for providing a PhD scholarship. This research was supported by Riddet Institute, a New Zealand Centre of Research Excellence funded by the New Zealand Tertiary Education Commission.

Conflicts of Interest: The authors declare no conflict of interest.

References

1. Faridnia, F.; Burritt, D.J.; Bremer, P.J.; Oey, I. Innovative approach to determine the effect of pulsed electric fields on the microstructure of whole potato tubers: Use of cell viability, microscopic images and ionic leakage measurements. *Food Res. Int.* **2015**, *77*, 556–564. [CrossRef]
2. Ignat, A.; Manzocco, L.; Brunton, N.P.; Nicoli, M.C.; Lyng, J.G. The effect of pulsed electric field pre-treatments prior to deep-fat frying on quality aspects of potato fries. *Innov. Food Sci. Emerg. Technol.* **2015**, *29*, 65–69. [CrossRef]
3. Liu, T.; Dodds, E.; Leong, S.Y.; Eyres, G.T.; Burritt, D.J.; Oey, I. Effect of pulsed electric fields on the structure and frying quality of "kumara" sweet potato tubers. *Innov. Food Sci. Emerg. Technol.* **2017**, *39*, 197–208. [CrossRef]
4. Shayanfar, S.; Chauhan, O.; Toepfl, S.; Heinz, V. The interaction of pulsed electric fields and texturizing—Antifreezing agents in quality retention of defrosted potato strips. *Int. J. Food Sci. Technol.* **2013**, *48*, 1289–1295. [CrossRef]
5. Fauster, T.; Schlossnikl, D.; Rath, F.; Ostermeier, R.; Teufel, F.; Toepfl, S.; Jaeger, H. Impact of pulsed electric field (PEF) pretreatment on process performance of industrial French fries production. *J. Food Eng.* **2018**, *235*, 16–22. [CrossRef]
6. Giteru, S.G.; Oey, I.; Ali, M.A. Feasibility of using pulsed electric fields to modify biomacromolecules: A review. *Trends Food Sci. Technol.* **2018**, *72*, 91–113. [CrossRef]
7. Zeng, F.; Gao, Q.Y.; Han, Z.; Zeng, X.A.; Yu, S.J. Structural properties and digestibility of pulsed electric field treated waxy rice starch. *Food Chem.* **2016**, *194*, 1313–1319. [CrossRef]
8. Han, Z.; Zeng, X.A.; Fu, N.; Yu, S.J.; Chen, X.D.; Kennedy, J.F. Effects of pulsed electric field treatments on some properties of tapioca starch. *Carbohydr. Polym.* **2012**, *89*, 1012–1017. [CrossRef] [PubMed]
9. Han, Z.; Yu, Q.; Zeng, X.A.; Luo, D.H.; Yu, S.J.; Zhang, B.S.; Chen, X.D. Studies on the microstructure and thermal properties of pulsed electric fields (PEF)-treated maize starch. *Int. J. Food Eng.* **2012**, *8*. [CrossRef]
10. Han, Z.; Zeng, X.A.; Yu, S.J.; Zhang, B.S.; Chen, X.D. Effects of pulsed electric fields (PEF) treatment on physicochemical properties of potato starch. *Innov. Food Sci. Emerg. Technol.* **2009**, *10*, 481–485. [CrossRef]

11. Han, Z.; Zeng, X.; Zhang, B.; Yu, S. Effects of pulsed electric fields (PEF) treatment on the properties of corn starch. *J. Food Eng.* **2009**, *93*, 318–323. [CrossRef]
12. Bertoft, E.; Blennow, A. Structure of potato starch. In *Advances in Potato Chemistry and Technology*, 2nd ed.; Academic Press: Cambridge, MA, USA, 2016; pp. 57–73.
13. Burlingame, B.; Mouillé, B.; Charrondière, R. Nutrients, bioactive non-nutrients and anti-nutrients in potatoes. *J. Food Compos. Anal.* **2009**, *22*, 494–502. [CrossRef]
14. Singh, J.; Kaur, L. Chemistry, processing, and nutritional attributes of potatoes—An introduction. In *Advances in Potato Chemistry and Technology*, 2nd ed.; Academic Press: Cambridge, MA, USA, 2016.
15. Yang, Y.; Achaerandio, I.; Pujolà, M. Effect of the intensity of cooking methods on the nutritional and physical properties of potato tubers. *Food Chem.* **2016**, *197*, 1301–1310. [CrossRef]
16. Tian, J.; Chen, S.; Wu, C.; Chen, J.; Du, X.; Chen, J.; Liu, D.; Ye, X. Effects of preparation methods on potato microstructure and digestibility: An in vitro study. *Food Chem.* **2016**, *211*, 564–569. [CrossRef]
17. Tian, J.; Chen, S.; Shi, J.; Chen, J.; Liu, D.; Cai, Y.; Ogawa, Y.; Ye, X. Microstructure and digestibility of potato strips produced by conventional frying and air-frying: An in vitro study. *Food Struct.* **2017**, *14*, 30–35. [CrossRef]
18. Nielsen, H.B.; Sonne, A.-M.; Grunert, K.G.; Banati, D.; Pollák-Tóth, A.; Lakner, Z.; Olsen, N.V.; Žontar, T.P.; Peterman, M. Consumer perception of the use of high-pressure processing and pulsed electric field technologies in food production. *Appetite* **2009**, *52*, 115–126. [CrossRef]
19. Lee, P.Y.; Lusk, K.; Mirosa, M.; Oey, I. Effect of information on Chinese consumers' perceptions and purchase intention for beverages processed by high pressure processing, pulsed-electric field and heat treatment. *Food Qual. Prefer.* **2015**, *40*, 16–23. [CrossRef]
20. Singh, J.; Singh, N. Studies on the morphological, thermal and rheological properties of starch separated from some Indian potato cultivars. *Food Chem.* **2001**, *75*, 67–77. [CrossRef]
21. Bordoloi, A.; Singh, J.; Kaur, L. In vitro digestibility of starch in cooked potatoes as affected by guar gum: Microstructural and rheological characteristics. *Food Chem.* **2012**, *133*, 1206–1213. [CrossRef]
22. Goñi, I.; Garcia-Alonso, A.; Saura-Calixto, F. A starch hydrolysis procedure to estimate glycemic index. *Nutr. Res.* **1997**, *17*, 427–437. [CrossRef]
23. Minekus, M.; Alminger, M.; Alvito, P.; Ballance, S.; Bohn, T.; Bourlieu, C.; Carrière, F.; Boutrou, R.; Corredig, M.; Dupont, D.; et al. A standardised static in vitro digestion method suitable for food—An international consensus. *Food Funct.* **2014**, *5*, 1113–1124. [CrossRef]
24. Leong, S.Y.; Oey, I. Effect of pulsed electric field treatment on enzyme kinetics and thermostability of endogenous ascorbic acid oxidase in carrots (*Daucus carota* cv. Nantes). *Food Chem.* **2014**, *146*, 538–547. [CrossRef]
25. Oey, I.; Faridnia, F.; Leong, S.Y.; Burritt, D.J.; Liu, T. Determination of Pulsed Electric Fields Effect on the Structure of Potato Tubers. In *Handbook of Electroporation*; Springer International Publishing: New York, NY, USA, 2017; pp. 1–20.
26. Busch, J.M. Enzymic browning in potatoes: A simple assay for a polyphenol oxidase catalysed reaction. *Biochem. Educ.* **1999**, *27*, 171–173. [CrossRef]
27. Bertoft, E. Understanding starch structure: Recent progress. *Agronomy* **2017**, *7*, 56. [CrossRef]
28. Liu, Q.; Charlet, G.; Yelle, S.; Arul, J. Phase transition in potato starch-water system I. Starch gelatinization at high moisture level. *Food Res. Int.* **2002**, *35*, 397–407. [CrossRef]
29. Maache-Rezzoug, Z.; Zarguili, I.; Loisel, C.; Doublier, J.L.; Buléon, A. Investigation on structural and physicochemical modifications of standard maize, waxy maize, wheat and potato starches after DIC treatment. *Carbohydr. Polym.* **2011**, *86*, 328–336. [CrossRef]
30. Svensson, E.; Eliasson, A.C. Crystalline changes in native wheat and potato starches at intermediate water levels during gelatinization. *Carbohydr. Polym.* **1995**, *26*, 171–176. [CrossRef]
31. Parada, J.; Aguilera, J.M. Effect of native crystalline structure of isolated potato starch on gelatinization behavior and consequently on glycemic response. *Food Res. Int.* **2012**, *45*, 238–243. [CrossRef]
32. Jarvis, M.C.; Hall, M.A.; Threlfall, D.R.; Friend, J. The polysaccharide structure of potato cell walls: Chemical fractionation. *Planta* **1981**, *152*, 93–100. [CrossRef] [PubMed]
33. Chakraborty, M.; McDonald, A.G.; Nindo, C.; Chen, S. An α-glucan isolated as a co-product of biofuel by hydrothermal liquefaction of Chlorella sorokiniana biomass. *Algal Res.* **2013**, *2*, 230–236. [CrossRef]

34. Liu, X.; Yu, L.; Liu, H.; Chen, L.; Li, L. In situ thermal decomposition of starch with constant moisture in a sealed system. *Polym. Degrad. Stab.* **2008**, *93*, 260–262. [CrossRef]
35. Srichuwong, S.; Sunarti, T.C.; Mishima, T.; Isono, N.; Hisamatsu, M. Starches from different botanical sources I: Contribution of amylopectin fine structure to thermal properties and enzyme digestibility. *Carbohydr. Polym.* **2005**, *60*, 529–538. [CrossRef]
36. Fang, J.; Fowler, P.; Tomkinson, J.; Hill, C.A. The preparation and characterisation of a series of chemically modified potato starches. *Carbohydr. Polym.* **2002**, *47*, 245–252. [CrossRef]
37. Leivas, C.L.; da Costa, F.J.O.G.; de Almeida, R.R.; de Freitas, R.J.S.; Stertz, S.C.; Schnitzler, E. Structural, physico-chemical, thermal and pasting properties of potato (*Solanum tuberosum* L.) flour. *J. Therm. Anal. Calorim.* **2013**, *111*, 2211–2216. [CrossRef]
38. Pravisani, C.I.; Califano, A.N.; Calvelo, A. Kinetics of starch gelatinization in potato. *J. Food Sci.* **1985**, *50*, 657–660. [CrossRef]
39. Karlsson, M.E.; Eliasson, A.C. Gelatinization and retrogradation of potato (*Solanum tuberosum*) starch in situ as assessed by differential scanning calorimetry (DSC). *LWT—Food Sci. Technol.* **2003**, *36*, 735–741. [CrossRef]
40. Maache-Rezzoug, Z.; Zarguili, I.; Loisel, C.; Queveau, D.; Buléon, A. Structural modifications and thermal transitions of standard maize starch after DIC hydrothermal treatment. *Carbohydr. Polym.* **2008**, *74*, 802–812. [CrossRef]
41. Marchal, L.M.; Jonkers, J.; Franke, G.T.; de Gooijer, C.D.; Tramper, J. The effect of process conditions on the α-amylolytic hydrolysis of amylopectin potato starch: An experimental design approach. *Biotechnol. Bioeng.* **1999**, *62*, 348–357. [CrossRef]
42. Morgan, J.E.; Williams, P.C. Starch damage in wheat flours: A comparison of enzymatic, iodometric, and near-infrared reflectance techniques. *Cereal Chem.* **1995**, *72*, 209–212.
43. Jung, Y.S.; Lee, B.H.; Yoo, S.H. Physical structure and absorption properties of tailor-made porous starch granules produced by selected amylolytic enzymes. *PLoS ONE* **2017**, *12*, e0181372. [CrossRef] [PubMed]
44. Vallons, K.J.R.; Arendt, E.K. Effects of high pressure and temperature on the structural and rheological properties of sorghum starch. *Innov. Food Sci. Emerg. Technol.* **2009**, *10*, 449–456. [CrossRef]
45. Dhital, S.; Shrestha, A.K.; Gidley, M.J. Relationship between granule size and in vitro digestibility of maize and potato starches. *Carbohydr. Polym.* **2010**, *82*, 480–488. [CrossRef]
46. Sujka, M.; Jamroz, J. α-Amylolysis of native potato and corn starches—SEM, AFM, nitrogen and iodine sorption investigations. *LWT—Food Sci. Technol.* **2009**, *42*, 1219–1224. [CrossRef]
47. Pinhero, R.G.; Waduge, R.N.; Liu, Q.; Sullivan, J.A.; Tsao, R.; Bizimungu, B.; Yada, R.Y. Evaluation of nutritional profiles of starch and dry matter from early potato varieties and its estimated glycemic impact. *Food Chem.* **2016**, *203*, 356–366. [CrossRef] [PubMed]
48. Li, Q.; Wu, Q.-Y.; Jiang, W.; Qian, J.-Y.; Zhang, L.; Wu, M.; Rao, S.-Q.; Wu, C.-S. Effect of pulsed electric field on structural properties and digestibility of starches with different crystalline type in solid state. *Carbohydr. Polym.* **2019**, *207*, 362–370. [CrossRef] [PubMed]
49. Lehmann, U.; Robin, F. Slowly digestible starch—Its structure and health implications: A review. *Trends Food Sci. Technol.* **2007**, *18*, 346–355. [CrossRef]
50. Nugent, A.P. Health properties of resistant starch. *Nutr. Bull.* **2005**, *30*, 27–54. [CrossRef]
51. Topping, D.L.; Fukushima, M.; Bird, A.R. Resistant starch as a prebiotic and synbiotic: State of the art. *Proc. Nutr. Soc.* **2003**, *62*, 171–176. [CrossRef]
52. Fuentes-Zaragoza, E.; Riquelme-Navarrete, M.J.; Sánchez-Zapata, E.; Pérez-Álvarez, J.A. Resistant starch as functional ingredient: A review. *Food Res. Int.* **2010**, *43*, 931–942. [CrossRef]
53. Thompson, D.B. Strategies for the manufacture of resistant starch. *Trends Food Sci. Technol.* **2001**, *11*, 245–253. [CrossRef]
54. Lee, J.S.; Ee, M.L.; Chung, K.H.; Othman, Z. Formation of resistant corn starches induced by gamma-irradiation. *Carbohydr. Polym.* **2013**, *97*, 614–617. [CrossRef] [PubMed]
55. Papathanasiou, M.M.; Reineke, K.; Gogou, E.; Taoukis, P.S.; Knorr, D. Impact of high pressure treatment on the available glucose content of various starch types: A case study on wheat, tapioca, potato, corn, waxy corn and resistant starch (RS3). *Innov. Food Sci. Emerg. Technol.* **2015**, *30*, 24–30. [CrossRef]
56. Li, S.; Ward, R.; Gao, Q. Effect of heat-moisture treatment on the formation and physicochemical properties of resistant starch from mung bean (*Phaseolus radiatus*) starch. *Food Hydrocoll.* **2011**, *25*, 1702–1709. [CrossRef]

57. Van Hung, P.; Vien, N.L.; Lan Phi, N.T. Resistant starch improvement of rice starches under a combination of acid and heat-moisture treatments. *Food Chem.* **2016**, *191*, 67–73. [CrossRef] [PubMed]
58. Ashwar, B.A.; Gani, A.; Wani, I.A.; Shah, A.; Masoodi, F.A.; Saxena, D.C. Production of resistant starch from rice by dual autoclaving-retrogradation treatment: In vitro digestibility, thermal and structural characterization. *Food Hydrocoll.* **2016**, *56*, 108–117. [CrossRef]
59. Liu, H.; Guo, X.; Li, W.; Wang, X.; Lv, M.; Peng, Q.; Wang, M. Changes in physicochemical properties and in vitro digestibility of common buckwheat starch by heat-moisture treatment and annealing. *Carbohydr. Polym.* **2015**, *132*, 237–244. [CrossRef]

© 2019 by the authors. Licensee MDPI, Basel, Switzerland. This article is an open access article distributed under the terms and conditions of the Creative Commons Attribution (CC BY) license (http://creativecommons.org/licenses/by/4.0/).

Article

Selection and Evaluation of 21 Potato (*Solanum Tuberosum*) Breeding Clones for Cold Chip Processing

Benjamin Opuko Wayumba [1,2], Hyung Sic Choi [1,2] and Lim Young Seok [1,2,*]

1. Department of Bio-Health Technology, Kangwon National University, Chuncheon 200-701, Korea; bwayumba@gmail.com (B.O.W.); luckychs@naver.com (H.S.C.)
2. KPBR (Korea Potato Breeding Resource Bank), Kangwon National University, Chuncheon 200-701, Korea
* Correspondence: potatoschool@gmail.com

Received: 11 February 2019; Accepted: 11 March 2019; Published: 14 March 2019

Abstract: Quality evaluations in potatoes are of necessity to meet the strict demands of the chip processing industry. Important parameters assessed include specific gravity, dry matter content, chip color, reducing sugars, and glycoalkaloids. This study was designed with the purpose of identifying specialized potato clones with acceptable qualities for processing chips, in comparison with two selected control varieties, Dubaek and Superior. As a result, high dry matter and specific gravity were observed for three potato clones, and the quantified ά-solanine levels ranged from 0.15 to 15.54 mg·100 g^{-1} fresh weight (FW). Significant variations ($p < 0.05$) in reducing sugar levels were observed in clones stored at different temperature conditions. After reconditioning of the tubers at 22 °C for 21 days, a significant drop in reducing sugar levels was recorded. In addition, fried chips for each potato clone were evaluated, and the color measured on the basis of the Snack Food Association (SFA) chip color score standard. Reconditioned tubers exhibited much lighter and better chip color compared to their counterparts cold-stored at 4 °C. This study observed that for quality processing of potato chips, clones with combined traits of high dry matter, low levels of glycoalkaloids and reducing sugars, and acceptable chip color should be used as raw materials.

Keywords: chip processing; cold storage; reconditioning; reducing sugar; potato

---

1. Introduction

The importance of potatoes as food includes their use as processed products, since they can be processed into many value-added food items like chips (crisps), dried flakes, French fries, and various snacks. Processing of potatoes into chips and other products has a great potential in ensuring reduced loss and waste post-harvesting, in handling and storage.

The industry requirements for processing chips in terms of tuber appearance should be of shallow eyes, appropriate weight, and round-oval shape [1], whereas long-oval-shaped tubers are preferred for processing French fries. In addition, tubers should be free from cracks, hollow heart, secondary damage, rusty spots, and greening. The peel color, flesh color, and flour content should also satisfy the national consumer preference of any given country.

In the recent years, there has been increasing demand for food convenience, and potato chips meet these requirements. Potato chips are one of the most convenient ways to serve potatoes, with minimal preparation needed. Potato chips can be made flavored, plain, with chili, cheese, or seasoning. Their slicing can be plain, regular, wavy, or waffle-cut. They are a very popular food item for picnics and parties and can be served conveniently any time as a snack. This withstanding, the quality variation of the raw materials used to produce chips has led to large differences in their culinary value and consumer acceptance. Therefore, it is important to address this issue prior to processing in order to maintain taste and flavor. Potato flavor results from the combination of taste, aroma, and texture [2]. However, the resultant flavor after processing is sometimes altered because of the effects of cold storage. Although the storage of potato tubers at low temperature (4 °C) minimizes tuber respiration and sprouting [3], low-temperature storage also activates a process known as low-temperature sweetening that results in the conversion of starch to reducing sugars. High levels of reducing sugars (glucose and fructose) lead to undesirable non-enzymatic browning reactions and the formation of amine groups of free amino acids [4] that cause off-flavors and darkening of the processed potato products [5] during frying.

In order to mitigate this problem, it is necessary to develop and identify cultivars with first-rate qualities that can be a game changer in the chip processing industry. In this regard, this study was designed with the purpose of identifying specialized potato clones with acceptable tuber qualities for processing chips with respect to tuber specific gravity, dry matter content, chip frying tests, solanine content, and reducing sugar profiles under different temperature storage regimes.

2. Materials and Methods

2.1. Plant Materials

Among potato breeding clones grown in 2017 spring season in South Korea, Gangwon-do Province, 21 breeding clones and 2 control varieties (Dubaek & Superior) were finally selected for further evaluation for chip processing. A criterion of fully mature and healthy tubers of high yield were used to select the tubers. The selected tubers from each clone were stored at different temperatures: 22 °C at harvest, 4 °C during cold storage, and 10 °C for 3 weeks before experimental analysis. The cold-stored tubers were then reconditioned at 22 °C for additional three weeks and re-evaluated.

2.2. Dry Matter (DM) Content

Clean, dry, standard-size aluminum foil boats were used as crucibles. Sliced samples, 3 mm thin, (10 g) of each potato clone were weighed in the foil boat, and the initial weight measured in grams. The samples were dried in an electric oven overnight for 16 h at 105 °C to a constant weight. The total solid content of each clone was calculated as a percentage. Four 10 g samples were measured for each clone. (DM % = (final dried weight/initial weight) × 100).

2.3. Specific Gravity (SG)

Six (80–130 g) tubers from each potato clone were weighed in air and under water. Tubers were all weight-matched to ensure uniformity per clone. Average underwater weights were used to calculate the specific gravity.

$$SG = (\text{weight in air}/ \text{weight in water} \times \text{density water} (g \cdot cm^{-3}))$$

2.4. Reducing Sugar

The reducing sugar was quantified by the dinitrosalicylic acid (DNS) method [6]. For the procedure, 1 g of freeze-dried powdered sample was added to 50 mL of distilled water and shaken for 1 h. The supernatant was centrifuged at 3000 rpm for 30 min. In the experiment, 2 mL of DNS was added to 1 mL of sample in a glass tube (15 × 100) and incubated at 99 °C in a water bath for 10 min. After cooling, the absorbance was measured at 550 nm in a microplate reader. The experiment was replicated twice. The samples were analyzed against glucose standards of known concentrations.

2.5. Chip Frying Test

Potato tubers of different clones were washed clean and sliced from apical to basal ends. Thin slices (1 mm) were cut using a hand-held slicer. The slices were then rinsed in water to remove excess starch and blot-dried on paper towels. A deep-fryer machine containing vegetable oil was used to prepare the chips. At a constant temperature of 180 °C, the chips were fried for 2 min and then placed on a paper towel to drain off excess oil. The chips were then photographed after cooling, using a high-resolution Canon EOS 5D Digital SLR Camera (Chuncheon, Korea), and their color measured on the basis of the Snack Food Association (SFA) chip color measurement standard.

2.6. Glycoalkaloid Analysis

Tubers of 21 clones and 2 control varieties were used. The samples were sliced whole with the skin intact and freeze-dried (ilshin Lab Co., LTD, Chuncheon, Korea) prior to glycoalkaloid content analysis to determine ά-solanine. All reagents for the analysis were prepared in HPLC-grade deionized water. The ά-solanine standard and acetonitrile HPLC-grade were purchased form Sigma Aldrich (Chuncheon, Korea). All other chemicals used were of standard analytical grade.

Analysis was carried out according to Hellenas et al., 1995a [7], with some modifications. In the experiment, 10 g powder extracted from a whole tuber was mixed with 20 mL water/acetic acid/sodium bisulphate (N_aHSO_3) in the ratio 95:5:0.5 (v/v/w) in a blender. The mixture was diluted up to 50 mL using the same extraction solvent and vacuum-filtered through Whatman filter paper No.1. The filtrate was further cleaned up by centrifugation at 6500 rpm for 10 min. A volume of 5 mL of supernatant was obtained which was further cleaned up by extraction with 1mL acetonitrile, 1 mL water/acetic acid/N_aHSO_3 solvent, 0.8 mL water/acetonitrile, 0.8 mL acetonitrile/0.022 M potassium phosphate buffer, pH 7.6, 55:45 (v/v), all preconditioned and filtered.

ά-solanine was quantified using High-performance liquid chromatography (HPLC) apparatus (Shimadzu Corp, Tokyo, Japan) consisting of a 250 × 4.6 mm NH_2 analytical column; mobile phase of acetonitrile/0.022 M potassium dihydrogenphosphate (KH_2PO_4) buffer, pH 4.7, 75:25 (v/v), at a flow rate of 1.5 mL/min, UV absorption of 200 nm wavelength, and 0.05 Absorbance units full scale (AUFS) sensitivity. The injection volume used was 20 µL. The retention time for ά-solanine was approximately 5.7 min, and sample extracts were quantified by comparing their corresponding peak areas with those of known amounts of standard. All samples were replicated twice and read against solanine standards of known concentrations.

2.7. Statistical Analysis

The samples' mean variations were analyzed by analysis of variance (ANOVA) software using IBM Corp. IBM SPSS Statistics for Windows, Version 23.0 Armonk, NY, USA. Significant differences were evaluated using Tukey's test at a 95% confidence interval.

3. Results and Discussion

3.1. Dry Matter Content and Specific Gravity

The potato clones varied with respect to dry matter content and specific gravity (Table 1). Tuber dry matter ranged from 18.37 ± 1.08 to 25.10 ± 0.88%, and specific gravity from 1.079 ± 0.006 to 1.096 ± 0.005. For both parameters, four specific clones (V50, V48, N109-35, N357) performed better than the control variety Dubaek (DM 23.22%, SG 1.088) with $p < 0.05$, while other 11 cultivars showed a similar trend, with significantly higher total solids and specific gravity levels compared to Superior (DM 21.92%, SG 1.078).

Table 1. Tuber dry matter content and specific gravity of potato clones.

Clones	Dry Matter (%)	Specific Gravity
V50	25.10 ± 0.88 [a]	1.096 ± 0.005 [a]
V48	24.84 ± 0.30 [a]	1.093 ± 0.008 [ab]
N109-35	24.75 ± 0.63 [ab]	1.092 ± 0.006 [abc]
N357	24.02 ± 0.42 [abc]	1.091 ± 0.005 [a–d]
Dubaek	23.22 ± 0.41 [a–d]	1.088 ± 0.006 [a–e]
V93	23.14 ± 0.51 [a–d]	1.082 ± 0.005 [a–g]
A165-4	23.06 ± 0.74 [a–d]	1.076 ± 0.010 [d–g]
N2-6	22.77 ± 0.44 [a–d]	1.086 ± 0.010 [a–f]
N96-29	22.43 ± 1.19 [b–e]	1.074 ± 0.008 [efg]
Gogu	22.36 ± 0.57 [cde]	1.073 ± 0.003 [fg]
V18	22.27 ± 0.25 [cde]	1.079 ± 0.008 [b–g]
N189	22.26 ± 0.72 [cde]	1.079 ± 0.004 [b–g]
V17	22.22 ± 1.60 [cde]	1.086 ± 0.008 [a–f]
V16	21.97 ± 0.85 [cde]	1.081 ± 0.004 [b–g]
A188-10	21.84 ± 0.74 [cde]	1.069 ± 0.008 [g]
A9	21.77 ± 0.24 [cde]	1.082 ± 0.007 [a–g]
Superior	21.92 ± 0.69 [cde]	1.078 ± 0.005 [c–g]
N69-2	21.67 ± 2.50 [cde]	1.083 ± 0.008 [a–g]
V63	21.64 ± 0.36 [cde]	1.082 ± 0.008 [a–g]
A27	21.60 ± 0.74 [de]	1.080 ± 0.009 [b–g]
A186-23	20.33 ± 0.51 [ef]	1.074 ± 0.006 [efg]
A12-5	20.20 ± 0.93 [ef]	1.071 ± 0.005 [g]
A35-6	18.37 ± 1.08 [g]	1.079 ± 0.006 [b–g]

Table results expressed as mean ± standard deviation. The means with different letters in each column are significantly different with $p < 0.05$ in Tukey's test.

The observed differences in dry matter and specific gravity among potato clones may be mainly due to genetic constitution, since all clones were grown and tested in one location with similar management. Tuber dry matter content is the most important attribute that determines the quality and yield of fried and dehydrated products. Higher dry matter or total solids result in higher recovery of the processed products, lower oil absorption, less energy consumption, and crispier texture [8,9].

Based on Pearson correlation test (r = 0.696) performed in the study, it was observed that clones with high specific gravity exhibited significantly ($p < 0.01$) high dry matter content, which is an important feature in the selection for processing chips. Tuber specific gravity is often used in the processing industry as a means for a quick estimation of total solids, as the two parameters are highly correlated [10,11].

3.2. Reducing Sugar

Significant ($p < 0.05$) variations in the reducing sugar content of the potato clones were observed under different temperature storage regimes. Overall, the reducing sugar content across the three temperature storage conditions ranged from 2.24 to 40.84 mg·g^{-1} dry weight (DW) (Table 2).

Table 2. Tuber reducing sugar contents in different cold storage conditions.

Clones	Reducing Sugar (mg·g^{-1} dry weight (DW))		
	At Harvest (22 °C)	Cold Storage (4 °C)	Cold Storage (10 °C)
V50	2.43 ± 0.07 [a]	7.02 ± 0.10 [c]	2.64 ± 0.01 [ab]
V93	2.69 ± 0.04 [ab]	3.43 ± 0.04 [b]	2.83 ± 0.04 [abc]
Dubaek	2.77 ± 0.03 [ab]	2.56 ± 0.17 [a]	2.24 ± 0.02 [a]
V48	2.79 ± 0.06 [abc]	2.61 ± 0.03 [a]	2.98 ± 0.01 [bc]
Superior	3.37 ± 0.03 [cd]	7.01 ± 0.15 [c]	6.75 ± 0.07 [gh]
N109-35	2.63 ± 0.01 [ab]	17.36 ± 0.26 [j]	3.33 ± 0.10 [c]
N2-6	5.58 ± 0.26 [f]	10.27 ± 0.23 [e]	8.70 ± 0.23 [k]
A9	4.66 ± 0.10 [e]	12.73 ± 0.30 [g]	3.37 ± 0.14 [c]
A165-4	9.86 ± 0.02 [i]	9.96 ± 0.20 [e]	7.24 ± 0.46 [hij]
V17	3.13 ± 0.02 [bcd]	8.99 ± 0.24 [d]	2.52 ± 0.10 [ab]
A12-5	5.15 ± 0.06 [ef]	37.92 ± 0.12 [o]	4.97 ± 0.03 [d]
V63	2.36 ± 0.04 [a]	11.19 ± 0.04 [f]	2.81 ± 0.08 [abc]
N357	2.42 ± 0.05 [a]	6.94 ± 0.07 [c]	2.32 ± 0.27 [a]
V18	2.85 ± 0.10 [abc]	21.96 ± 0.04 [m]	6.93 ± 0.09 [ghi]
N69-2	7.05 ± 0.10 [g]	28.49 ± 0.31 [n]	6.46 ± 0.09 [fg]
N96-29	8.58 ± 0.27 [h]	16.42 ± 0.10 [i]	13.93 ± 0.23 [m]
A35-6	6.65 ± 0.00 [g]	12.41 ± 0.08 [g]	6.07 ± 0.04 [ef]
Gogu	3.71 ± 0.10 [d]	18.71 ± 0.08 [k]	9.68 ± 0.12 [l]
N189	2.65 ± 0.44 [ab]	15.40 ± 0.24 [h]	4.61 ± 0.01 [d]
A186-23	12.82 ± 0.28 [k]	28.75 ± 0.34 [n]	19.18 ± 0.04 [n]
A188-10	11.36 ± 0.00 [j]	37.87 ± 0.00 [o]	7.47 ± 0.10 [ij]
V16	2.82 ± 0.02 [abc]	40.84 ± 0.47 [p]	5.78 ± 0.27 [e]
A27	5.70 ± 0.03 [f]	21.11 ± 0.06 [l]	7.62 ± 0.22 [j]
Mean	**4.96 ± 0.09**	**16.52 ± 0.16**	**6.11 ± 0.11**

Table results expressed as mean ± standard deviation. The means with different letters in each column are significantly different with $p < 0.05$ in Tukey's test. The Bold is used to contrast the average mean against individual values

For the tubers analyzed at harvest, three clones (V63, N357, V50) exhibited significantly lower levels of total reducing sugars compared to the main control variety 'Dubaek', while another set of genotypes (N109-35, N189, V93, V48, V16, V18) showed just as great processing aptitude potential at harvest, with similar sugar levels as the control.

At 4 °C cold storage, the clones (V93 & V48) exhibited the desired low reducing sugar levels alongside the control variety 'Dubaek'. Such cultivars with cold chipping properties are of great importance to the chip processing industry as they can provide high-quality material for processing all year round and thus increase industry efficiency.

At 10 °C cold storage, the overall mean reducing sugars of the genotypes were considerably low (Table 2), and many clones (N357, V17, V50, V63, V93, V48, A9) showed great aptitude for processing in terms of the desired total reducing sugar levels. This could be attributed to the fact that, in general, potato tubers for chip processing are stored at temperatures of about 8 to 12 °C in order to avoid an increase in sugar content [12].

At harvest, the potato clones' mean levels of reducing sugars (Table 2) were lower than those of their counterparts cold-stored at 4° and 10 °C. This has also been noted by Hayes and Thill [13] and Meena et al. [14] who observed that 'tubers have low levels of reducing sugars and produce light-colored chips directly after harvest'. The high levels of sugars in 4 °C stored tubers occurred as a result of low-temperature sweetening (LTS). During LTS, potato tubers accumulate sucrose and reducing sugars (glucose and fructose). This accumulation of sugars at low temperature causes a reduction in starch content and ultimately affects the quality of the fried products [15].

Among other parameters, the reducing sugar level is considered one of the most limiting factors in cold chip processing. The sugars in tubers react with proteins and free amino acids in the cytoplasm during the frying process, producing dark pigments with a bitter taste that devalue the final product [16]. Therefore, the importance of using material with low sugar contents is further emphasized by our study results.

3.3. Reconditioning Effect on Potato Clones

After cold storage at 4 °C, the potato clones were put at ambient temperature of 22 °C for 21 days in order to evaluate their reconditioning ability. During reconditioning, accumulated sugars are eliminated in the increased storage temperature. The decrease in sugar levels results from respiration as well as reconversion of sugars to starch [17].

From the results (Table 3), it was seen that two genotypes (V50, V93) that had performed relatively poorly in 4 °C cold storage were able to recondition well enough alongside the main control variety. Another set of genotypes (V48, N109-35, N2-6, A9) also showed high reconditioning ability, with the clone V48 showing the lowest and safest reconditioning ratio (Table 3) among the top performers. A genotype with safe reconditioning ratio would be desirable for the chip processing industry as it exhibits consistent reducing sugar levels both in cold storage and in reconditioned states. The reconditioning ratios were calculated and expressed as a percentage of reconditioned tubers against non-reconditioned ones.

$$\text{Recondition Ratio (\%)} = 100 - (\text{Reconditioned state}/\text{cold-storage state} \times 100)$$

Table 3. Reducing sugar levels after cold storage and after reconditioning.

Clones	Reducing Sugar (mg·g^{-1} DW)		Recondition Ratio (%)
	Cold Storage (4 °C)	Reconditioned (22 °C)	
V50	7.02 ± 0.10 [c]	1.73 ± 0.06 [a]	75.4
V93	3.43 ± 0.04 [b]	1.85 ± 0.03 [a]	46.1
Dubaek	2.56 ± 0.17 [a]	1.92 ± 0.08 [a]	25.0
V48	2.61 ± 0.03 [a]	2.18 ± 0.01 [ab]	16.7
Superior	7.01 ± 0.15 [c]	2.23 ± 0.13 [ab]	68.2
N109-35	17.36 ± 0.26 [j]	2.42 ± 0.00 [abc]	86.1
N2-6	10.27 ± 0.23 [e]	2.43 ± 0.15 [abc]	76.3
A9	12.73 ± 0.30 [g]	2.72 ± 0.03 [abc]	78.6
A165-4	9.96 ± 0.20 [e]	3.34 ± 0.01 [abc]	66.4
V17	8.99 ± 0.24 [d]	3.58 ± 0.08 [abc]	60.2
A12-5	37.92 ± 0.12 [o]	3.94 ± 0.04 [bc]	89.6
V63	11.19 ± 0.04 [f]	4.22 ± 0.01 [cd]	62.3
N357	6.94 ± 0.07 [c]	6.05 ± 0.09 [de]	12.8
V18	21.96 ± 0.04 [m]	6.60 ± 0.11 [e]	69.9
N69-2	28.49 ± 0.31 [n]	6.96 ± 0.06 [e]	75.6
N96-29	16.42 ± 0.10 [i]	7.27 ± 0.06 [e]	55.8
A35-6	12.41 ± 0.08 [g]	7.38 ± 0.24 [e]	40.5
Gogu	18.71 ± 0.08 [k]	13.99 ± 0.16 [f]	25.2
N189	15.40 ± 0.24 [h]	15.05 ± 0.24 [f]	2.2
A186-23	28.75 ± 0.34 [n]	17.01 ± 2.11 [g]	40.8
A188-10	37.87 ± 0.00 [o]	18.45 ± 0.13 [g]	51.3
V16	40.84 ± 0.47 [p]	20.53 ± 0.23 [h]	49.7
A27	21.11 ± 0.06 [l]	21.97 ± 0.74 [h]	−4.0

Table results expressed as mean ± standard deviation. The means with different letters in each column are significantly different with $p < 0.05$ in Tukey's test.

On the basis of the reconditioning trends, we observed that not all potato cultivars reconditioned positively, and although some clones (A165-4, V17, A12-5) showed a great come-back ability, their reducing sugar levels were still not within the acceptable range. This is a confirmation that the degree to which quality can be restored with reconditioning varies with cultivar [18].

3.4. Chip Frying Test

A major indicator of chip quality after frying is the resultant chip color. This attribute can be adequately measured on the SFA color scale (Figure 1), with the lighter color (score 1.0) being highly desirable, and the darker color (score 5.0) considered as unacceptable. The SFA chip color scores of the fried potato clones stored at different temperature conditions were evaluated and recorded (Table 4).

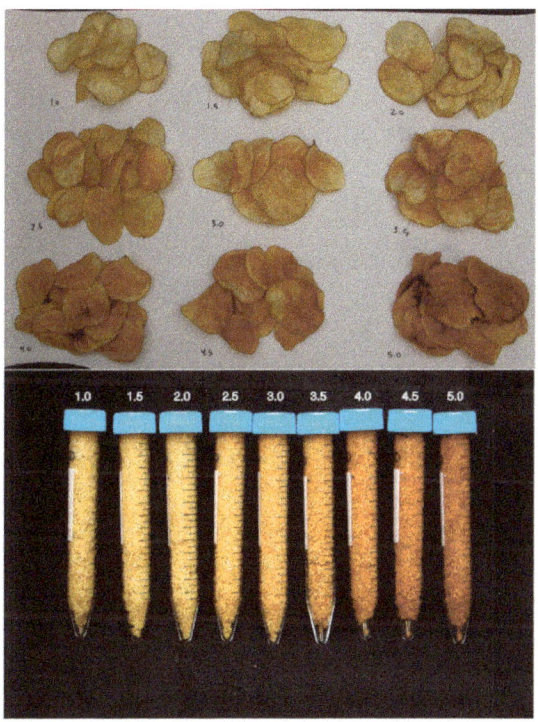

Figure 1. Snack Food Association (SFA) chip color measurement standard.

For tubers fried after 4 °C cold storage, four clones (A9, V93, V48, A165-4) exhibited acceptable light chip color alongside the control 'Dubaek'. Among these, only two clones (V93, V48) showed consistency in chip color and their corresponding reducing sugar content (Table 2, Figure 2).

Figure 2. Fry chip color of samples prepared from clones cold-stored at 4 °C: V50, V93, Dubeak, V48, Superior, N109-35, N2-6, A9, A165-4 and V17 from left to right respectively.

Table 4. Potato chip color scores after frying tests in relation to different storage conditions.

Clones	Snack Food Association (SFA) Chip Color measurement [z]		
	Cold Storage (4 °C)	Cold Storage (10 °C)	Reconditioned (22 °C)
A9	2.0	2.5	2.5
A188-10	4.5	4.0	4.0
V16	3.0	3.5	4.0
A12-5	4.0	3.5	3.0
V17	3.0	2.5	2.5
V93	1.0	2.0	1.5
N357	3.0	1.5	2.5
V48	2.5	1.5	1.5
Superior	2.5	2.5	1.5
Dubaek	1.5	1.5	1.0
V63	3.0	3.0	3.0
V18	3.5	3.5	3.5
A165-4	2.5	4.5	2.0
V50	3.0	1.5	1.5
N2-6	3.0	2.5	2.0
N189	3.5	3.0	3.5
N69-2	3.0	3.0	3.0
N96-29	4.0	3.5	3.5
A27	3.5	3.5	4.5
Gogu	4.5	4.5	4.5
N109-35	3.0	1.5	3.5
A35-6	3.0	3.0	3.0
A186-23	4.5	5.0	4.5

[z] SFA scores measured on a scale of 1–5, with 1 = light color, and 5 = dark color.

The study observed that reconditioning tubers at 22 °C for 21 days had a positive effect on many clones (A9, V93, N357, V48, A165-4, V50, N2-6, N69-2) alongside the experimental controls Dubaek and Superior. Among these, five clones (A9, V93, V48, V50, N2-6) were graded as acceptable on the basis of the corresponding light chip color and low reducing sugar levels (Table 2, Figure 3).

Figure 3. Fry chip color of clone samples reconditioned at 22 °C: V50, V93, Dubeak, V48, Superior, N109-35, N2-6, A9, A165-4 and V17, from left to right.

3.5. Glycoalkaloids

The isolation of glycoalkaloids from tubers is based on the fact that the major alkaloids are soluble in slightly acidified water [19]. The glycoalkaloid (ά-solanine) contents in tubers showed wide variation among the potato genotypes, ranging from 0.15 to 15.54 mg·100 g^{-1} fresh weight (FW). In the study, one cultivar (V50) exhibited undetectable levels of ά-solanine alongside the control variety, and although only one of the major glycoalkaloids (ά-solanine) was quantified, the levels were generally considered acceptable with regards to potato chip processing, on the basis that the total glycoalkaloid content considered safe for consumption should not exceed 20 mg·100 g^{-1} FW [20]. The variations observed in solanine contents (Figure 4) could be attributed to genetically inherited differences of the potato clones and, on a broader perspective, to environmental factors during growth and storage.

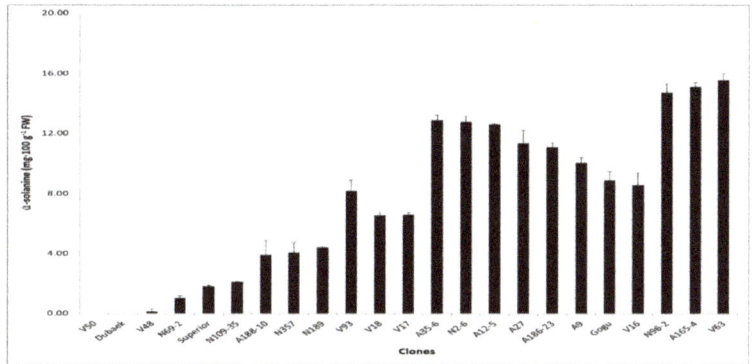

Figure 4. Solanine content in potato clones. Significant mean differences shown with $p < 0.05$ in Tukey's test.

While thermal processing of tubers such as boiling potatoes in water and frying partly eliminate glycoalkaloids, processes such as baking, rinsing, cooking, slicing, and pulsed electric field have no significant effect on glycoalkaloid reduction [21]. Therefore, it is of utmost importance in processing chains that glycoalkaloid levels be kept low, especially when introducing new cultivars in the market.

4. Conclusions

The reconditioning of potato clones at 22 °C for up to 21 days was seen to be effective for eight potato clones. We observed that for quality processing of potato chips, clones that possess combined traits of high dry matter, low levels of reducing sugars, and low levels of glycoalkaloids with acceptable light chip color should be used as raw materials. On the basis of these parameters, we identified three clones (V93, V48, A9) that exhibited an outstanding chip processing aptitude both in cold storage and in reconditioned states. These clones were finally selected for recommendation to the chip processing industry in Korea and beyond.

Author Contributions: B.O.W.: designed the experiments, performed analysis and interpretation of data, and authored the manuscript. H.S.C.: support in experimental design, supervised the study, and edited the manuscript. L.Y.S.: proofreading, supervision of the study, and revision of the manuscript.

Funding: This study was funded by a 2015 research grant from Kangwon National University (Fund #: 520150312) in collaboration with the Korea Potato Breeding Resource Bank (KPBRB), Kangwon National University.

Conflicts of Interest: The authors declare no conflict of interest.

References

1. Nacheva, E.; Pevicharova, G. Potato breeding lines for processing. *Genet. Breed.* **2008**, *37*, 3–4.
2. John Wiley and Sons, Inc. *Handbook of Fruit and Vegetable Flavors*; Hui, Y.H., Ed.; Lebensmittelchemie: Garching, Germany, 2010; Volume 64, pp. 171–172.
3. Blenkinsop, R.; Copp, L.; Yada, R.; Marangoni, A. Changes in Compositional Parameters of Tubers of Potato (*Solanum tuberosum*) during Low-Temperature Storage and Their Relationship to Chip Processing Quality. *J. Agric. Food Chem.* **2002**, *50*, 4545–4553. [CrossRef] [PubMed]
4. Habib, A.T.; Brown, H.D. Role of reducing sugars and amino acids in the browning of potato chips. *Food Technol.* **1957**, *11*, 85–89.
5. Marquez, G.; Añon, M.C. Influence of reducing sugars and amino acids in the color development of fried potatoes. *J. Food Sci.* **1986**, *51*, 157–160. [CrossRef]
6. Miller, G. Use of dinitrisalicylic acid reagent for determination of reducing sugar. *Anal. Chem.* **1959**, *31*, 426–428. [CrossRef]

7. Hellenas, K.E.; Branzell, C.; Johnsson, H.; Slanina, P. High levels of glycoalkaloids in the established Swedish potato variety Magnum Bonum. *J. Sci. Food Agric.* **1995**, *68*, 249–255. [CrossRef]
8. Marwaha, R.S.; Pandey, S.K.; Singh, S.V.; Khurana, S.M.P. Processing and nutritional qualities of Indian and exotic potato cultivars as influenced by harvest date, tuber curing, pre-storage holding period, storage and reconditioning under short days. *Adv. Hortic. Sci.* **2005**, *19*, 130–140.
9. Marwaha, R.S.; Kumar, D.; Singh, S.V.; Pandey, S.K. Nutritional and qualitative changes in potatoes during storage and processing. *Process. Food Ind.* **2008**, *11*, 22–30.
10. Grewal, S.S.; Uppal, D.S. Effect of dry matter and specific gravity on yield, color and oil content of potato chips. *Indian Food Pack.* **1989**, *43*, 17–20.
11. Killick, R.; Simmonds, N. Specific gravity of potato tubers as a character showing small genotype-environment interactions. *Heredity* **1974**, *32*, 109–112. [CrossRef]
12. Matsuura-Endo, C.; Ohara-Takada, A.; Chuda, Y.; Ono, H.; Yada, H.; Yoshida, M.; Kobayashi, A.; Tsuda, S.; Takigawa, S.; Noda, T.; et al. Effects of Storage Temperature on the Contents of Sugars and Free Amino Acids in Tubers from Different Potato Cultivars and Acrylamide in Chips. *Biosci. Biotechnol. Biochem.* **2006**, *70*, 1173–1180. [CrossRef] [PubMed]
13. Hayes, R.; Thill, C. Selection for Potato Genotypes from Diverse Progenies that Combine 4 °C Chipping with Acceptable Yields, Specific Gravity, and Tuber Appearance. *Crop Sci.* **2002**, *42*, 1343–1349. [CrossRef]
14. Meena, R.; Manivel, P.; Bharadwaj, V.; Gopal, J. Screening potato wild species for low accumulation of reducing sugars during cold storage. *Electron. J. Plant Breed.* **2009**, *56*, 89–92.
15. Satish Datir, S. Researcharchive.lincoln.ac.nz. Available online: http://researcharchive.lincoln.ac.nz/bitstream/handle/10182/4266/datir_phd.pdf?sequence=5&isAllowed=y (accessed on 24 November 2017).
16. Araújo, T.; Pádua, J.; Spoto, M.; Ortiz, V.; Margossian, P.; Dias, C.; Melo, P. Productivity and quality of potato cultivars for processing as shoestrings and chips. *Hortic. Bras.* **2016**, *34*, 554–560. [CrossRef]
17. Iritani, W.; Weller, L. Factors influencing reconditioning of Russet Burbank potatoes. *Am. Potato J.* **1978**, *55*, 425–430. [CrossRef]
18. Kumar, D.; Singh, B.; Kumar, P. An overview of the factors affecting sugar content of potatoes. *Ann. Appl. Biol.* **2004**, *145*, 247–256. [CrossRef]
19. Rodriguez-Saona, L.; Wrolstad, R.; Pereira, C. Glycoalkaloid Content and Anthocyanin Stability to Alkaline Treatment of Red-Fleshed Potato Extracts. *J. Food Sci.* **1999**, *64*, 445–450. [CrossRef]
20. Sarquís, J.; Coria, N.; Aguilar, I.; Rivera, A. Glycoalkaloid Content in Solanum species and hybrids from a breeding program for resistance to late blight (*Phytophthora infestans*). *Am. J. Potato Res.* **2000**, *77*, 295–302. [CrossRef]
21. Zarins, R.; Kruma, Z. Glycoalkaloids in Potatoes: A Review. Available online: http://llufb.llu.lv/conference/foodbalt/2017/Zarins_Kruma_FoodBalt2017.pdf (accessed on 1 February 2018).

© 2019 by the authors. Licensee MDPI, Basel, Switzerland. This article is an open access article distributed under the terms and conditions of the Creative Commons Attribution (CC BY) license (http://creativecommons.org/licenses/by/4.0/).

MDPI
St. Alban-Anlage 66
4052 Basel
Switzerland
Tel. +41 61 683 77 34
Fax +41 61 302 89 18
www.mdpi.com

Foods Editorial Office
E-mail: foods@mdpi.com
www.mdpi.com/journal/foods

www.ingramcontent.com/pod-product-compliance
Lightning Source LLC
LaVergne TN
LVHW071948080526
838202LV00064B/6706